Dedicated to the 55th Anniversary of G.B. Elyakov Pacific Institute of Bioorganic Chemistry of the Far Eastern Branch of the Russian Academy of Sciences

Dedicated to the 55th Anniversary of G.B. Elyakov Pacific Institute of Bioorganic Chemistry of the Far Eastern Branch of the Russian Academy of Sciences

Editor

Pavel S. Dmitrenok

MDPI • Basel • Beijing • Wuhan • Barcelona • Belgrade • Manchester • Tokyo • Cluj • Tianjin

MDPI

Editor
Pavel S. Dmitrenok
G.B. Elyakov Pacific Institute of Bioorganic Chemistry
Russia

Editorial Office
MDPI
St. Alban-Anlage 66
4052 Basel, Switzerland

This is a reprint of articles from the Special Issue published online in the open access journal *Molecules* (ISSN 1420-3049) (available at: https://www.mdpi.com/journal/molecules/special_issues/molecules_Russian_Academy_Sciences).

For citation purposes, cite each article independently as indicated on the article page online and as indicated below:

LastName, A.A.; LastName, B.B.; LastName, C.C. Article Title. *Journal Name* **Year**, *Volume Number*, Page Range.

ISBN 978-3-0365-2504-4 (Hbk)
ISBN 978-3-0365-2505-1 (PDF)

Cover image courtesy of Dr. Boris B. Grebnev, G. B. Elyakov Pacific Institute of Bioorganic Chemistry, FEB RAS (PIBOC, Vladivostok, Russia)

Contents

molecules

MDPI

Editorial

Editorial to the Special Issue: "Dedicated to the 55th Anniversary of G.B. Elyakov Pacific Institute of Bioorganic Chemistry of the Far Eastern Branch of the Russian Academy of Sciences"

Pavel S. Dmitrenok

G.B. Elyakov Pacific Institute of Bioorganic Chemistry, Far Eastern Branch, Russian Academy of Sciences, Pr. 100 Let Vladivostoku, 159, 690022 Vladivostok, Russia; paveldmt@piboc.dvo.ru; Tel.: +7-423-231-14-30

check for updates

Citation: Dmitrenok, P.S. Editorial to the Special Issue: "Dedicated to the 55th Anniversary of G.B. Elyakov Pacific Institute of Bioorganic Chemistry of the Far Eastern Branch of the Russian Academy of Sciences". *Molecules* **2021**, *26*, 4971. https://doi.org/10.3390/molecules26164971

Received: 4 August 2021
Accepted: 7 August 2021
Published: 17 August 2021

The G.B. Elyakov Pacific Institute of Bioorganic Chemistry of the Far-Eastern Branch of the Russian Academy of Sciences (PIBOC FEB RAS) was founded in 1964 in Vladivostok in the Far East of Russia. Over many years, we have been carrying out studies on the natural products of both marine and terrestrial origin. In collaboration with many Russian and foreign scientists, we have investigated many hundreds of diverse biomolecules, including steroids and terpenoids, quinoid compounds and alkaloids, polysaccharides and lipids, enzymes and lectins, proteins, and peptides. The Institute has a collection of marine microorganisms (KMM) PIBOC, which includes more than 4000 strains of marine bacteria and more than 1000 strains of marine fungi. The biological activity of natural compounds is also being studied.

This part of the Special Issue is devoted to the investigation of structures and biological activities of low molecular weight secondary metabolites of marine origin. These studies began in the 60s of the last century, but they remain relevant to this day [1]. In the review written by Valentin Stonik and Inna Stonik [2], the information concerning biosynthesis, environmental roles, biological action, and syntheses of marine excitatory amino acids, including kainic acid, domoic acid, dysiherbaine, and neodysiherbaine A, is summarized.

Several articles have described the structural studies of marine secondary metabolites. The structures and absolute configurations of all new compounds have been established by extensive NMR, MS, and ECD analyses, sometimes together with quantum-chemical modeling. Kicha et al. [3] reported about the isolation of four new polyhydroxylated steroids from the Vietnamese starfish *Anthenoides laevigatus*, which is able to inhibit the formation of colonies of human colorectal carcinoma HT-29 and breast cancer MDA-MB-231 cells in non-toxic concentrations. Two compounds have the rare starfish steroid compounds 5β-cholestane skeleton. The studies on a Vietnamese marine sponge *Stelletta* sp. by Kolesnikova et al. [4] led to the isolation of two new isomalabaricane triterpenoids and four new isomalabaricane-derived nor-terpenoids. One of these compounds contains an acetylenic fragment, unprecedented in the isomalabaricane family and extremely rare in other marine sponge terpenoids.

In their research article, Santalova et al. [5] analyzed rare minor oxidized cerebrosides isolated from the extract of the Far-Eastern deep-sea glass sponge *Aulosaccus* sp. Along with NMR spectroscopy and mass spectrometry, GC analysis and chemical transformations were used for structural elucidation of components. The additional instrumental and chemical methods used made it possible, for the first time, to carry out a detailed structural analysis of a complex mixture of glycosphingolipids containing allyl oxygenated monoene acyl chains.

Mishchenko et al. [6] reported on the products formed during the oxidation of Echinochrome A (Ech A). Ech A is one of the main pigments of several species of sea urchins and is registered in the Russian Pharmacopoeia as an active medicinal substance

(Histochrome®) used in cardiology and ophthalmology. The importance of this work is due to the need to characterize the products formed during the destruction of this pharmaceutical compound and to evaluate the toxic properties of obtaining products.

Sabutsky et al. [7] described the synthesis of new tetracyclic oxathiine-fused quinone-thioglycoside conjugates based on biologically active 1,4-naphthoquinones and 1-mercapto derivatives of per-O-acetyl D-glucose, D-galactose, D-xylose, and L-arabinose. The cytotoxic and antimicrobial activities of the obtained compounds were studied, and the positive effect of heterocyclization with mercaptosugars on the cytotoxic and antimicrobial properties for 1,4-naphthoquinones was shown.

In recent years, studies of the anticancer activity of marine natural compounds have become increasingly important [1,8,9]. Malyarenko et al. [10] investigated the anticancer and radiosensitizing effects of high molecular weight floretols CcPh from the brown algae *Costaria costata* on human colorectal carcinoma cells HCT 116 and HT-29. It was shown that CcPh at non-toxic concentrations suppressed the colony formation of colon cancer cells and significantly increased their sensitivity to low, non-toxic X-ray irradiation, showing a synergistic effect, which can be used to improve the radiation therapy scheme. Kvetkina et al. [11] studied the anticancer activity of the recombinant analog of actinoporin Hct-S3 (rHct-S3) from the sea anemone *Heteractis crispa*. The mechanisms of the anti-migration activity of rHct-S3 and its effect on the programmed death of cancer cells were described. Kaluzhskiy et al. [12], using both a surface plasmon resonance optical biosensor and spectral titration assays, showed that the natural flavonoid luteolin 7,3′-disulfate inhibits the activity of lanosterol 14-alpha demethylase CYP51A1, which may be important for further study of natural flavonoids as cholesterol-lowering and anticancer compounds.

The study of biopolymers of marine origin remains extremely relevant. In order to study the channels of the outer membranes that determine cell permeability, Novikova et al. [13] isolated and characterized the porin MpOmp from the extreme living marine bacterium *Marinomonas primoryensis* KMM 3633^T. It was concluded that lipid–protein interactions could be a factor that stabilizes the trimeric structure of MpOmp. Bakholdina et al. [14] investigated the effect of cultivation temperatures on the conformational quality of *Yersinia pseudotuberculosis* phospholipase A1 in inclusion bodies (IBs) using green fluorescent protein (GFP) as a folding reporter. Obtained data showed that the GFP-marker could be useful for studying the molecular organization of IBs, their morphology, and localization in *E. coli*, as well as for visualization of IBs interactions with eukaryotic cells.

The study of the genomes of marine bacteria is extremely important, not only for species identification but also for the search for bacteria that are promising sources of unique enzymes and secondary metabolites. Noskova et al. [15] proposed to use the gene of alkaline phosphatase as an additional marker in the description of strains of the genus *Cobetia* due to difficulty in their identification, both in terms of phenotypic parameters and because of the high homology of their 16S rRNA. Bystritskaya et al. [16] investigated general porin regulation in the Far-Eastern strain of *Y. pseudotuberculosis* in response to sublethal concentrations of antibiotics. As a result, the phenotypic heterogeneity of the *Y. pseudotuberculosis* population was found, manifested in the variable expression of the porin gene under the influence of carbenicillin. Chernysheva et al. [17] carried out a detailed study of PL7 alginate lyase in the representatives of the genus *Zobellia*. PL7 was found to belong to subfamilies 3, 5, and 6, undergoing local and horizontal gene transfer and gene duplication processes.

Overall, 14 manuscripts published in the current SI cover almost all aspects of PIBOC research activity in the fields of bioorganic chemistry, biochemistry, organic synthesis of natural compounds, marine microbiology, and genetic engineering, and, we hope, provide interesting new information for scientists working in these fields.

Funding: This article received no external funding.

Acknowledgments: The guest editor wishes to thank all authors for their contribution to this Special Issue. MDPI staff and, personally, Larry Li are also kindly acknowledged for their kind assistance in developing this Special Issue.

Conflicts of Interest: The author declares no conflict of interest.

References

1. Carroll, A.R.; Copp, B.R.; Davis, R.A.; Keyzers, R.A.; Prinsep, M.R. Marine natural products. *Nat. Prod. Rep.* **2021**, *38*, 362–413. [CrossRef] [PubMed]
2. Stonik, V.A.; Stonik, I.V. Marine excitatory amino acids: Structure, properties, biosynthesis and recent approaches to their syntheses. *Molecules* **2020**, *25*, 3049. [CrossRef] [PubMed]
3. Kicha, A.A.; Ha, D.T.; Malyarenko, T.V.; Kalinovsky, A.I.; Popov, R.S.; Malyarenko, O.S.; Thuy, T.T.T.; Long, P.Q.; Ha, N.T.T.; Ivanchina, N.V. Unusual polyhydroxylated steroids from the starfish *Anthenoides laevigatus*, collected off the coastal waters of Vietnam. *Molecules* **2020**, *25*, 1440. [CrossRef] [PubMed]
4. Kolesnikova, S.A.; Lyakhova, E.G.; Kozhushnaya, A.B.; Kalinovsky, A.I.; Berdyshev, D.V.; Popov, R.S.; Stonik, V.A. New isomalabaricane-derived metabolites from a *Stelletta* sp. marine sponge. *Molecules* **2021**, *26*, 678. [CrossRef] [PubMed]
5. Santalova, E.A.; Denisenko, V.A.; Dmitrenok, P.S. Structural analysis of oxidized cerebrosides from the extract of deep-sea sponge *Aulosaccus* sp.: Occurrence of amide-linked allylically oxygenated fatty acids. *Molecules* **2020**, *25*, 6047. [CrossRef] [PubMed]
6. Mishchenko, N.P.; Vasileva, E.A.; Gerasimenko, A.V.; Grigorchuk, V.P.; Dmitrenok, P.S.; Fedoreyev, S.A. Isolation and structure determination of Echinochrome A oxidative degradation products. *Molecules* **2020**, *25*, 4778. [CrossRef] [PubMed]
7. Sabutski, Y.E.; Menchinskaya, E.S.; Shevchenko, L.S.; Chingizova, E.A.; Chingizov, A.R.; Popov, R.S.; Denisenko, V.A.; Mikhailov, V.V.; Aminin, D.L.; Polonik, S.G. Synthesis and evaluation of antimicrobial and cytotoxic activity of oxathiine-fused quinone-thioglucoside conjugates of substituted 1,4-naphthoquinones. *Molecules* **2020**, *25*, 3577. [CrossRef] [PubMed]
8. Khalifa, S.A.M.; Elias, N.; Farag, M.A.; Chen, L.; Saeed, A.; Hegazy, M.F.; Moustafa, M.S.; Abd El-Wahed, A.; Al-Mousawi, S.M.; Musharraf, S.G.; et al. Marine natural products: A source of novel anticancer drugs. *Mar. Drugs* **2019**, *17*, 491. [CrossRef] [PubMed]
9. Katanaev, V.L.; Di Falco, S.; Khotimchenko, Y. The anticancer drug discovery potential of marine invertebrates from Russian Pacific. *Mar. Drugs* **2019**, *17*, 474. [CrossRef] [PubMed]
10. Malyarenko, O.S.; Imbs, T.I.; Ermakova, S.P. In vitro anticancer and radiosensitizing activities of phlorethols from the brown alga *Costaria costata*. *Molecules* **2020**, *25*, 3208. [CrossRef] [PubMed]
11. Kvetkina, A.; Malyarenko, O.; Pavlenko, A.; Dyshlovoy, S.; von Amsberg, G.; Ermakova, S.; Leychenko, E. Sea anemone *Heteractis crispa* actinoporin demonstrates in vitro anticancer activities and prevents HT-29 colorectal cancer cell migration. *Molecules* **2020**, *25*, 5979. [CrossRef] [PubMed]
12. Kaluzhskiy, L.; Ershov, P.; Yablokov, E.; Shkel, T.; Grabovec, I.; Mezentsev, Y.; Gnedenko, O.; Usanov, S.; Shabunya, P.; Fatykhava, S.; et al. Human lanosterol 14-alpha demethylase (CYP51A1) Is a putative target for natural flavonoid luteolin 7,3′-disulfate. *Molecules* **2021**, *26*, 2237. [CrossRef] [PubMed]
13. Novikova, O.D.; Khomenko, V.A.; Kim, N.Y.; Likhatskaya, G.N.; Romanenko, L.A.; Aksenova, E.I.; Kunda, M.S.; Ryzhova, N.N.; Portnyagina, O.Y.; Solov'eva, T.F.; et al. Porin from marine bacterium *Marinomonas primoryensis* KMM 3633T: Isolation, physico-chemical properties, and functional activity. *Molecules* **2020**, *25*, 3131. [CrossRef] [PubMed]
14. Bakholdina, S.I.; Stenkova, A.M.; Bystritskaya, E.P.; Sidorin, E.V.; Kim, N.Y.; Menchinskaya, E.S.; Gorpenchenko, T.Y.; Aminin, D.L.; Shved, N.A.; Solov'eva, T.F. Studies on the structure and properties of membrane phospholipase a1 inclusion bodies formed at low growth temperatures using gfp fusion strategy. *Molecules* **2021**, *26*, 3936. [CrossRef] [PubMed]
15. Noskova, Y.; Seitkalieva, A.; Nedashkovskaya, O.; Shevchenko, L.; Tekutyeva, L.; Son, O.; Balabanova, L. Are the closely related *Cobetia* strains of different species? *Molecules* **2021**, *26*, 690. [CrossRef] [PubMed]
16. Bystritskaya, E.; Chernysheva, N.; Stenkova, A.; Guzev, K.; Rakin, A.; Isaeva, M. Differential expression of *Yersinia pseudotuberculosis* general porin genes during short-and long-term antibiotic stresses. *Molecules* **2021**, *26*, 3956. [CrossRef] [PubMed]
17. Chernysheva, N.; Bystritskaya, E.; Likhatskaya, G.; Nedashkovskaya, O.; Isaeva, M. Genome-wide analysis of PL7 alginate lyases in the genus *Zobellia*. *Molecules* **2021**, *26*, 2387. [CrossRef] [PubMed]

molecules

MDPI

Review

Marine Excitatory Amino Acids: Structure, Properties, Biosynthesis and Recent Approaches to Their Syntheses

Valentin A. Stonik [1] and Inna V. Stonik [2,*]

1 G.B. Elyakov Pacific Institute of Bioorganic Chemistry, Far Eastern Branch, Russian Academy of Sciences, prospect 100 let Vladivostoku, 159, 690022 Vladivostok, Russia; stonik@piboc.dvo.ru

2 National Scientific Center of Marine Biology, Far Eastern Branch, Russian Academy of Sciences, ul. Palchevskogo, 17, 690041 Vladivostok, Russia

* Correspondence: innast2004@mail.ru; Tel.: +7-84232317107

Academic Editor: Derek J. McPhee

Received: 16 June 2020; Accepted: 1 July 2020; Published: 3 July 2020

Abstract: This review considers the results of recent studies on marine excitatory amino acids, including kainic acid, domoic acid, dysiherbaine, and neodysiherbaine A, known as potent agonists of one of subtypes of glutamate receptors, the so-called kainate receptors. Novel information, particularly concerning biosynthesis, environmental roles, biological action, and syntheses of these marine metabolites, obtained mainly in last 10–15 years, is summarized. The goal of the review was not only to discuss recently obtained data, but also to provide a brief introduction to the field of marine excitatory amino acid research.

Keywords: kainic acid; domoic acid; dysiherbaine; neodysiherbaine A

1. Introduction

Throughout the 55-year history of the G.B. Elyakov Pacific Institute of Bioorganic Chemistry (PIBOC), Far Eastern Branch, Russian Academy of Science, a vast variety of new marine-derived natural compounds including different glycosides, polar steroids, alkaloids, lipids, and other metabolites have been isolated from echinoderms and sponges, ascidians, algae, fungi, and marine bacteria and thoroughly studied. Many of the discovered new natural compounds proved to have a protective action and were repeatedly mentioned in well-known series of review articles published in Natural Products Reports [1]. To protect themselves against predators and pathogens, marine organisms use a variety of structural groups of toxins. The latter are synthesized by lower plants, including microalgae [2–4], and then accumulated by some invertebrates that transfer them up to higher trophic levels. Marine excitatory amino acids (MEAAs) that belong to this category and play an important role in marine communities have found application in pharmacology. To date, there are only three known groups of marine-derived excitatory amino acids: (**1**) kainic acid and related compounds, (2) domoic acid and some its derivatives, (3) sponge-derived dysiherbaine and neodysiherbaine A, which specifically bind and activate the kainate-type of glutamate receptors (GluRs).

MEAAs, including the main compounds shown in Figure 1, are a chemical group of particular research interest to us, although many of them were discovered long ago. This is explained by our studies on highly polar marine metabolites from northwestern Pacific invertebrates, including sponges, and marine bacteria and search for new excitatory compounds or their analogs. Another explanation is the long-term interest of one of the coauthors of the present review (S.I.) [5] in domoic acid (DA), and in DA-producers from the northern Pacific Ocean.

In general, these compounds are known as bioactive agents, particularly some of them as biotoxicants, contaminating marine waters. The application of these compounds in experimental pharmacology has expanded the knowledge about transmission of nerve impulses in higher animals. In fact, excitatory amino acids have a structure similarity with that of the principal excitatory neurotransmitter glutamic acid (**1**) and act as central nervous system (CNS) excitants. In the past decade, MEAAs, sometimes called kainoids, have become an important research field, with several discoveries made.

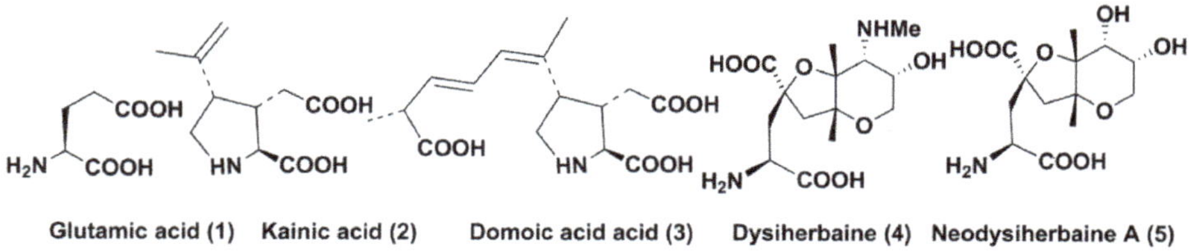

Figure 1. Chemical structures of glutamic acid (**1**) and marine excitatory acids (**2–5**).

Despite the successes of the most recent studies on biosynthesis of these natural products, a number of questions about their appearance and disappearance in producers and consumers, transfer via food chains, action on own producers, and origin of some of them still remain unresolved.

The last review [6] considering excitatory amino acids was published more than 20 years ago. Herein, we discuss the recent results of studies on marine-derived natural products belonging to this group, including their discovery, structure, biosynthesis, recent approaches to syntheses, origin, biological action in producers and recipients, and environmental roles.

2. Kainic Acid and Related Metabolites

2.1. Discovery, Structure, and Some Properties

The key member of the kainoid family, kainic acid (**2**) [(−)α-kainic acid], sometimes referred to as digenic acid, was isolated by Japanese chemists in the 1950 s from aqueous extracts of the red macroalga Diginea simplex (Floridiophyceae, Ceramiales, Rhodomelaceae). Compound **2** in these extracts was used as a vermifuge in the traditional medicine of East Asian countries for over a thousand years [7]. In 1953, Murakami and co-authors [8] showed this acid to be the most active anthehelmintic agent from this alga and named it kainic acid after Kaininso, the Japanese name of this alga. The structures of this compound and its isomer, allo-kainic acid (**6**), were established by X-ray analysis of the zinc salt of kainic acid and, later of kainic and allokainic acids themselves [9,10]. Compound (**2**) was also detected in several other species of lower macrophytes such as the red alga Centroceras clavulatum [9] and others [11]. The Corsican moss Alsidium helminthochorton [12] contains kainic acid (**2**), allokainic acid (**6**), and α-kainic acid lactone (**7**) (Figure 2) along with a peptide of 37 amino acids including two kainic acid residues. A mutant strain of the alga Palmaria palmata was found to produce high levels of kainic acid [13]. Chemical transformation of **2** provided several additional compounds of this series such as isokainic acid (**8**) [12].

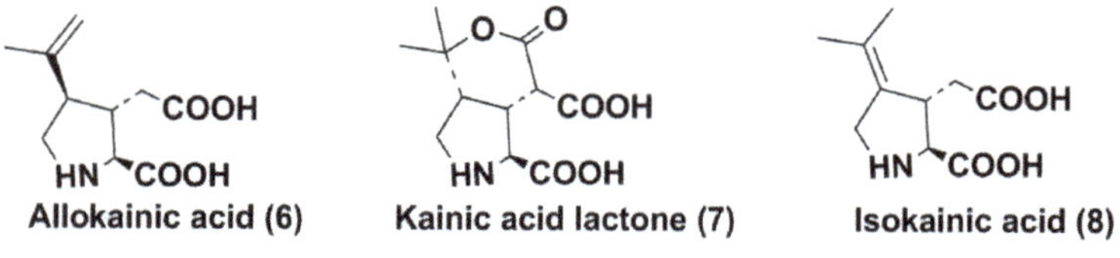

Figure 2. Some compounds closely related to kainic acid (**6–8**).

A new stage in the fundamental research of kainic acid came when its most interesting property, the ability to specifically activate a subgroup of glutamic acid receptors, was discovered in the early

1970s. Later, these receptors were named kainate receptors (KARs) [14–16]. Kainic acid is one of the best natural agonists of KARs. It causes the influx of cellular Ca^{2+} ions, production of reactive oxygen species, and mitochondrial dysfunctions that leads to neuronal apoptosis and necrosis. Hyperstimulation of KARs is involved in the pathogenesis of various neurodegenerative disorders such as epilepsy, Hantington's chorea, and stroke. These properties of **2** and its application in experimental neurobiology and pharmacology have attracted much attention and stimulated development of different approaches to synthesize this excitatory acid and related compounds. In 1995, after the discovery of better anthelmintics than **2**, the production of this acid as anthehelmintic in Taiwan was stopped. However, application in experimental pharmacology continued to require increasing amounts of this compound, and it began to be produced not only from algae, but also by synthesis. In the early 2000s, its price increased more than 50 times, to $100 per 10 mg [17]. In 2012, Evens and Inglesby [18] noted that kainic acid had an estimated market value of 1 billion USA $ per annum. The price of synthetic (−)-α-kainic acid from Sigma-Aldrich in January 2012 was $750/50 mg and increased up to $1749/50 mg by May 2020. The deficit and the exceptionally high price of this product necessitated development of optimal schemes for its complete syntheses, the number of which has increased significantly in the recent decade.

2.2. Recent Syntheses

Thus, kainic acid raised great interest in synthetics due its activity, wide use in experimental pharmacology, and probability to synthesize analogs and derivatives which could be applicable in medicine to treat schizophrenia and other brain diseases. Taking into account that configurations of its three stereogenic centers are crucial in binding to receptors and functional activities of this compound, the stereoselective synthesis of **2** was directed first to the optically active form of **2**, identical to the natural product. Different synthetic strategies, based on C2–C3 or C3–C4 bond formations (the latter approach was used in majority of syntheses), C4–C5 bond formation and C-*N* bond formation pathways, as well as on the use of the existing pyrrolidine ring and cycloaddition reactions, were applied to synthesize this excitatory amino acid. As a result, about 40 total multi-step syntheses of **2** were reported in literature and reviewed in 2012 [19].

In our review, we discuss only some of syntheses reported later. Most of them are based on novel schemes. We omit listing all the stages of these syntheses here and, instead, pay attention to main ideas and results achieved in comparison with the already known approaches.

Poison et al. [20] elaborated an efficient synthesis of (–)-kainic acid (**2**) using just two of the above-mentioned approaches: application of a pyrrolidine precursor and cycloaddition. A high-pressure Diels-Alder cycloaddition of the obtained from 4-hydroxy-L-prolin (**9**) 3,4-unsaturated pyrrolidine derivative **10** with Danishefsky's diene at high pressure (15 kbar) and room temperature gave bicyclic product **11**. In this case, high pressure provided the gain in reactivity and led, after some additional treatment and 82 h of exposure, to a 96% conversion into **11** that contained *N*-protected trisubstituted pyrrolidine cycle. The subsequent decarboxylation and transformation of its six-membered ring to isopropenyl and carboxymethyl groups gave **12** converted into target product via pyrrolidine derivative **13** with almost total stereocontrol and an approximately 10% yield (Scheme 1). At the same time, the key intermediate product, enone **11**, and related compounds were partly racemic in their first synthesis attempts. However, preliminary purification of the initial Danishevsky's diene by rapid distillation and removal of trace triethylamine from it made it possible to synthesize **11** with excellent yield and high enantiomeric excess.

New approaches, based on the formation of C3–C4 bond and involving the Ireland-Claisen rearrangement of allylic esters, were used by Indian and Japanese groups. Reddy and Chandraseker [21] utilized this rearrangement along with the Sharpless asymmetric epoxidation. The first of these reactions was used to create C3 and C4 *cis* stereocenters, while the Sharpless oxidation allowed the designing of chirality at C2. Authors constructed the target compound 2 from ynone **14**. At the first stages, this compound was converted into alcohol **15** by Noyori reduction and Red-Al transformation of triple bond into double one. Reaction of **15** with 3-methyl-3-butenoate catalyzed by *N,N′*-dicyclohexylcarbodiimide (DCC) and 4-dimethylaminopyridine (DMAP) gave the allylic ester **16**, which was converted into **17** by the Ireland-Claisen rearrangement with lithium bis(trimethylsilyl)amide (LiHMDS) and trimethylsilyl chloride (TMSCl). The subsequent four steps of transformations led to allyl alcohol **18** with removal of p-methoxybenzyloxy (PMBO) protective group. The obtained **18** was epoxidated by asymmetric Sharpless epoxidation using (–)-diethyltratrate (DET), titanium isopropoxide $Ti(iOPr)_4$, and *tert*-butyl hydroperoxide (TBHP) to give **19**. Further conversion provided azide **20**. Reduction of **20** allowed the formation of pyrrolidine ring, and then NH in this ring was protected by reaction with *tert*-butyloxycarbonyl anhydryde ($(tBoc)_2O$) to obtain **21** in several steps. Removal of the protective group and oxidation with Jones reagent (CrO_3 in aqueous sulfuric acid) led to kainic acid (**2**) (Scheme 2). Thus, a new strategy for constructing the core system of **2** was realized in this synthesis.

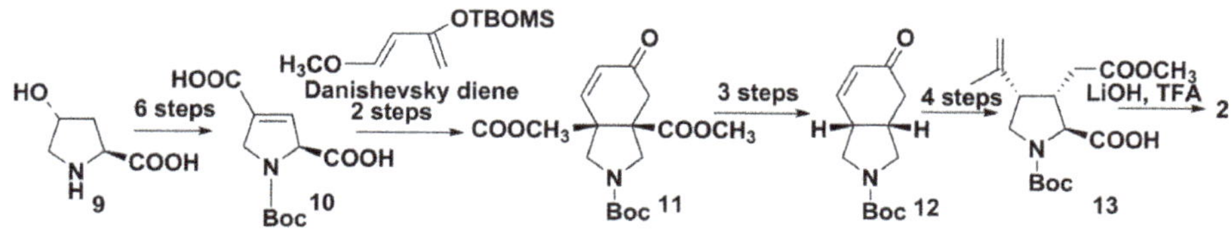

Scheme 1. Synthesis of kainic acid by Diels-Alder reaction.

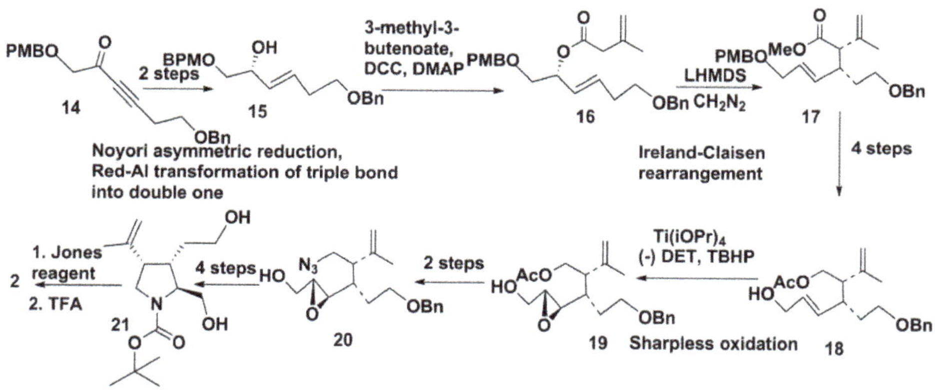

Scheme 2. A shortcut scheme of synthesis of kainic acid via chirality transfer through the Ireland-Claisen rearrangement.

A similar strategy, also based on Ireland-Claisen rearrangement, was proposed by Japanese scientists as a unified approach to synthesize not only kainic acid proper, but also bioactive 4-substituted kainoids [22]. A source compound **22** was obtained from L-tartaric acid through four steps. Condensation of 22 with 3-methyl-3-butenoic acid gave ester **23** which, through the Claisen-Ireland rearrangement with LHMDS in the presence of TMSCl, was converted with high diastereoselectivity into carbonic acid **24**. The subsequent reduction of a carboxy group and other transformations gave aminomethyl derivative **25**. The resulting product was cyclized exceptionally easily through the palladium-mediated pyrrolidine-ring formation into the product **26** that had the required stereochemistry for transformation into kainic acid or its derivatives. After protection of the isopropenyl group, the corresponding conversion into **2** was achieved by oxidative cleavage of the vinyl group by ozonolysis followed by the Jones oxidation (Scheme 3).

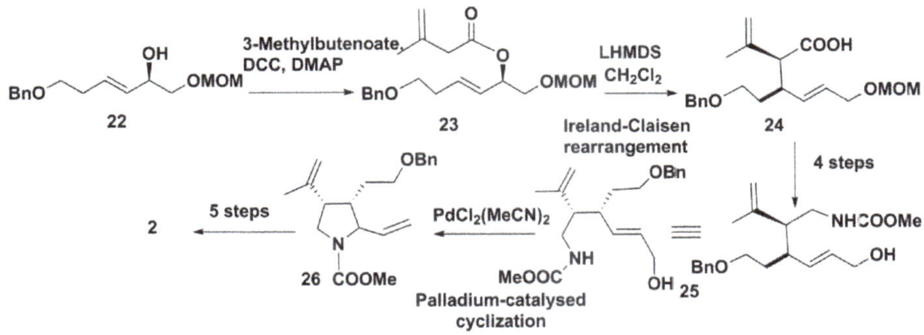

Scheme 3. Synthesis of kainic acid using the Ireland-Claisen rearrangement and palladium catalyzed formation of pyrrolidine ring.

The advantage of the recent synthesis of kainic acid, developed by Japanese scientists [23], unlike most other multi-stage syntheses, consisted in only nine stages. This short total synthesis was carried out on the basis of the Cu-catalyzed Michael addition-cyclization reaction between the bearing chiral auxiliary of camphorsultam-type isonitrile **27** and ester of unsaturated ketoacid **28** (in mixture with an isomeric ester) with Cu(t-ButSal)$_2$ as catalyst (Scheme 4). The additional ester was isomerized into **28** in the conditions of this reaction. The treatment of the obtained adduct **29** by sodium methoxide with the loss of chiral substituent, followed by protection with (Boc)$_2$O in the presence of triethylamine and DMAP, gave unsaturated pyrrolidine **30**. The alkaline hydrolysis of the amide group at the position 2 and the selective reduction of a double bond by boron-organic reagent L-selectride led to the protected analog of kainic acid **31** which was converted into the target compound via intermediates **32–34**. This synthesis was carried out with a 16.8% overall yield on chiral isonitrile and was conducted on a 300 mg scale, although usually syntheses of **2** had given smaller amounts of this product.

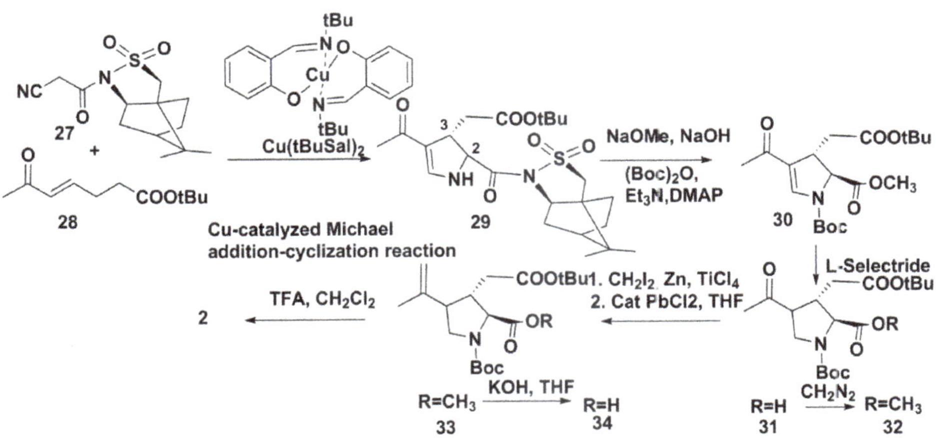

Scheme 4. Short total synthesis of (–)-kainic acid.

A gram-scale synthesis of (–)-kainic acid through six steps and with a 34% overall yield was elaborated using the Pt-catalyzed direct allylic amination [24]. This amination was catalyzed by the combination of dichloro(1,5-cyclooctadiene) platinum (Pt(cod)Cl$_2$) with bis[2-di-phenylphosphino) ether (PPEphos). Such new approach allows catalyzing the direct introduction of amino group into allylic alcohols. The chosen scheme provided the minimum number of stages in this synthesis (Scheme 5). Amination of **35** under microwave heating conditions gave monoallylamine **36**. As a protective group

(Pg), authors preferred to use 2,4-dimethoxybenzyl (DMB) or p-methoxybenzyl (PMB) with almost no loss of enantiopurity. The unsaturated ester **38** was obtained by one-pot oxidation of epoxide **37** with 2,2,6,6-tetramethyl-1-piperidinyloxy (TEMPO) followed by the Wittig reaction. It was found that the transformation of **36** and **38** into diallylamine derivative **39** at a higher concentration inhibited the epimerization, and the desired product was obtained in 95% yield and 98% ee (enantiomeric excess). The following stage of heating in a sealed tube with xylenes and a catalytic amount of iPr$_2$N-Et gave pyrrolidine **40** in 75% yield and 14:1 diastereoselectivity. The Jones oxidation with application of H_5IO_6 as terminal oxidant gave satisfactory results at conversion of **40** into **2** (Scheme 5). Finally, 1.11 g pure kainic acid was obtained by this method in 34% overall yield. Taking into account its yield and the minimal number of steps, this proved to be the best chemical scheme elaborated to synthesize **2**.

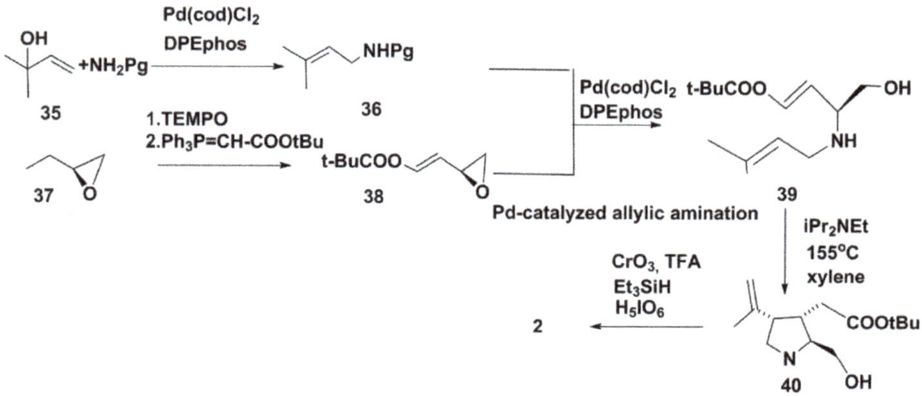

Scheme 5. Synthesis of kainic acid using the Pd-catalyzed direct allylic amination.

Kainic and allo-kainic acids (**2** and **3**, respectively) were also synthesized using SmI_2, a unique single-electron reducing agent that allows the induction of reductive coupling with the formation of C–C bond under mild conditions [25]. The key initial products for this new synthesis were obtained from D-serine methyl ester hydrochloride using the known procedure and a mixture of E- and Z-isomers of α,β-unsaturated esters, designated as E-14 and Z-14, with a general formula of **41**. The cyclization of **41** gave derivatives of kainic and/or allo-kainic acids (**42** and **43**) with different yields depending on ratio of the reagents (SmI_2, hexamethylphosphoramide (HMPA), NiI_2 and H_2O) and additional ligands such as ethylenediamine, 2,2′-bipyridine, 2,2′-bipyridylamine, triphenylphosphine, etc. Pure compounds were isolated in optimal conditions of this reaction for each of them followed by separation and purification of the products through HPLC. The obtained compounds were converted into kainic acid **2** or allo-kainic acid **3** through a 3-step transformation by conversion ester groups into acids and deprotection of nitrogen in pyrrolidine ring (Scheme 6).

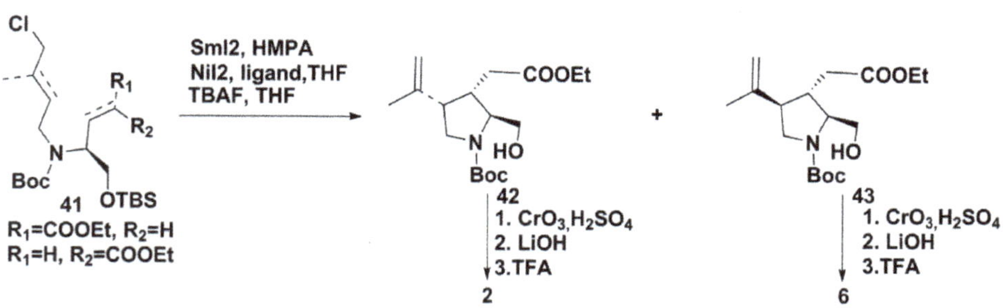

Scheme 6. Syntheses of kainic and *allo*-kainic acids using SmI_2-induced cyclization.

The formal total synthesis of (–)-kainic acid, recently described by Chinese chemists [26], was based on a new approach to synthesis of the same important intermediate, which was already used for synthesizing kainic acid by different authors [19,20].

In general, there have been several syntheses of kainic acid published in recent years, which confirms the constant attention to this excitatory amino acid and the rapid progress of organic synthesis. These syntheses are shorter and can provide better production of kainic acid as a pharmacological probe for application in experimental pharmacology, although the costs and availability of the synthetic precursors and reagents used were not compared with previously known syntheses. All the studies have discovered a large number of kainoids, a total of 70 [26], including those obtained through recent chemical syntheses.

2.3. *Biosynthesis and Application of Kainoids*

Recently, Moore's team from the USA studied the biosynthesis of two related excitatory amino acids: domoic acid from diatoms and kainic acid from red macrophytes. After elucidating the origin of kainic acid **2**, they performed the whole genome sequencing of a kainic acid producer, *Digenea simplex*, to identify the gene clusters involved in the biosynthesis of **2** [27]. They used the new single-molecule, long-read sequencing platforms such as Oxford Nanopore Technologies. As a result, the long reads that included genes encoding the kainic acid biosynthesis were sequenced; however, due to the presence of impurity of microbial DNA, they were not assembled into the whole genome of this alga. After the analysis of *D. simpex* on a higher-throughput platform, Prometh ION, 47 Gp of sequence was generated, with the longest read being 1.2 Mb. A series of *Digenea simplex* kainic acid biosynthesis (Dskab) genes were uncovered. They included genes of annotated *N*-prenyltransferase (DsKabA), α-ketoglutarate-dependent dioxygenase (DskabC), and several retrotransposable elements such as integrase, reverse transcriptases, and RNA H domains. Both sequences, DsKabA and DskabC, were successfully expressed in *Echerichia coli* and purified. Incubation of recombinant DskabA with L-glutamate **1** and dimethylallylpyrophosphate **44** gave *N*-dimethylallyl-L-glutamic acid **45** (the so-called prekainic acid). This compound was converted into kainic acid in an aqueous extract from *D. simplex*. Another enzyme of kainic acid biosynthesis, the so-called DsKabC, was incubated with prekainic acid, α-keto-glutarate, L-ascorbate, and Fe^{2+}, and transformed prekainic acid into kainic acid (**2**) and kainic acid lactone (**7**) (Scheme 7). The main product was confirmed to be kainic acid using isolation from the culture medium and NMR elucidation. The obtained kainic acid lactone **7** has been shown to be antagonist of iGluR in contrast to kainic acid proper.

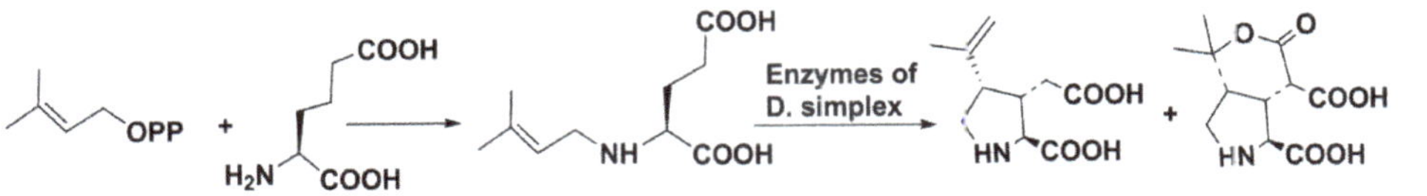

44 **1** **Prekainic acid (45)** **2** **7**

Scheme 7. Biosynthesis of kainic acid.

To show the power of both chemical synthesis and biocatalysis, prekainic acid was synthesized through the reductive amination of L-glutamate with 3-methyl-2-butenal, by the method reminiscent of the cobalt-mediated strategy of Baldwin et al. [28]. For the next step, DsKabC was used without additional purification, but in a medium with E. coli expressing this enzyme. The employment of this combined approach and the purification procedure using activated carbon followed by preparative reverse phase HPLC provided 1.1 g of kainic acid with a 32% overall yield and 95% purity. It proved to be better compared to any chemical synthesis. Thus, this work has not only discovered enzymes of biosynthesis of kainic acid, but also developed a new method combining chemical and enzymatic transformations to obtain **2** for pharmacological research.

Analyses of genomes of other red algae revealed similar genes. For example, genes named as PpkabA and PpkabC were identified in the red alga Palmara palmata, the second known producer of **2**. In comparison with Digenea simplix, the kab genes in the edible P. palmata were tightly clustered, and the retrotransposable elements were absent in the intergenic region. The corresponding kab genes were also found in four other red algae species [27].

As it is well known, **2** has found a wide range of applications to model epilepsy, to determine molecular events indicating human neurodegenerative disorders, and to evaluate efficiency of various therapeutic interventions on laboratory animals with simulated diseases [29–31].

The specific interaction of kainic acid with glutamic acid receptors (GARs) in CNS was studied. Most of ionotropic kainate receptors have ligand-binding domains which form a clamshell-like structure consisting of lobes. Kainic acid, like glutamic acid, is bound between these lobes, but their influence on activities of these receptors remains to be explained [32]. As a neurotransmitter, kainic acid acts also on another class of receptors, metabotropic glutamate receptors (mGluRs), which respond slower than iGluRs, but their interaction with **2** shows an effect on the learning and memory functions.

The biological role of kainic acid in its producers was more poorly studied than distribution, biological activities, and biosynthesis. Using rabbit polyclonal antibodies for kainic acid, which did not cross-react with other amino acids including glutamate, Japanese scientists immunochemically localized this toxin in the fine cylindrical thallus of D. simpex [33]. They concluded that the presence of kainic acid on the surface of the alga can be related to its role as a protection against grazers.

Little is currently known about the functions of algal glutamate receptors-like compounds or the probable influence of kainic acid on these receptors in algae. However, the highest variability in glutamate receptors sequences was recorded from algae [34]. These receptors are associated with many plant-specific physiological functions, such as sperm signaling in moss, pollen tube growth, root meristem proliferation, and innate immune and wound responses. It should be taken into account that the main physiological roles and modes of action of plant GluRs, in contrast to those in animals, are performed in peripheral, non-neuronal tissues [35].

Further studies on some synthetic derivatives of kainic acid and analogs (Figure 3) are aimed, first, at discovering the earlier unknown events in the brain depending on kainate receptors. Second, analogous compounds can be used to create novel drugs. Some of these compounds have reached clinical trials. For example, the compound LY404039 (**46**) is a selective agonist for the metabotropic glutamate 2/3 (MGlu2/3) receptor. This preparation, when dosed as prodrug (**47**) (LY2140023), showed efficacy in schizophrenia. A series of similar compounds was studied as potential psychiatric medicines by such pharmaceutical companies as Lilly and Taisho [36].

Figure 3. Structures of synthetic derivatives of glutamic acid (**46**,**47**).

3. Domoic Acid and Related Compounds

3.1. Discovery, Structure, and Some Properties

Domoic acid (DA) **3** was first found and isolated from the red alga *Chondria armata* by Japanese scientists in the 1960s [37]. After three years of studies, its structure was determined and then refined through organic synthesis. Like kainic acid, **3** is a derivative of dicarboxylated pyrrolidine containing glutamate moiety, but in its another moiety, domoic acid bears octadienoic substituent derived from a monoterpenoid acid (Figure 1). Closely related metabolites, isodomoic acids **48–55**, were identified for more than 25 subsequent years in the red alga *C. armata* and contaminated mussels [38–40], as reviewed in [41,42]. All these metabolites, including isodomoic acids A-H along with domoilactones and some related compounds, also contain a glutamic acid residue and should be considered as effectors of GARs. However, as a rule, their effect is much weaker (Table 1). The number of known analogues of DA has increased, particularly in recent years [43].

Table 1. Metabolites related to domoic acid.

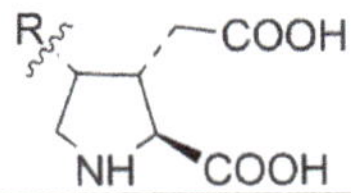

Compound	R (Side Chain)=	Name	KARs, μM	References
3		Domoic acid	2.2	[38,44,45]
48	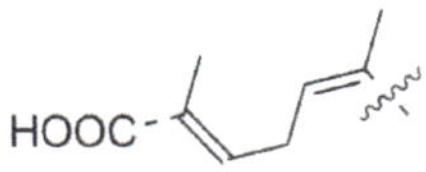	Isodomoic acid A	4.4	[39,42]
49	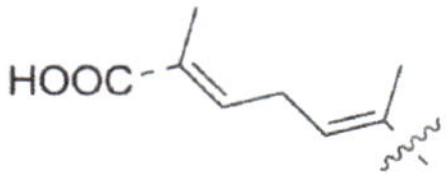	Isodomoic acid B	4990	[39,42]
50	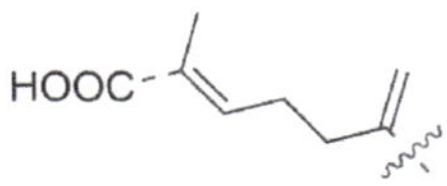	Isodomoic acid C	171	[39,42]
51	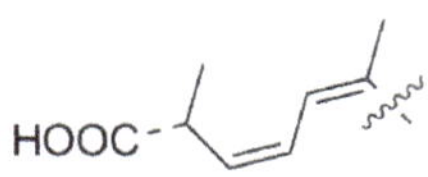	Isodomoic acid D	600	[40,41]
52	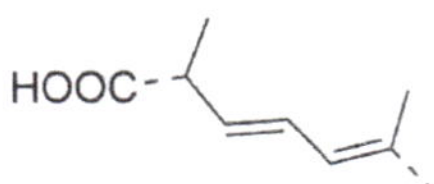	Isodomoic acid E	600	[42]
53	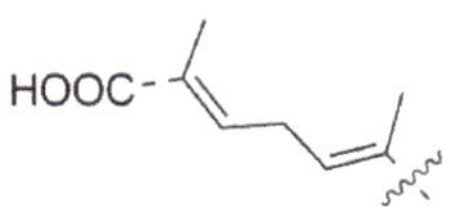	Isodomoic acid F	67	[42]
54	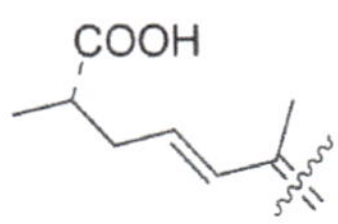	Isodomoic acid G		[41,42]
55	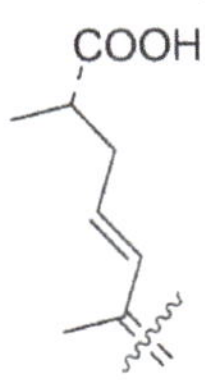	Isodomoic acid H		[41,42]

KARs, μM = displacement of kainic acid from the complex with kainate receptors.

Several compounds related to domoic acid (dainic acids) have recently been described and named as 7′-methyl-isodomoic acid A (**56**), 7′-methyl-isodomoic acid B (**57**), 7′-hydroxymethyl-isodomoic A (**58**), and 7′-hydroxymethyl-isodomoic acid B (**59**). Closely related minor metabolites **60**, **61**, which do not contain pyrrolidine ring, have also been isolated from the same alga *C. armata* (Figure 4) [43].

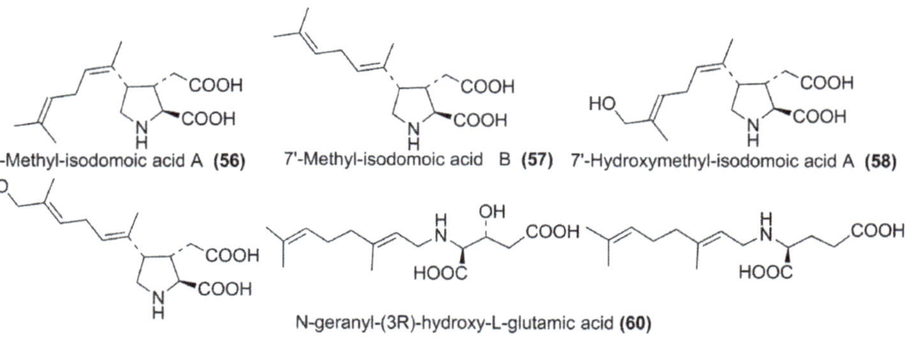

7'-Methyl-isodomoic acid A **(56)** 7'-Methyl-isodomoic acid B **(57)** 7'-Hydroxymethyl-isodomoic acid A **(58)**

7'-Hydroxymethyl-isodomoic acid B **(59)** N-geranyl-(3R)-hydroxy-L-glutamic acid **(60)** N-geranyl-L-glutamic acid **(61)**

Figure 4. Dainic acids (**56–59**) and related metabolites (**60,61**) from *C. armata*.

The main turn in the fate of domoic acid and related compounds occurred after the discovery that these metabolites are produced not only by several species of red algae, but also by planktonic diatoms [43]. As a result of subsequent transfer of domoic acid (DA) (**3**) up the food chain from microalgae into edible mollusks [44–46], it may cause the so-called Amnesic Shellfish Poisoning (ASP) in humans that consume edible mollusks contaminated by DA [47–49]. The first case of ASP was described in 1987 from Canada, where 107 people, who had consumed cultivated mussel *Mytilus edulis*, showed symptoms of this disease and three of them died [45]. The diatom *Pseudo-nitzshia multiseries* was identified as a phytoplanktonic producer of DA, which also produces isodomoic acids E and F. Since 1987, it was found that this toxin and related compounds are responsible not only for human poisoning, but also for numerous cases of poisoning and mortality of birds and marine mammals [46–49].

Symptoms of ASP in humans include gastrointestinal alterations (abdominal cramp, nausea, vomiting), and/or neurological disorders (dizziness, short-term memory loss, seizure, epilepsy or coma in the acute cases) [47]. DA binds to glutamate receptors in the central nervous system (CNS) and myocardium [50] causing overexcitation and, as a consequence, neuro-excitatory behavior in humans, marine mammals, and fish [51–55]. The binding of DA to kainate or AMPA (α-amino-3-hydroxy-5-methyl-4-isoxazolepropionic acid) receptors, subclasses of glutamate receptors, results in the intracellular accumulation of Ca^{2+} [47,55]. In contrast to glutamic acid, DA causes long-lasting depolarization, neuronal swelling induced by intraneuronal accumulation of excess Ca^{2+}, production of reactive oxygen species, DNA and mitochondrial damages, energy depletion, and cell death [55,56]. Along with the CNS, the heart is often also considered as a potential target site for adverse effects of DA. DA-induced cardiotoxicity has been confirmed not only in people suffering from ASP, but also in sea lions and zebrafish that died as a result of DA-poisoning [51,52].

Isodomoic acids (iso-DAs), which have an insecticidal effect against the American cockroach *Periplaneta americana* [38], were also isolated from marine diatoms including *P. australis*, *P. seriata*, and *Nitzschia navis-varingica* [57]. Some diatom species (e.g., *Nitzschia navis-varingica*) produce isodomoic acids A and B as the major toxin components [49]. However, iso-DAs are often present in the environment at lower concentration than DA [57]. The affinity of iso-DAs to the glutamate receptors is lower than that of DA (Table 1) [58]. Therefore, parent toxin is a major threat for humans and animals, in contrast to iso-DAs [59].

Thus, a series of DA isomers have been identified, while data on occurrence of DA and epidomoic acid (referred to as total DAs) and their toxicity have been used to establish the current regulation limits for DA of ≤ 20 mg·kg^{-1} [54]. The discovery that exposure to low levels of DA leads to long-lasting neurological effects in mammalian species [53] may suggest the need to reduce its permissible level in tissues of consumed finfish and shellfish [57].

3.2. Synthesis

Synthesis of the domoic acid **3** and correction of the earlier proposed geometry of the double bonds in its side chain have been carried out [60]. The bicyclic core was constructed from the compound **62**, obtained by multistep synthesis from **1** and diene **63** using the thermic Diels-Alder reaction to give bicyclic adduct **64** with *cis*-fused ring junction. This adduct was ozonized, treated with diazomethane and 2-methyl-2-ethyl-1,3-dioxolane in the presence of p-TsOH and yielded 1,3-dioxolane **65**. The employment of borane-dimethylsulfide complex reduced the carbonyl group. Deprotection of the silyl ester and subsequent oxidation of intermediate diol with pyridinium dichromate (PDC) provided the diester **66**. Selective removal of ketal group by 60% AcOH gave the aldehyde **67**. The Wittig reaction led to **68**, the subsequent reaction with PhSeCl to the selenium-containing aldehyde **69**. The key conversions of this synthesis consisted in the obtaining of *trans*- and *cis*-enal systems from **69**. For example, the treatment with *N*-bromosuccinimide (NBS) in tetrahydrofuran and then with NaOAc led to *cis*-enal **70** as a major product. Transformation of this particular compound, but not its trans-isomer by the Wittig reaction followed by the Jones oxidation gave the product **71**. The subsequent removal of protective groups converted it into domoic acid **3**, thus confirming the *E*,*Z*,*R*-configurations in the side chain of this toxin (Scheme 8).

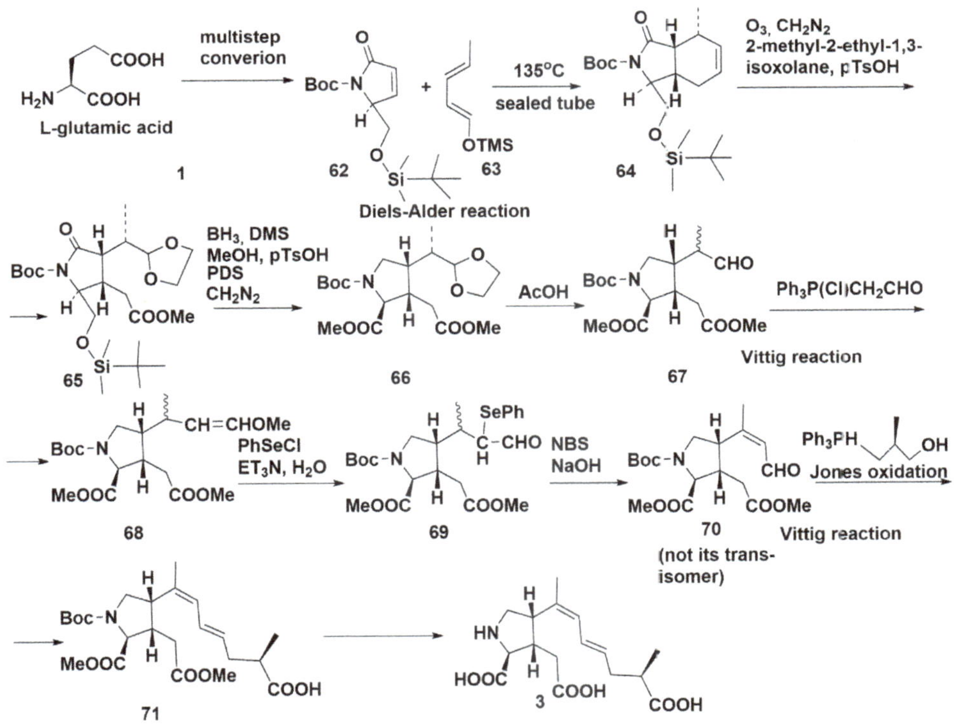

Scheme 8. Synthesis of domoic acid.

3.3. Biosynthesis, Producers, Biological Action, and Environmental Role

Several attempts were made to study the process of biosynthesis of this toxin in order to understand how and why environmental factors trigger the biosynthesis leading to high levels of DA accumulated in diatoms. Using feeding experiments, Savage et al. [61] established that DA arises as a result of condensation of geranyl diphosphate (**72**) with glutamate. Labeled by a stable isotope (deuterium), geranyl diphosphate was incorporated from culture medium into DA that was confirmed by gas chromatography–mass spectrometry (GC-MS). It was suggested that the condensation of geranyl diphosphate with amino group glutamic acid occurred via nucleophilic substitution of the diphosphate group with the amino group of glutamic acid (**1**).

A recent publication in Science [62] provided new important data concerning the DA biosynthesis. It was shown that these genetically programmed biosynthetic pathways are encoded by a four-gene cluster which is involved in this biosynthesis and exists in genomes of toxic microalgae belonging to the genus *Pseudo-nitzschia*. In the conditions of phosphate limitation and elevated CO_2 level, probably characteristic of the final stage of microalgal bloom, *Pseudo-nitzschia* spp. express a series of enzymes involved in construction of the pyrrolidine skeleton system of DA. The up-regulation of CYP450 gene, encoding an enzyme which is responsible for the formation of 7′-carboxylic acid from geranyl diphosphate **72** at its reaction with glutamic acid (**1**), was revealed. It was shown that this gene is clustered with several other genes of DA biosynthesis (the so-called dab genes). The DA biosynthetic gene cluster was established to consist of terpene cyclase (dabA), hypothetical protein Hypo (dabB), dioxygenase (dabC), CYP450 (dabD), and probably other genes. The expression of recombinant dabA without *N*-terminal transit peptide in *E. coli* and its use as a catalyst in the reaction between **72** and glutamic acid **1** confirmed this enzyme to catalyze *N*-geranylation of L-glutamic acid in Mg^{2+}-dependent manner. This is the first and key stage in the DA biosynthesis, which yields *N*-geranyl-L-glutamic acid **61**, as was earlier suggested by Savage et al. [61]. The use of recombinant dabC in the presence of Fe^{3+}, oxygen, and L-ascorbic acid, and α-ketoglutarate-dependent dioxygenase, or both dabC and dabD in one put, gave one of so-called dainic acids (**56**) with a non-oxidized geranyl side chain. When the primary product of this biosynthesis **61** was incubated with *Saccharamyses cerevisae* microsomes, which expressed the transmembrane dabD, *N*-geranyl-L-glutamic acid in the presence of Fe^{3+} and oxygen was converted into small, but reproductible amounts of 7′-carboxy-*N*-geranyl-L-glutamic acid (**73**) and its 7′-hydroxy analog. Both products were identified by being compared with synthetic fractions. The product **73**, in the same conditions as were used for the transformation of **61** into **56**, was converted into an isomer of domoic acid **48**. However, no isomerase activity necessary for the conversion of **48** into the end product of this biosynthesis was found (Scheme 9). All the above results show that this biosynthesis pathway begins with the dab A-catalyzed geranylation of L-glutamic acid in chloroplasts. Therefore, the presence of dab genes is a character that allows the recognition of toxic strains of this genus. It can help identify environmental conditions, which may potentially cause neurotoxicity in microalgae and in their consumers such as mollusks and diatom-feeding invertebrates.

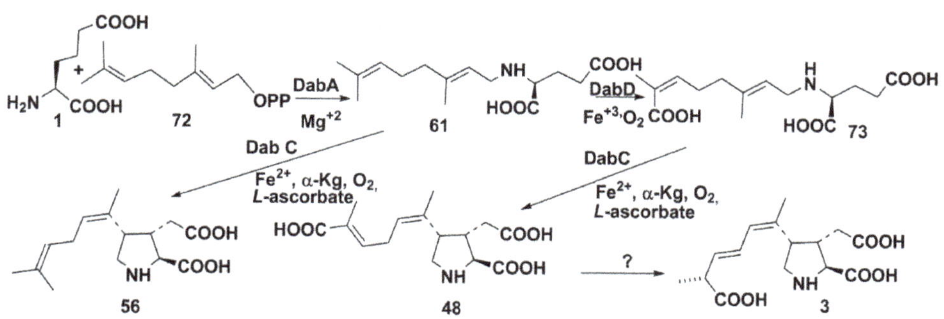

Scheme 9. Biosynthesis of domoic acid.

Thus, it has been confirmed that DA, as a potent excitatory amino acid, is produced primarily by *Pseudo-nitzschia* and *Nitzschia* diatoms distributed across the world and is naturally accumulated in filter-feeding marine organisms. Although many *Pseudo-nitzschia* species have the potential ability to produce this neurotoxin [49], *P. multiseries*, *P. seriata*, and *P. australis* were established as the most toxic species [57].

There have been numerous bloom events caused by highly toxic DA-producing microalgae along the USA and Canadian coasts over the past 15 years. In many cases, it led to mass poisoning of marine mammals and birds [48]. For instance, in the spring of 2015, an unusually intense and long-lasting toxigenic bloom of *Pseudo-nitzschia* microalgae associated with an abnormal warming of water, was observed in a vast area from California to Alaska. Besides the mass poisoning of marine animals, it caused significant economic losses due to the closure of farms that cultivated bivalves and crabs [63]. Thus, DA exposure has become more widespread due to the higher intensity of toxigenic *Pseudo-nitzschia* blooms and related consumption of DA-contaminated mollusks.

A suggestion can be made that the isomerase activity necessary for the transformation of less active biosynthetic precursors into DA may be present not only in these microalgae but also in consumers, or in symbionts and/or epiphytic microorganisms. It is possible that bacteria play an important role in accumulation of DA in microalgae, but the details of this are still elusive [47]. It was established that axenic diatom cultures produce less DA than xenic ones and that bacteria can enhance DA production [64]. It was shown, for example, that after the addition of the gamma-proteobacterium *Alteromonas macleodii*, isolated from the Russian clone of *P. multiseries*, to an axenic culture of another clone of this species, the amount of DA produced by the latter significantly increased [65]. The stimulating effect of bacteria on the DA production has also been found for another known DA-producer, *Nitzschia* sp. [66]. However, mechanisms of bacteria's influence on intracellular DA levels in microalgae remain insufficiently studied [49].

There are a few other hypotheses that discuss environmental factors influencing toxicity of diatoms. It was suggested that DA production may arise as a defense against grazing. However, the effect of predatory copepods on toxicity of microalgae has not been definitively confirmed [67]. The hypothesis that DA may be involved in a high-affinity iron uptake system seems likely [68]. Trick et al. [69] demonstrated that the addition of iron stimulated the growth of toxigenic *Pseudo-nitzschia* spp., providing a competitive advantage over other phytoplankton. Each of these hypotheses requires further validation.

4. Dysiherbaine and Related Compounds

4.1. Discovery and Structures

In 1997, Japanese scientists described a new potent marine excitatory amino acid **74** [70], isolated through homogenization of the sponge *Dyidea herbacea* in water, centrifugation of the extract, and precipitation of high-molecular-weight compounds with 2-propanol. Further purification of water-soluble materials using several types of column chromatography yielded this toxin known as disyherbaine (Figure 5). Its structure was identified as a novel diamino dicarboxylic acid using NMR, FABMS, ISIMS, and other methods of molecular structure analysis. Structurally, the acid contains the bicyclic core consisting of *cis*-fused tetrahydropyran and tetrahydrofuran rings with a glutamic acid fragment attached to the tetrahydrofuran moiety. Dysiherbaine core contains four contiguous stereogenic centers with an additional quaternary stereocenter in the tetrahydrofuran moiety at the site of amino acid section attachment. This excitatory acid stimulates binding of kainic acid and 1-amino-3-hydroxy-5-methyl-4-isooxazolepropionic acid, but not *N*-methyl-D-aspartic acid to the rat brain synaptic membranes, suggesting that **74** is an agonist of KARs in CNS.

Radioligand binding assay showed that dysiherbaine binds kainic acid receptors at significantly lesser concentrations than other agonists of KARs. Intraperitoneal injection into mice (20 μg/kg) caused the same behavior as that observed after injection of domoic acid. In general, dysiherbaine exhibits the most potent epileptogenic activity among so far known amino acids. ED_{50} for substitution of kainic acid, labeled by tritium in kainate receptors by dysiherbaine is 0.00074 μM. This excitatory amino acid is useful for evaluating the physiological roles of KARs in the central nervous system [71].

The above-mentioned findings explained the increased attention to synthesis **74** and related compounds. Another excitatory amino acid, neodysiherbaine A **75** (Figure 5), was isolated by a bioassay-guided chromatographic procedure as a minor constituent of the aqueous extract from the same sponge species. Its structure was deduced by NMR and MS methods and unambiguously confirmed by the total synthesis. After HPLC purification, Japanese scientists obtained as small amount of **75** as 0.26 mg [72]. Neodysiherbaine A is also potent agonist of KARs which induces characteristic epilepsy-like seizures in mice. Like dysiherbaine, it activates neuronal glutamate receptors with a considerable preference toward kainic acid receptors. However, neodysiherbaine is less active in comparison with dysiherbaine: the same activity associated with crowding out kainic acid from kainate receptors is induced by 0.052 μM of neodysiherbaine [72].

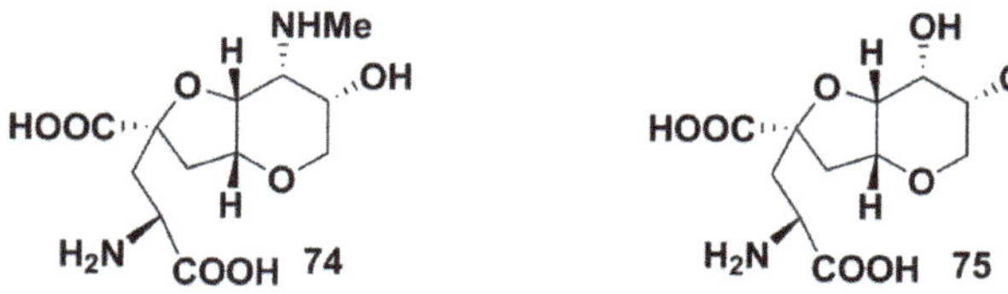

Figure 5. Structures of dysiherbaine (**74**) and neodysiherbaine (**75**).

After the discovery of dysiherbaine, it was established that various fragments of its structure play different roles in interaction with receptors. The glutamate section of this molecule is responsible for binding to a receptor, while structural and stereochemical features in the tetrahydropyran ring determine the profile of its activity, in particular, agonistic or antagonistic properties [70,71].

4.2. Some Recent Approaches to Syntheses

These compounds represent a new structural group of amino acids containing a unique hexahydrofuro[3,3–b]pyrane core system. It was shown that **74** and **75**, as well as many their synthetic analogs, exhibit preferential binding to GluK1 kainate receptors and almost no activity towards AMPA receptors. In total more 35 syntheses of dysiherbaine and neodysiherbaine were reported. Most of them were reviewed in 2013 by Cachet and Poree [73]. In our review, we will only consider several additional syntheses which were developed later and published in 2013–2018.

Taking into account that several syntheses of dysiherbaine, reported earlier, were based on a known intermediate, obtained through 13 to 16 steps, a formal synthesis of **74** was carried out [74]. In this synthesis, D-mannitol diacetonide, an available initial compound, was converted into the same compound by a shorter synthetic pathway.

Gilbeston and co-authors [75] from the University of Houston, USA, proposed a new variant of enantioselective synthesis of **74** which differed in a later design of functionalities in the tetrahydropyran ring than in other known pathways. The scheme of this synthesis suggested, first, the creation of tetrahydrofuran derivative and then the formation of tetrahydropyran ring with the use of metathesis reaction. The following introduction of the necessary substituents and, finally, the attachment of glutamic acid moiety to the furan fragment completed the synthesis. The furan fragment of **74** was constructed from methylglycidate **76**. This compound was transformed by a series of reactions into α,β-unsaturated epoxide **77**. The palladium catalyzed opening of this epoxide yielded the furan derivative **78**. Then **78** was converted by allylation of free alcohol with allyl bromide and silver oxide to provide the diene metathesis substrate **79**, in which oxymethyl group was introduced by reaction with LDA and formaldehyde. The derivative **79** was obtained as a mixture of stereoisomers and one of them, suitable for further transformations, was isolated by column chromatography. Metathesis of **79** with protection of hydroxy group gave **80**, containing the bicyclic core system of dysiherbaine. The oxidation of the double bond in **80** with OSO_4 and *N*-oxide of methylmorpholine (MMO) led to diol **81**. On the following stages of introduction of mesyl and azide groups in the pyrane cycle was carried out to obtain **82**. The treatment of **82** with PPh3 and *N,N*-diisopropylethylamine (DIPEA) successfully provided aziridine **83**, protected with Boc. The consequent reaction with copper (II) triflate, methylation, and deprotection of oxymethyl group provided the rearranged production of **84**. Further steps were the transformation into aldehyde **85**, introduction of glutamic acid residue to obtain **86**. Asymmetric hydrogenation with rhodium DuPhos (DuPhos is a class of organophosphorus compounds that are used as ligands for asymmetric syntheses), followed by deprotection of amino group and alkaline hydrolysis gave the target compound **74** (Scheme 10).

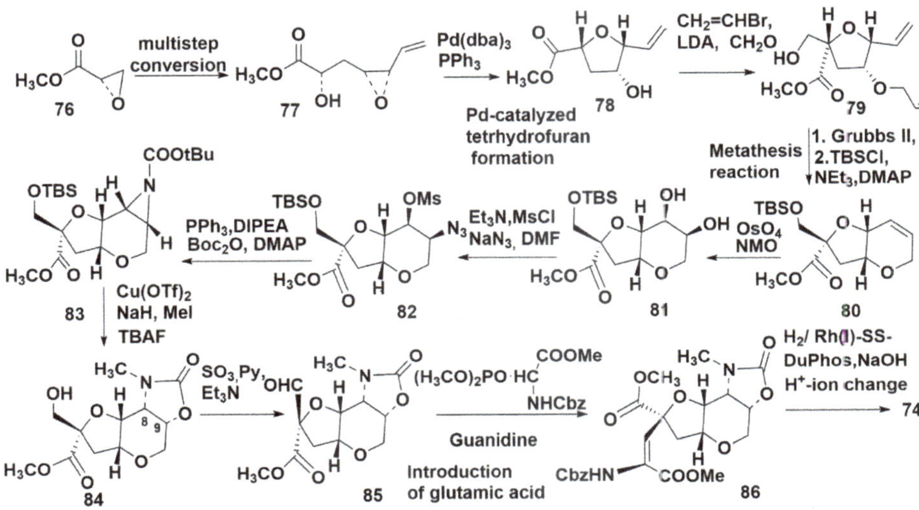

Scheme 10. Enanthioselective synthesis of (–)-dysiherbaine.

A total synthesis of neodysiherbaine A (**75**) based on 1,3-dipolar cycloaddition of a chiral nitrone to sugar-derived product, containing tetrahydropyran moiety, has also been reported recently [76]. The pyran allylic alcohol was obtained from methyl-α-*D*-mannopyranoside (**87**) through conversion into derivative **88** by the previously known multistep procedure, and, after the Swern oxidation and Wittig reaction, gave **89**. The further hydrogenation and Mannich reaction with Eschenmoser's salt yielded aldehyde, which, after DIBAL-H reduction, gave the compound **90**. The bipolar addition of a nitrone, catalyzed by etherate of $MgBr_2$, led to **91**. This cycloaddition constructed the C2 and C4 asymmetric centers in a single step. Further transformations resulting in reductive cleavage of O-benzyl, *N*–O, and *N*-benzyl bond, which occurred simultaneously, led to intermediate **92** which was converted by multistep transformations into another key product (**93**). The intramolecular SN2 reaction in **93** led to the compound **94** containing the required configurations of all six stereogenic centers of neodysiherbaine A. The multistep substitution and removal of protective groups and formation of tricyclic intermediate **95**, followed by hydrolysis with 6M HCl, completed this synthesis (Scheme 11).

4.3. Probable Origin in Sponges

The cellular origin of **74** and **75** has remained a mystery for a long time, since excitatory acids are usually present in plants or microorganisms, but not in animals such as sponges. In 2008, using dysiherbaine antibodies, Japanese scientists from Kitasato University [77] have found dysiherbaine-to show immunoreactivity in spherical cells harbored in mesohyl of the sponge *Lendenfeldia chondrodes*. These spherical cells were identified as *Synechocystis* cyanobacteria by a combination of ribosomal RNA gene sequencing and cell morphology analysis. Therefore, these excitatory amino acids are probably formed in symbiotic cyanobacteria inhabiting some sponges.

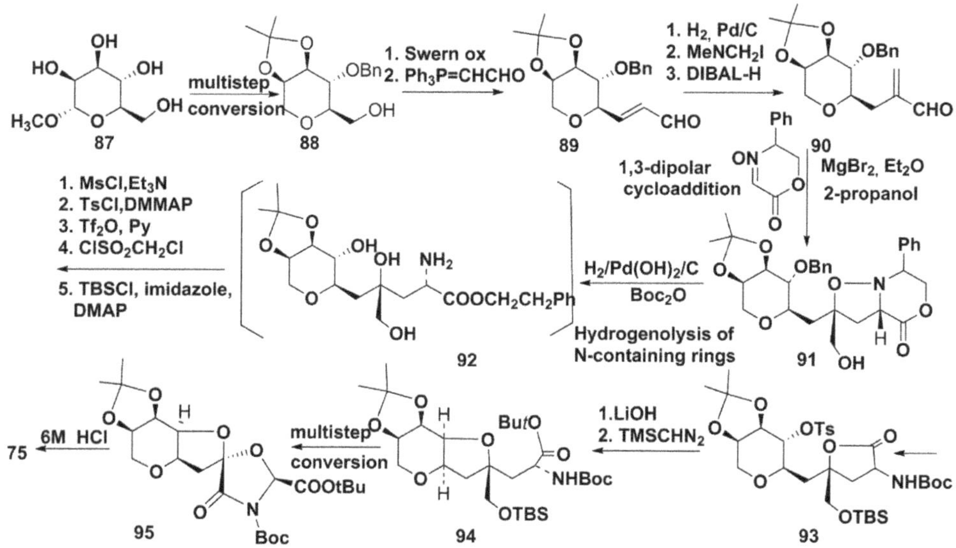

Scheme 11. Synthesis of neodysiherbaine A from methyl-α-*D*-mannopyranoside.

5. Conclusions

Excitatory amino acids, found in marine organisms a few dozen years ago, continue to attract increasing attention by their unusual properties. Some of them, such as domoic acid and its analogues, are potent excitotoxicants harmful for humans and different marine mammals and birds. These substances may cause significant damage to both natural ecosystems and aquaculture farms. Kainic acid and some similar compounds, being specific agonists of kainate receptors, are used in experimental pharmacology to model epilepsy and other neurodegenerative diseases in animals. Application of these compounds have become an important approach to understanding of the biological roles, classification and modes of action of synaptic receptors. Other excitatory amino acids isolated from some sponges exert even more potent action on neurons of animals and humans than kainic and domoic acids do. The promising pharmacological properties of marine excitatory amino acids were an incentive for developing numerous syntheses of these natural products and their analogs.

The most interesting results of very recent research concerns the biosynthesis of kainic acid and chemico-enzymatic procedure, providing a more efficient obtaining of this compound than chemical syntheses. Another important discovery is the decoding of biosynthesis processes leading to domoic acid and isolation of a number of new natural derivatives of this excitant, so-called dainic acids. In the recent decade, the chiral organic synthesis allowed obtaining of kainic acid by short and effective pathways using different synthetic strategies. New hypotheses have been proposed that explain the biological origin of dysiherbaine and neodysiherbaine A in sponges from symbiotic cyanobacteria and the biological significance of kainic acid in algae producing this excitatory amino acid for their defense against grazers.

Author Contributions: V.A.S. analyzed chemical data and schemes of syntheses of excitatory amino acids, and wrote chemical parts of the review. I.V.S. analyzed information concerning biological aspects, including properties and biological roles of compounds, and wrote other parts of the review. All authors have read and agreed to the published version of the manuscript.

Funding: This research received no external funding beside Governmental.

Acknowledgments: Authors thank Natalya Ivanchina (PIBOC, FEB RAS) for critical reading of the manuscript and useful comments.

Conflicts of Interest: The authors declare no conflict of interest.

References

1. Caroll, A.R.; Copp, B.R.; Davis, R.A.; Keyzers, R.A.; Prinsep, M.R. Marine natural products. *Nat. Prod. Rep.* **2019**, *36*, 122–173. [CrossRef] [PubMed]
2. Shimizu, Y.; Gupta, S.; Masuda, K.; Moranda, L.; Walker, C.R.; Wang, R. Dinoflagellate and other microalgal toxins; chemistry and biochemistry. *Pure Appl. Chem.* **1989**, *61*, 513–516. [CrossRef]
3. Visciano, P.; Schirone, M.; Berti, M.M.; Milandri, A.; Tofalo, R.; Suzzi, G. Marine biotoxins: Occurrence, toxicity, regulatory limits and reference methods. *Front. Microbiol.* **2016**, *7*, 1051. [CrossRef] [PubMed]
4. Stonik, V.A.; Stonik, I.V. Toxins produced by marine microorganisms: A short review. In *Marine and Freshwater Toxins*; Gopalakrishnakone, P., Haddad, J.V., Tubaro, A., Kim, E., Kem, W., Eds.; Springer: Dordrecht, The Netherlands, 2016; pp. 3–21. [CrossRef]
5. Stonik, I.V.; Orlova, T.Y. Domoic acid-producing diatoms of the genus *Pseudo-nitzschia* H. Peragallo, 1900 (Bacillariophyta) from the North Pacific. *Russ. J. Mar. Biol.* **2018**, *44*, 347–354. [CrossRef]
6. Moloney, M.G. Excitatory amino acids. *Nat. Prod. Rep.* **1998**, *15*, 205–219. [CrossRef]
7. Pei-Gen, X.; Shan-Lin, F. Traditional antiparasitic drugs in China. *Parasitol. Today* **1986**, *2*, 353–355. [CrossRef]
8. Murakami, S.; Takemoto, T.; Shimizu, Z. Studies on the effective principle of *Digenea simplex* Aq.1, separation of the effective fraction by liquid chromatography. *J. Pharm. Soc. Jpn.* **1953**, *73*, 1026–1028. [CrossRef]
9. Watase, H.; Nitta, I. The structure of kainic acid, the most active component of *Digenea simplex* Aq. I. The molecular and crystal structure of zinc kainite dihydrate. *Bull. Chem. Soc. Jpn.* **1957**, *30*, 889–900. [CrossRef]
10. Nitta, I.; Watase, H.; Tomiie, Y. Structure of kainic acid and its isomer, allokainic acid. *Nature* **1958**, *181*, 761–762. [CrossRef]
11. Impellizezeri, G.; Mangiafico, S.; Oriente, G.; Piatelli, M.; Scuit, S.; Fatorusso, E.; Magno, S.; Santacroce, C.; Sica, D. Amino acid and low-molecular-weight carbohydrates of some marine red algae. *Phytochemistry* **1975**, *7*, 1549–1557. [CrossRef]
12. Calaf, R.; Barlatier, A.; Garson, D.; Balansard, G.; Pellegrini, M.; Reynaud, J. Isolation of an unknown kainic peptide from the red alga *Alsidium helminthocorton*. *J. Appl. Phycol.* **1989**, *1*, 257–266. [CrossRef]
13. Laycook, M.V.; de Freitas, A.S.W.; Wright, J.L.C. Glutamate agonists from marine algae. *J. Appl. Phycol.* **1989**, *1*, 113–122. [CrossRef]
14. Shinozaki, H.; Konishi, S. Action of several anthelmintics and insecticides on rat cortical neurons. *Brain Res.* **1970**, *24*, 368–371. [CrossRef]
15. Ferkany, J.W.; Zaczek, R.; Colyle, J.T. Kainic acid stimulates excitatory amino acid neurotransmitter release at presynaptic receptors. *Nature* **1982**, *298*, 757–759. [CrossRef]
16. Garthwaite, J.; Garthwaite, G. The mechanism of kainic acid neurotoxicity. *Nature* **1983**, *305*, 138–140. [CrossRef]
17. Tremblay, J.-F. Producers strive to bring kainic acid back on the market. *Chem. Eng. News* **2000**, *78*, 31. [CrossRef]
18. Evans, P.A.; Inglesby, P.A. Diastereoselective rhodium-ctalyzed ene-cycloisomerization reactions of alkenylidenecyclopropanes: Total synthesis of (-)-alpha-kainic acid. *J. Am. Chem. Soc.* **2012**, *134*, 3635–3638. [CrossRef]
19. Stathakis, C.I.; Yioti, E.; Gallos, J.K. Total syntheses of (-)-α-kainic acid. *Eur. J. Org. Chem.* **2012**, *25*, 4661–4673. [CrossRef]
20. Orellana, A.; Pandey, S.K.; Carret, S.; Greene, A.E.; Poison, J.-F. A Diels-Alder-based total synthesis of (-) kainic acid. *J. Org. Chem.* **2012**, *77*, 5286–5296. [CrossRef]
21. Reddy, N.K.; Chandrasekhar, S. Total synthesis of (-)-α-kainic acid via chirality transfer through Ireland-Claisen rearrangement. *J. Org. Chem.* **2013**, *78*, 3355–3360. [CrossRef]
22. Fujii, M.; Yokoshima, S.; Fukuyama, T. A unified strategy for kainoid syntheses. *Eur. J. Org. Chem.* **2014**, *22*, 4823–4836. [CrossRef]
23. Oe, K.; Ohfune, Y.; Shinada, T. Short total synthesis of (-)-kainic acid. *Org. Lett.* **2014**, *16*, 2550–2553. [CrossRef] [PubMed]
24. Zhang, M.; Watanabe, K.; Tsukamoto, M.; Shibuya, R.; Morimoto, H.; Ohshima, T. A short scalable route to (-)-α-kainic acid using Pt-catalyzed direct allylic amination. *Chem. Eur. J.* **2015**, *21*, 3937–3941. [CrossRef] [PubMed]

25. Suzuki, J.; Miyano, M.; Yashiro, S.; Umezawa, T.; Matsuda, F. Total synthesis of (-)-kainic acid and (+)-*allo*-kainic acid through SmI_2-mediated intramolecular coupling between allyl chloride and α,β-unsaturated ester. *Org. Biomol. Chem.* **2017**, *15*, 6557–6566. [CrossRef] [PubMed]
26. Hong, W.S.; Li, J.L.; Lui, Y.M.; Du, Z.L.; Huang, Z.Y.; Zhao, N.; Li, N.; Yang, J. Formal total synthesis of (-)-kainic acid. *Tetrahedron* **2016**, *72*, 5502–5506. [CrossRef]
27. Chekan, J.R.; McKinnie, S.M.K.; Moore, M.L.; Poplawski, S.G.; Michael, T.P.; Moore, B.S. Scalable biosynthesis of the seaweed neurochemical, kainic acid. *Angew. Chem. Int. Ed. Engl.* **2019**, *58*, 8454–8457. [CrossRef]
28. Baldwin, J.E.; Moloney, M.G.; Parsons, A.F. Enantioselective synthesis of kainoid analysis by cobalt-mediated cyclization of an amino acid derivative. *Tetrahedron* **1990**, *46*, 7263–7282. [CrossRef]
29. Wang, Q.; Yu, S.; Simonyi, A.; Sun, G.Y.; Sun, A.Y. Kainic acid-mediated excitotoxicity as a model for neurodegeneration. *Mol. Neurobiol.* **2005**, *31*, 3–16. [CrossRef]
30. Zheng, X.Y.; Zhang, H.; Liqang, L.; Luo, Q.; Zhu, J. Kainic acid-induced neurodegenerative model: Potentials and limitations. *Biomed. Res. Int.* **2011**, *2011*, 1–10. [CrossRef]
31. Bertoglio, D.; Amhaoul, H.; Van Eetveldt, A.; Houbrechts, R.; Van De Viver, S.; Ali, I.; Dedeurwaerdere, S. Kainic-acid-induced post-status epilepticus models of temporal lobe epilepsy with diverging seizure phenotype and neuropathology. *Front. Neurobiol.* **2017**, *8*, 588. [CrossRef]
32. Møllerud, S.; Frydenvang, K.; Pickering, D.S.; Kastrup, J.S. Lessons from the crystal structures of kainoid receptors. *Neuropharmacol.* **2017**, *112*, 16–28. [CrossRef]
33. Sakai, R.; Minato, S.; Koike, K.; Koike, K.; Jimbo, M.; Kamiya, H. Cellular and subcellular localization of kainic acid in the marine red alga *Digenea simplex*. *Cell Tissue Res.* **2005**, *322*, 491–502. [CrossRef] [PubMed]
34. De Bortoli, S.; Teardo, E.; Scabo, I.; Morosinotto, T.; Alboresi, A. Evolutionary insight into the ionotropic glutamate receptor superfamily of photosynthetic organisms. *Biophys. Chem.* **2016**, *218*, 14–26. [CrossRef] [PubMed]
35. Wudick, M.M.; Michard, E.; Nunes, C.O.; Feijo, J.A. Comparing plant and animal glutamate receptors: Common traits but different fates? *J. Exp. Bot.* **2018**, *69*, 4151–4163. [CrossRef] [PubMed]
36. Blaskovich, M.A.T. Unusual amino acids in medicinal chemistry. *J. Med. Chem.* **2016**, *59*, 10807–10836. [CrossRef] [PubMed]
37. Takemoto, T.; Daigo, K.; Kondo, Y.; Kondo, K. Studies on the constituents of *Chondria armata*. VIII. On the structure of domoic acid. *Yakugakku Zasshi* **1966**, *86*, 874–877. [CrossRef]
38. Meda, M.; Kodama, T.; Tanaka, T.; Yoshizumi, H.; Takemoto, T.; Nomoto, K.; Fujita, T. Structures of isodomoic acids A and B, novel insecticidal amino acids from the red alga Chondrus armata. *Chem. Parm. Bull.* **1986**, *34*, 4892–4895. [CrossRef]
39. Wright, J.L.C.; Falk, M.; McInnes, A.G.; Walter, J.A. Identification of isodomoic acid D and two new geometrical isomers of domoic acid in toxic mussels. *Can. J. Chem.* **1990**, *68*, 22–25. [CrossRef]
40. Zaman, L.; Arakawa, O.; Shimosu, A.; Onoue, Y.; Nishio, S.; Shida, Y.; Noguchi, T. Two new isomers of domoic acid from a red alga, Chondria armata. *Toxicon* **1997**, *35*, 205–212. [CrossRef]
41. Clayden, J.; Read, B.; Hebditch, K.R. Chemistry of domoic acid, isodomoic sacid, and their analogues. *Tetrahedron* **2005**, *61*, 5713–5724. [CrossRef]
42. Stonik, V.; Stonik, I. Low-molecular-weight metabolites from diatoms: Structures, biological roles, and biosynthesis. *Mar. Drugs* **2015**, *13*, 3672–3709. [CrossRef] [PubMed]
43. Maeno, Y.; Kotaki, Y.; Terada, R.; Cho, Y.; Konoki, K.; Yotsu-Yamashita, M. Six domoic acids related compounds from the red alga, *Chondria armata*, and domoic acid biosynthesis by the diatom, *Pseudo-nitzchia multiseries*. *Sci. Rep.* **2018**, *8*, 356. [CrossRef]
44. Wright, J.L.C.; Boyd, R.K.; de Freitas, A.S.W.; Falk, M.; Foxall, R.A.; Jamieson, W.; Pathak, V.P. Identification of domoic acid, a neuroexcitatory amino acids, in toxic mussels from eastern Prince Edward Island. *Can. J. Chem.* **1989**, *67*, 481–490. [CrossRef]
45. Bates, S.; Bird, C.; de Freitas, A.; Foxall, R.; Gilgan, M.; Hanic, L.; Johnson, G.; McCulloch, A.; Odense, P.; Pocklington, R.; et al. Pennate diatom *Nitzschia pungens* as the primary source of domoic acid, a toxin in shellfish from eastern Prince Edward Island, Canada. *Can. J. Fish. Aquat. Sci.* **1989**, *46*, 1203–1215. [CrossRef]
46. Lefebvre, K.; Bargu, S.; Kieckhefer, T.; Silver, M. From sanddabs to blue whales: The pervasiveness of domoic acid. *Toxicon* **2002**, *40*, 971–977. [CrossRef]
47. Lelong, A.; Hegaret, H.; Soudant, P.; Bates, S. *Pseudo-nitzschia* (Bacillariophyceae) species, domoic acid and amnesic shellfish poisoning: Revisiting previous paradigms. *Phycologia* **2012**, *51*, 168–216. [CrossRef]

48. Trainer, V.; Bates, S.; Lundholm, N.; Thessen, A.; Cochlan, W.; Adams, N.; Trick, C. *Pseudo-nitzschia* physiological ecology, phylogeny, toxicity, monitoring and impacts on ecosystem health. *Harmful Algae* **2012**, *14*, 271–300. [CrossRef]
49. Bates, S.; Hubbard, K.; Lundholm, N.; Montresor, M.; Leaw, C. *Pseudo-nitzschia*, *Nitzschia*, and domoic acid: New research since 2011. *Harmful Algae* **2018**, *79*, 3–43. [CrossRef]
50. Kumar, K.P.; Kumar, S.P.; Nair, G.A. Risk assessment of the amnesic shellfish poison, domoic acid, on animals and humans. *J. Environ. Biol.* **2009**, *30*, 319–325. [PubMed]
51. Gulland, F.; Haulena, M.; Fauquier, D.; Langlois, G.; Lander, M.; Zabka, T.; Duerr, R. Domoic acid toxicity in Californian sea lions (*Zalophus californianus*): Clinical signs, treatment and survival. *Vet. Rec.* **2002**, *150*, 475–480. [CrossRef]
52. Hong, Z.; Zhang, Y.; Zuo, Z.; Zhu, R.; Gao, Y. Influences of domoic acid exposure on cardiac development and the expression of cardiovascular relative genes in zebrafish (*Danio rerio*) embryos. *J. Biochem. Mol. Toxicol.* **2015**, *29*, 254–260. [CrossRef] [PubMed]
53. Lefebvre, K.A.; Frame, E.R.; Gulland, F.; Hansen, J.D.; Kendrick, P.S.; Beyer, R.P.; Bammler, T.K.; Farin, F.M.; Hiolski, E.M.; Smith, D.R.; et al. A novel antibody-based biomarker for chronic algal toxin exposure and sub-acute neurotoxicity. *PLoS ONE* **2012**, *7*, e36213. [CrossRef] [PubMed]
54. Vale, C. Domoic acid: Chemistry and pharmacology. In *Seafood and Freshwater Toxins: Pharmacology, Physiology and Detection*, 3rd ed.; Botana, L.M., Ed.; CRC Press: Boca Raton, FL, USA, 2014; pp. 875–894.
55. Pulido, O.M. Domoic acid toxicologic pathology: A review. *Mar. Drugs* **2008**, *6*, 180–219. [CrossRef] [PubMed]
56. Stewart, G.; Zorumski, C.; Price, M.; Olney, J. Domoic acid: A dementia-inducing excitotoxic food poison with kainic acid receptor specificity. *Exp. Neurol.* **1990**, *110*, 127–138. [CrossRef]
57. Zabaglo, K.; Chrapusta, E.; Bober, B.; Kaminski, A.; Adamski, M.; Bialczyk, J. Environmental roles and biological activity of domoic acid: A review. *Algal Res.* **2016**, *13*, 94–101. [CrossRef]
58. Holland, P.; Selwood, A.; Mounfort, D.; Wilkins, A.; McNabb, P.; Rhodes, L.; Doucette, G.; Mikulski, C.; King, K. Isodomoic acid C, an unusual amnesic shellfish poisoning toxin from *Pseudo-nitzschia australis*. *Chem. Res. Toxicol.* **2005**, *18*, 814–816. [CrossRef]
59. Munday, R.; Holland, P.; McNabb, P.; Selwood, A.; Rhodes, L. Comparative toxicity to mice of domoic acid and isodomoic acids A, B and C. *Toxicon* **2008**, *52*, 954–956. [CrossRef]
60. Ohfune, Y.; Tomita, M. Total synthesis of (-)-domoic acid. A revision of the original structure. *J. Am. Chem. Soc.* **1982**, *104*, 3511–3513. [CrossRef]
61. Savage, T.J.; Smith, G.J.; Clark, A.T.; Saucedo, P.N. Condensation of the isoprenoid and amino precursors in the biosynthesis of domoic acid. *Toxicon* **2012**, *59*, 25–33. [CrossRef]
62. Brunson, J.K.; McKinnie, S.M.K.; Chekan, J.R.; McGrow, J.P.; Miles, Z.D.; Bertrand, E.M.; Bielinsky, V.A.; Moore, B.S.; Luhavaya, H.; Obornik, M.; et al. Biosynthesis of the neurotoxin domoic acid in a bloom-forming diatom. *Science* **2018**, *361*, 1356–1358. [CrossRef]
63. McCabe, R.; Hickey, B.; Kudela, R.; Lefebvre, K.; Adams, N.; Bill, B.; Gulland, F.; Thomson, R.; Cochlan, W.; Trainer, V. An unprecedented coastwide toxic algal bloom linked to anomalous ocean conditions. *Geophys. Res. Lett.* **2016**, *43*, 10366–10376. [CrossRef] [PubMed]
64. Bates, S.; Douglas, D.; Doucette, G.; Leger, C. Enhancement of domoic acid production by reintroducing bacteria to axenic cultures of the diatom *Pseudo-nitzschia multiseries*. *Nat. Toxins* **1995**, *3*, 428–435. [CrossRef] [PubMed]
65. Orlova, T.; Stonik, I.; Aizdaicher, N.; Bates, S.; Leger, C.; Fehling, J. Toxicity, morphology and distribution of *Pseudo-nitzschia calliantha*, *P. multistriata* and *P. multiseries* (Bacillariophyta) from the northwestern Sea of Japan. *Bot. Mar.* **2008**, *51*, 297–306. [CrossRef]
66. Kotaki, Y.; Koike, K.; Yoshida, M.; Thuoc, C.; Huyen, N.; Hoi, N.; Fukuyo, Y.; Kodama, M. Domoic acid production in *Nitzschia* sp. (Bacillariophyceae) isolated from a shrimp-culture pond in Do Son, Vietnam. *J. Phycol.* **2000**, *36*, 1057–1060. [CrossRef]
67. Olson, M.; Lessard, E.; Wong, C.; Bernhardt, M. Copepod feeding selectivity on microplankton, including the toxigenic diatom, *Pseudo-nitzschia* spp., in the coastal Pacific Northwest. *Mar. Ecol. Prog. Ser.* **2006**, *326*, 207–220. [CrossRef]
68. Maldonado, M.; Hughes, M.; Rue, E.; Wells, M. The effect of Fe and Cu on growth and domoic acid production by *Pseudo-nitzschia multiseries* and Pseudo-nitzschia australis. *Limnol. Oceanogr.* **2002**, *47*, 515–526. [CrossRef]

69. Trick, C.; Bill, B.; Cochlan, W.; Wells, M.; Trainer, V.; Pickell, L. Iron enrichment stimulates toxic diatom production in high-nitrate, low-chlorophyll areas. *Proc. Natl. Acad. Sci. USA* **2010**, *107*, 5887–5892. [CrossRef]
70. Sakai, R.; Kamiya, H.; Murata, M.; Shimamoto, K. Dysiherbaine: A new neurotoxic amino acid from the Micronesian marine sponge Dysidea herbaceae. *J. Am. Chem. Soc.* **1997**, *119*, 4112–4116. [CrossRef]
71. Sakai, R.; Swanson, G.T.; Shimamoto, K.; Green, T.; Contractor, A.; Ghetti, A.; Tamura-Horikawa, Y.; Oiwa, C.; Kamiya, H. Pharmacological properties of the potent epileptogenic amino acid dysiherbaine, a novel glutamate receptor agonist isolated from the marine sponge Dysidea herbacea. *J. Pharm. Exp. Theurap.* **2001**, *296*, 650–658. [PubMed]
72. Sakai, R.; Koike, T.; Sasaki, M.; Shimamoto, K.; Oiwa, C.; Yano, A.; Suzuki, K.; Tachibana, K.; Kamiya, H. Isolation, structure determination, and syntheses of neodysiherbaine A, a new excitatory amino acid from a marine sponge. *Org. Lett.* **2001**, *3*, 1479–1482. [CrossRef]
73. Cachet, X.; Poree, F.-H. Total synthesis of dysiherbaine and neodysiherbaine A. *RSC Adv.* **2013**, *3*, 12466–12484. [CrossRef]
74. Rao, M.V.; Naresh, A.; Saketh, G.; Rao, B.V. Formal synthesis of dysiherbaine. *Tetr. Lett.* **2013**, *54*, 6931–6933. [CrossRef]
75. Do, H.; Kang, C.W.; Cho, J.H.C.; Gilbertson, S.R. Enantioselective synthesis of (-) -dysiherbaine. *Org. Lett.* **2015**, *17*, 3972–3974. [CrossRef] [PubMed]
76. Hirai, T.; Shibata, K.; Niwano, Y.; Shiozuki, M.; Hashimoto, Y.; Morita, N.; Ban, S.; Tamura, O. Total synthesis of neodysiherbaine A via 1,3-dipolar cycloaddition of a chiral nitrone template. *Org. Lett.* **2017**, *19*, 6320–6323. [CrossRef]
77. Sakai, R.; Yoshida, K.; Kimura, A.; Koike, K.; Jimbo, M.; Koike, K.; Kobiyama, A.; Kamiya, H. Cellular origin of dysiherbaine, an excitatory amino acid derived from a marine sponge. *ChemBioChe* **2008**, *9*, 543–551. [CrossRef]

Sample Availability: Not available.

molecules

MDPI

Article

Unusual Polyhydroxylated Steroids from the Starfish *Anthenoides laevigatus*, Collected off the Coastal Waters of Vietnam

Alla A. Kicha [1,*], Dinh T. Ha [2,3], Timofey V. Malyarenko [1,4], Anatoly I. Kalinovsky [1], Roman S. Popov [1], Olesya S. Malyarenko [1], Tran T. T. Thuy [3,*], Pham Q. Long [3], Nguyen T. T. Ha [5] and Natalia V. Ivanchina [1]

1 G.B. Elyakov Pacific Institute of Bioorganic Chemistry, Far Eastern Branch of Russian Academy of Sciences, Pr. 100-let Vladivostoku 159, 690022 Vladivostok, Russia; malyarenko-tv@mail.ru (T.V.M.); kaaniw@piboc.dvo.ru (A.I.K.); prs_90@mail.ru (R.S.P.); malyarenko.os@gmail.com (O.S.M.); ivanchina@piboc.dvo.ru (N.V.I.)

2 Institute of Natural Products Chemistry, Vietnam Academy of Science and Technology, 18 Hoang Quoc Viet, Cau Giay, Hanoi, Vietnam; dinhha.inpc@gmail.com

3 Graduate University of Science and Technology, Vietnam Academy of Science and Technology, 18 Hoang Quoc Viet, Cau Giay, Hanoi, Vietnam; mar.biochem@fpt.vn

4 School of Natural Sciences, Far Eastern Federal University, Sukhanova Str. 8, 690000 Vladivostok, Russia

5 Institute of Chemistry, Vietnam Academy of Science and Technology, 18 Hoang Quoc Viet, Cau Giay, Hanoi, Vietnam; thuha.vast@gmail.com

* Correspondence: kicha@piboc.dvo.ru or allakicha@mail.ru (A.A.K.); thuy.tran@inpc.vast.vn (T.T.T.T.); Tel.: +7-423-2312-360 (A.A.K.); Fax: +7-423-2314-050 (A.A.K.)

Academic Editor: Pavel S. Dmitrenok

Received: 13 February 2020; Accepted: 20 March 2020; Published: 23 March 2020

Abstract: Four new polyhydroxylated steroids **1–4** were isolated along with two previously known related steroids **5** and **6** from the methanolic extract of the starfish *Anthenoides laevigatus* collected off the coastal waters of Vietnam. Structures of new compounds were substantially elucidated by one-dimensional (1D) and two-dimensional (2D) NMR spectroscopy and HRESIMS techniques. Heptaol **1** and hexaol **2** contain the common 5α-cholestane skeleton, while hexaol **3** and heptaol **4** have the rare among starfish steroid compounds 5β-cholestane skeleton. Compounds **1**, **5**, and **6** do not show cytotoxic effects against normal JB6 Cl41 and cancer HT-29 and MDA-MB-231 cells, however they inhibit cell proliferation and colony formation of cancer HT-29 and MDA-MB-231 cells.

Keywords: polyhydroxylated steroids; NMR spectra; starfish; *Anthenoides laevigatus*; cytotoxicity; soft agar assay

1. Introduction

Polyhydroxylated steroids have been found in diverse marine species of ophiuroids, gorgonians, sponges, and other marine invertebrates. However, the class Asteroidea (also known as starfish or sea stars) is the richest source of these kind of steroids [1–7]. These substances generally contain from four to nine hydroxyl groups in the steroid nucleus and side chains and are characterized by a wide variety of chemical structures. Polyhydroxylated steroids are present in starfish in both free, sulfated, or glycosylated by one to three monosaccharide residues forms. These compounds usually occur in very complicated mixtures that are often difficult to separate into individual components by chromatographic methods. In addition to the original chemical structure, polyhydroxylated steroids attract attention with a wide diversity of biological effects including neuritogenic, neuroprotective, antiviral, anti-inflammatory, immunomodulatory, and other activities [7,8]. Recently, new knowledge about the cancer preventive activity and toxicity against different human tumor cell lines and molecular mechanisms of action of some starfish steroid compounds has been acquired [9]. Moreover, for the first time, it has been shown that starfish polyhydroxylated compounds demonstrated the radio sensitizing activity that was realized through apoptosis induction by the regulation of anti- and pro-apoptotic protein expression followed by activation of caspases and DNA degradation [10]. On the basis of the data obtained, we assume that the study of polyhydroxylated steroids from the starfish could lead to new promising results on their biological activities.

The search for new metabolites from starfish is a long-term direction of the studies of G.B. Elyakov Pacific Institute of Bioorganic Chemistry which celebrated its 55th anniversary in 2019. Continuing our research on biologically active steroid metabolites from the starfish species inhabiting the Vietnamese sea waters [11–15], herein, we describe the results of our investigation of steroid constituents from the methanolic extract of the starfish *Anthenoides laevigatus* Liao & A.M. Clark, 1989 (order Valvatida, family Goniasteridae), collected off the coastal waters of the Qui Nhon Province, Vietnam. We have isolated and structurally studied four new polyhydroxylated steroids **1–4**, along with two previously known related steroids **5** and **6**. Additionally, the action of compounds **1**, **5**, and **6** on cell viability and proliferation of normal and cancer cells, as well as colony formation of cancer cells in a soft agar clonogenic assay *in vitro*, have been investigated.

2. Results and Discussion

2.1. Structure Elucidation of Compounds 1–6

The concentrated methanol extract of *A. laevigatus* was subjected to sequential separation by column chromatography using Amberlite XAD-2 silica gel followed by reversed-phase HPLC on Discovery C18 and YMC-Pack Pro C18 columns to yield four new polyhydroxylated steroids (**1–4**), along with two known compounds **5** and **6** (Figure 1). The known polyhydroxylated steroids were identified by analysis of their ^{1}H-, ^{13}C-NMR, and ESIMS spectra and comparison with those reported earlier for (24*S*)-5α-cholestane-3β,5,6β,15α,24-pentaol (**5**) and related sulfated compounds previously isolated from the starfish *Luidia clathrata* and *Henricia leviuscula* [16,17], and (25*S*)-5α-cholestane-3β,5,6β,15α,16β,26-hexaol (**6**) from the starfish *L. clathrata* [18].

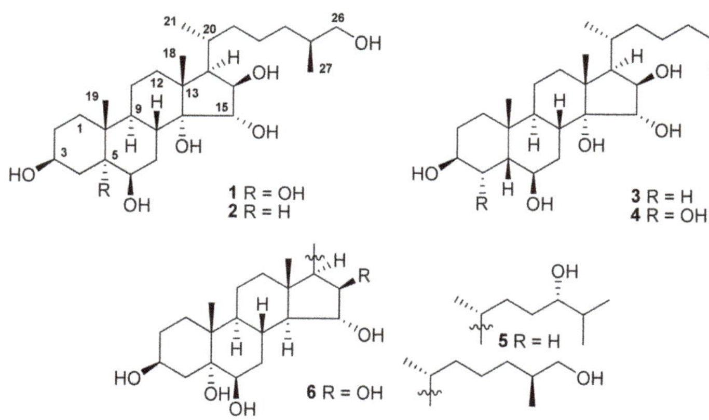

Figure 1. Structures of compounds **1–6** isolated from *A. laevigatus*.

The molecular formula of compound **1** was established to be of $C_{27}H_{48}O_7$ from the $[M + Na]^+$ sodium adduct ion peak at *m/z* 507.3292 in the (+)HRESIMS spectrum and the $[M - H]^-$ deprotonated ion peak at *m/z* 483.3321 in the (–)HRESIMS spectrum (Figure S1). The ^{1}H- and ^{13}C-NMR spectroscopic data of **1** displayed the proton and carbon chemical shifts of two angular methyl groups CH_3-18 (δ_H 1.00 s and δ_C 16.5) and CH_3-19 (δ_H 1.17 s and δ_C 17.3), one oxygenated methylene CH_2-26 (δ_H

3.42 dd (J = 10.5, 5.7), 3.30 m and δ_C 68.4), four oxygenated methines CH-3 (δ_H 4.00 m and δ_C 68.3), CH-6 (δ_H 3.49 t (J = 3.0) and δ_C 76.5), CH-15 (δ_H 3.86 d (J = 3.3) and δ_C 85.6), and CH-16 (δ_H 3.96 dd (J = 8.1, 3.3) and δ_C 82.1), and two oxygenated tertiary carbons C-5 (δ_C 76.6) and C-14 (δ_C 82.7) (Tables 1 and 2, Figures S2 and S3). These signals were similar to the corresponding resonances in the NMR spectra of co-occuring compound **6**, except for the proton and carbon resonances of C-14 and CH-15. Therefore, the doublet of doublets of H-15 at δ_H 3.73 (J = 10.4, 2.4) in the ^{1}H-NMR spectrum of **6** became doublet at δ_H 3.86 d (J = 3.3) in the ^{1}H-NMR spectrum of **1** suggesting the presence of an additional hydroxyl group at C-14 in the steroid nucleus of **1** as compared with **6**. In addition, the value of the $J_{16,17}$ = 8.1 Hz and the existence of the ROESY cross-peak H-16/H-17 indicated the α-orientation of the proton H-16 and, accordingly, 3β,5,6β,14,15α,16β,26-heptahydroxy substitution in **1**. An analysis of the COSY, HSQC, HMBC, and ROESY spectra ascertained all the proton and carbon signals in **1** (Tables 1 and 2, Figures S4). The COSY and HSQC experiments led to the assignment of the proton atom sequences at C-1 to C-4, C-6 to C-12 through C-11, C-15 to C-17, C-17 to C-21 through C-20, C-20 to the end of the side chain. The key HMBC correlations H-4/C-2; H-6/C-8, C-10; H-8/C-7, C-9, C-14; H-16/C-17; H_3-18/C-12, C-13, C-14, C-17; H_3-19/C-1, C-5, C-9, C-10; H_3-21/C-17, C-20, C-22; H_2-24/C-23, C-25, C-26; and H_3-27/C-24, C-25, C-26 confirmed the total structure of the molecule of **1** (Figure 2). The signal shape and coupling constants of protons H-3, H-6, H-15, and H-16 and the presence of the key ROESY cross-peaks Hα-4/H-6; H-8/H-15; H-16/H-17; H-17/H_3-21; H_3-18/H-8, H-15; and H_3-19/Hβ-2, Hβ-4, H-8 confirmed the 3β,6β,15α,16β relative configurations of the oxygenated carbons and common 5α-cholestane skeleton in **1** (Figure 2). The 20R-configuration was assigned based on the ROESY correlations of H_3-18/H-20 and H-16/H-22 and the downfield chemical shift of H_3-21 at δ_H 0.90 [19]. The absolute configuration of the asymmetric center C-25 was defined by examination of ^{1}H-NMR spectra of (R)- and (S)-MTPA derivatives obtained by reaction of **1** with S-(+)- and R-(−)-MTPA chlorides, respectively. The ^{1}H-NMR spectrum of 3,15,26-tri-(R)-MTPA ester of **1** showed H_2-26 signals as two close double doublets at δ_H 4.15 and 4.18, while that of 3,6,15,26-tetra-(S)-MTPA ester of **1** displayed two well-separated double doublets at δ_H 4.08 and 4.24 (Figure S8). These values were comparable with those of (R)- and (S)-MTPA derivatives obtained from other (25S)-26-hydroxy steroids [20]. Thus, the structure of compound **1** was determined to be the (25S)-5α-cholestane-3β,5,6β,14,15α,16β,26-heptaol.

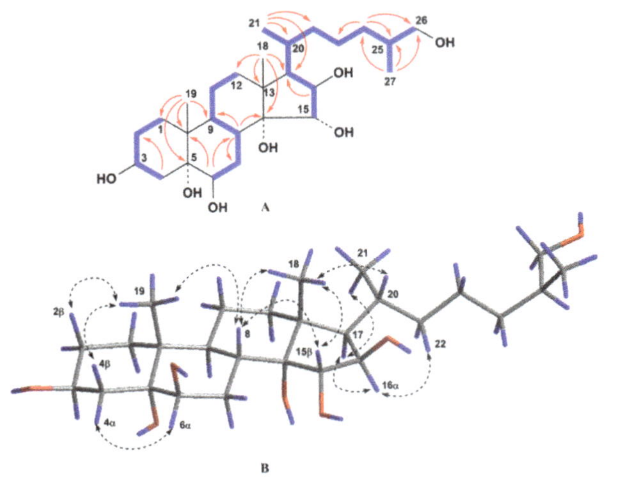

Figure 2. (**A**) COSY and key HMBC correlations for compound **1**; (**B**) Key ROESY correlations for compound **1**. Colors reveal the atoms of hydrogen (blue), oxygen (red) and carbon (grey) and their bonds.

Table 1. ^{1}H-NMR data of compounds **1–4** (CD_3OD, δ in ppm, J in Hz) [a].

Position	1	2	3	4
1	1.62, m	1.63, m	1.45, m	1.47, m
	1.32, m	0.98, m	1.39, m	
2	1.75, m	1.73, m	1.56, m	1.88, m
	1.47, m	1.40, m	1.44, m	1.32, m
3	4.00, m	3.53, m	3.99, br. q (2.7)	3.70, br. q (3.7)
4	2.04, dd (13.0, 11.3)	1.72, m	1.66, m	3.66, t (3.7)
	1.53, m	1.53, m	1.38, m	
5	-	1.09, dt (13.0, 2.6)	1.83, m	1.69, t (3.3)
6	3.49, t (3.0)	3.77, q (2.6)	3.66, q (2.7)	4.00, q (3.3)
7	1.69, ddd (14.2, 3.9, 2.6)	1.94, m	1.74, m	2.19, m
	2.03, m	1.50, m	1.66, m	1.74, m
8	2.22, td (12.4, 4.0)	2.22, td (12.4, 3.5)	2.25, td (11.7, 4.0)	2.21, m
9	1.91, m	1.28, m	1.98, m	2.11, m
10	-	-	-	-
11	1.33, m	1.44, m	1.35, m	1.37, m
12	1.81, m	1.78, m	1.80, m	1.77, m
		1.60, m	1.61, m	1.59, m
13	-	-	-	-
14	-	-	-	-
15	3.86, d (3.3)	3.89, d (3.3)	3.89, d (3.2)	3.87, d (3.2)
16	3.96, dd (8.1, 3.3)	3.96, dd (7.9, 3.3)	3.97, dd (8.0, 3.2)	3.96, dd (8.0, 3.2)
17	1.96, dd (11.3, 8.1)	1.96, t (10.9)	1.96, m	1.95, dd (11.0, 8.0)
18	1.00, s	1.01, s	1.00, s	1.00, s
19	1.17, s	1.04, s	1.13, s	1.10, s
20	1.80, m	1.80, m	1.80, m	1.79, m
21	0.90, d (6.7)	0.90, d (6.7)	0.90, d (6.8)	0.90, d (6.5)
22	1.58, m	1.58, m	1.58, m	1.58, m
	1.02, m	1.02, m	1.02, m	1.01, m
23	1.47, m	1.47, m	1.48, m	1.47, m
	1.22, m	1.21, m	1.22, m	1.21, m
24	1.41, m	1.41, m	1.42, m	1.41, m
	1.04, m	1.04, m	1.05, m	1.05, m
25	1.56, m	1.57, m	1.56, m	1.56 m
26	3.42, dd (10.5, 5.7)	3.42, dd (10.9, 6.0)	3.42, dd (10.7, 5.8)	3.41, dd (10.5, 6.0)
	3.30, m	3.30, m	3.30, m	3.30, m
27	0.90, d (6.6)	0.90, d (6.7)	0.90, d (6.8)	0.90, d (6.7)

[a] Assignments from 700 MHz COSY, HSQC, HMBC (8 Hz), and ROESY (250 ms) data.

The molecular formula of compound **2** was established to be of $C_{27}H_{48}O_6$ from the $[M + Na]^+$ sodium adduct ion peak at m/z 491.3337 in the (+)HRESIMS spectrum and the $[M - H]^-$ deprotonated ion peak at m/z 467.3379 in the (–)HRESIMS spectrum (Figure S9). Along with mass-spectra, the ^{1}H-, ^{13}C-, and DEPT NMR spectra revealed the presence of a hexahydroxy substitution in **2**. Similarity of the corresponding proton and carbon signals, as well as coupling constants in the NMR spectra of **2** and **1**, indicated that compound **2** has the same 14,15α,16β-trihydroxy substitution in the steroid C/D rings and 26-hydroxy cholestane side chain (Tables 1 and 2, Figures S10 and S11).

Table 2. ^{13}C-NMR data of compounds **1–4** (CD_3OD).

Position	1	2	3	4
1	33.7	40.0	31.5	31.3
2	31.7	32.2	28.1	24.4
3	68.3	72.4	67.1	72.2
4	41.4	36.4	34.4	75.3

Table 2. *Cont.*

Position	1	2	3	4
5	76.6	48.7	44.0	48.4
6	76.5	72.8	74.3	73.3
7	30.3	35.6	29.9	33.3
8	34.6	34.7	35.0	34.8
9	39.1	47.8	33.4	37.1
10	39.5	36.7	36.0	36.5
11	21.1	20.9	20.7	21.3
12	33.7	33.5	33.6	33.5
13	48.2	48.2	48.2	48.1
14	82.7	82.7	82.8	82.3
15	85.6	85.6	85.5	85.3
16	82.1	82.1	82.2	82.1
17	54.4	54.5	54.5	54.5
18	16.5	16.4	16.4	16.4
19	17.3	16.1	26.3	25.6
20	31.0	31.0	31.0	31.0
21	18.7	18.7	18.7	18.7
22	37.9	37.9	37.9	37.9
23	25.1	25.1	25.1	25.1
24	34.9	34.9	34.9	34.9
25	37.0	37.0	37.0	37.0
26	68.4	68.4	68.4	68.4
27	17.3	17.3	17.3	17.3

However, most of the proton and carbon chemical shifts of steroid A/B rings in the NMR spectra of **2** were quite different from those of **1**. The characteristic proton and carbon resonances of angular methyl group CH_3-19 (δ_H 1.04 s and δ_C 16.1), an oxygenated methine CH-3 (δ_H 3.53 m and δ_C 72.4), and an oxygenated methine CH-6 (δ_H 3.77 q (J = 2.6) and δ_C 72.8) observed in the NMR spectra of **2** testified to the 3β,6β-dihydroxy substitution in 5α-cholestane nucleus in **2** [21]. The COSY and HSQC spectroscopic data ascertained the proton sequences at C-1 to C-8, C-8 to C-12 through C-11, C-15 to C-17, C-17 to C-21 through C-20, and C-20 to the end of the side chain (Figures S12 and S13). The key HMBC cross-peaks H-6/C-8, C-10; H-16/C-13, C-15; H_3-18/C-12, C-13, C-14, C-17; H_3-19/C-1, C-5, C-9, C-10 and the key ROESY cross-peaks Hα-4/H-6; H-5/H-3, Hα-7, H-9; H-16/H-17, H-22; H-17/H_3-21; H_3-18/H-8, H-11β, H-15, H-20; and H_3-19/Hβ-1, Hβ-2, Hβ-4, H-8 confirmed the 3β,6β,15α,16β-tetrahydroxy pattern in 5α/9α/10β/13β steroid nucleus in **2** (Figures S14 and S15). The configuration at C-25 was determined as (*S*) by analogy with co-occurring compound **1** and similarity of the proton and carbon chemical shifts of the both side chains in the NMR spectra. Therefore, the structure of steroid **2** was established as the (25*S*)-5α-cholestane-3β,6β,14,15α,16β,26-hexaol.

According to the presence of the $[M + Na]^+$ sodium adduct ion peak at *m*/*z* 491.3341 in the (+)HRESIMS spectrum and the $[M - H]^-$ deprotonated ion peak at *m*/*z* 467.3380 in the (−)HRESIMS spectrum, the molecular formula $C_{27}H_{48}O_6$ of compound **3** has been found to be identical to that of **2** (Figure S16). The detailed comparison of the ^{1}H- and ^{13}C-NMR spectra of compounds **3** and **2** has revealed that the proton and carbon resonances belonging to the steroid C/D rings and side chain of **3** are close to those of **2** indicating the 14,15α,16β,26-tetrahydroxy substitution in **3**, while the proton and carbon signals of the steroid A/B rings of **3** substantially differed from those of **2** (Tables 1 and 2, Figures S17 and S18). The proton and carbon signals in the NMR spectroscopic data attributable to the A/B rings of **3** showed the presence of angular methyl group CH_3-19 (δ_H 1.13 s and δ_C 26.3) and two oxygenated methines CH-3 (δ_H 3.99 br. q (J = 2.7) and δ_C 67.1) and CH-6 (δ_H 3.66 q (J = 2.7) and δ_C 74.3). The signal of CH_3-19 in the ^{13}C-NMR spectrum of **3** was shifted from δ_C 16.1 to 26.3 in comparison with that of **2**. This fact strongly testified to *cis*-A/B ring fusion in **3** [16,20]. The coupling constant J = 2.7 of the broad quartet of H-3 corresponded well to the 3β-hydroxyl group in 5β-cholestane nucleus, and the coupling constant J = 2.7 of the quartet of H-6 indicated the 6β-hydroxyl group in **3** [16]. All the proton and carbon signals associated with the steroid nucleus and side chain were assigned by 2D experiments (Tables 1 and 2, Figure 3, Figures S19). Proton and carbon chemical shifts of the steroid A/B rings of **3** were similar to the corresponding data of (25*S*)-5β-cholestane-3β,6β,15α,16β,26-pentaol isolated from the starfish *L. clathrata* [16]. The key ROESY cross-peaks H_3-19/Hβ-1, H-5, H-8; H_3-18/H-8, H-15, H-20; Hα-4/H-6, Hα-7; Hβ-4/H-6; H-8/H-15; H-16/H-17; and H-17/H_3-21 and proton coupling constants confirmed the 3β,6β,15α,16β relative configurations of the oxygenated carbons and the 5β-cholestane skeleton of **3** (Figure 3). As a result, steroid **3** was proved to be the *cis*-A/B ring fusion isomer of steroid **2** and its structure was established as (25*S*)-5β-cholestane-3β,6β,14,15α,16β,26-hexaol.

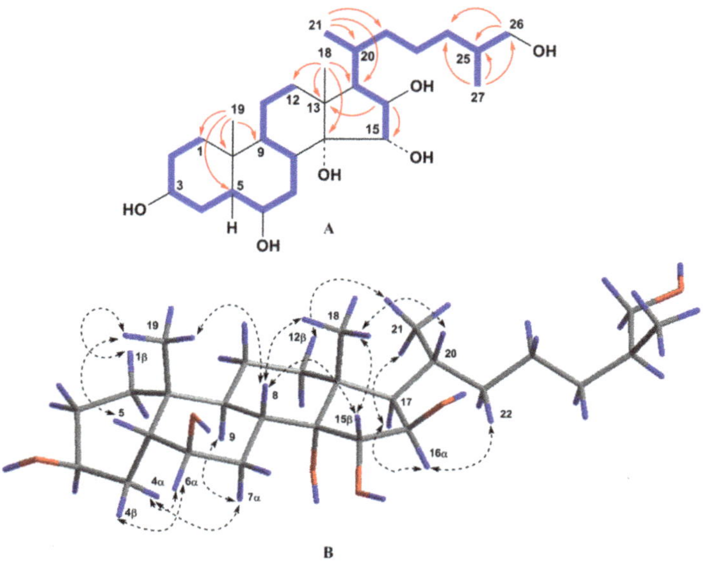

Figure 3. (**A**) COSY and key HMBC correlations for compound **3**; (**B**) Key ROESY correlations for compound **3**. Colors reveal the atoms of hydrogen (blue), oxygen (red) and carbon (grey) and their bonds.

The molecular formula of compound **4** was established to be of $C_{27}H_{48}O_7$ from the $[M + Na]^+$ sodium adduct ion peak at *m/z* 507.3290 in the (+)HRESIMS spectrum and the $[M - H]^-$ deprotonated ion peak at *m/z* 483.3327 in the (−)HRESIMS spectrum (Figure S23). Data of the mass-spectra and the ^{1}H- and ^{13}C-NMR spectra showed the presence of seven hydroxyl groups in **4**. The examination of the ^{1}H-, ^{13}C-, and 2D NMR spectra of steroids **4** and **3** revealed that both compounds have the identical 14,15α,16β-trihydroxy substitution and 26-hydroxy cholestane side chain, but the proton and carbon resonances of the steroid A/B rings of **4** differed from those of **3** (Tables 1 and 2, Figures S24). The deshielded shift of the signal of CH_3-19 at δ_C 25.6 in the ^{13}C-NMR spectrum and the existence of the ROESY cross-peak H_3-19/H-5 immediately showed a 5β-cholestane skeleton in **4**. The proton connectivities from C-1 to C-9 in A/B rings were ascertained using the COSY and HSQC experiments. The ^{1}H- and ^{13}C-NMR spectroscopic data, referred to the steroid A/B rings of **4**, revealed the proton and carbon chemical shifts of three oxygenated methines, including CH-3 (δ_H 3.70 br. q (J = 3.7) and δ_C 72.2), CH-4 (δ_H 3.66 t (J = 3.7) and δ_C 75.3), and CH-6 (δ_H 4.00 q (J =3.3) and δ_C 73.3). The irradiation of the proton H-5 in the 1D TOCSY experiment gave an enhancing signal of the neighboring proton H-4, that confirmed the presence of an additional hydroxyl group at C-4 in **4** as compared with **3** (Figure S30). Small values of the coupling constants of the protons H-3, H-4, and H-6 showed the absence of their axially axial interaction with neighboring protons. As a result, the 3β,4α,6β-trihydroxy pattern and the *cis*-A/B ring fusion were determined. Accordingly, the structure of **4** was established as (25*S*)-5β-cholestane-3β,4α,6β,14,15α,16β,26-heptaol.

Compounds **3** and **4** have the 5β-cholestane skeleton, which are rare among starfish steroids. Previously, only two steroid compounds with the *cis*-A/B ring junction, (25*S*)-5β-cholestane-3β,6β,15α,16β,26-pentaol from the starfish *L. clathrata* [17] and (25*S*)-5β-cholestane-3α,6β,15α,16β,26- pentaol from the starfish *Tremaster novaecaledoniae* [20], were found.

2.2. Biological Evaluation

2.2.1. The Effect of Compounds 1, 5, and 6 on Cancer Cells' Viability and Proliferation of Normal and Cancer Cells

In the first step of bioactivity investigations, the cytotoxicity of compounds **1**, **5**, and **6** was determined by measuring the metabolic activity of normal mouse epidermal JB6 Cl41, human colorectal carcinoma HT-29, and breast cancer MDA-MB-231 cells using MTS reagent. None of the tested compounds inhibited the viability of JB6 Cl41, HT-29, and MDA-MB-231 cells by 50% at concentrations up to 100 μM. The compounds **1**, **5**, and **6** decreased the cell viability by less than 20% at 100 μM (data not shown).

Next we determined the ability of the investigated compounds to affect cell proliferation of the tested cell lines. JB6 Cl41, HT-29, and MDA-MB-231 cells were treated with compounds **1**, **5**, and **6** at a non-toxic concentration of 20 μM for 24, 48, and 72 h. All tested compounds inhibited cell growth to a comparable degree (Figure 4). Compounds **1**, **5**, and **6** decreased proliferation of JB6 Cl41 cells by 20%, 22%, and 24%, respectively; HT-29 cells by 20%, 22%, and 26%, respectively; and MDA-MB-231 cells by 24%, 27%, and 29%, respectively, after 72 h of treatment.

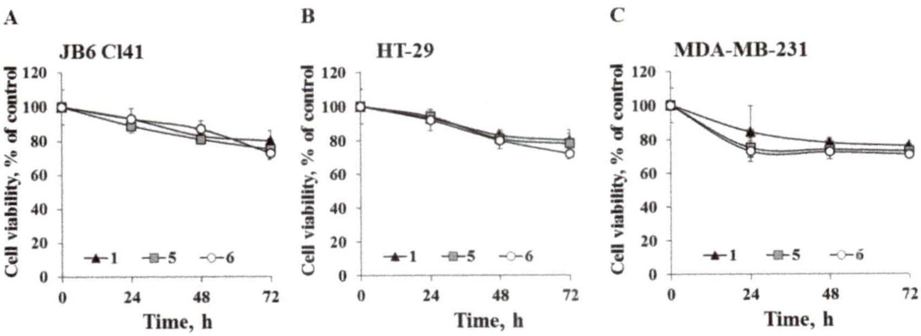

Figure 4. The effect of compounds **1**, **5**, and **6** on cell proliferation. (**A**) JB6 Cl41; (**B**) HT-29; or (**C**) MDA-MB-231 cells were treated with compounds **1**, **5**, and **6** at concentration of 20 μM for 24 h, 48 h, and 72 h. Cell viability was estimated using the MTS assay. Data are represented as the mean ± SD as determined from triplicate experiments. A Student's t-test was used to evaluate the data with the following significance levels: *$p < 0.05$, **$p < 0.01$, ***$p < 0.001$.

2.2.2. The Effect of Compounds 1, 5, and 6 on Colony Formation of Cancer Cells

The effect of compounds **1**, **5**, and **6** on the colony formation of human cancer HT-29 and MDA-MB-231 cells was investigated using the soft agar assay. Compounds **1**, **5**, and **6** (20 μM) were demonstrated to possess comparable inhibiting activity on colony formation of cancer cells and the decrease in the colony number of HT-29 cells was by 18%, 12%, and 18%, respectively, while MDA-MB-231 cells was by 35%, 30%, and 31%, respectively as compared with non-treated cells (control) (Figure 5). In the present study it was demonstrated that triple negative human breast cancer cells MDA-MB-231 were more sensitive to the treatment of compounds **1**, **5**, and **6** than colorectal carcinoma cells HT-29.

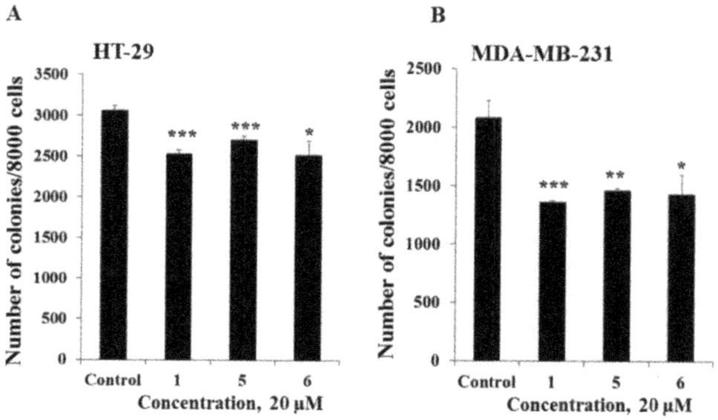

Figure 5. The effect of compounds **1**, **5**, and **6** on colony formation in human cancer cells. (**A**) HT-29; or (**B**) MDA-MB-231 cells (2.4×10^4) with or without investigated compounds (20 µM) treatment were subcultured onto 0.3% Basal Medium Eagle (BME) agar containing 10% FBS, 2 mM L-glutamine, and 25 µg/mL gentamicin. After 14 days of incubation, the number of colonies was evaluated under a microscope with the aid of the ImageJ software program. Results are expressed as the mean ± standard deviation (SD). The asterisk (*) indicates a significant decrease in colony number of cancer cells treated by compounds as compared with the control (*$p < 0.05$, **$p < 0.01$, ***$p < 0.001$).

In summary, compounds **1**, **5**, and **6** are non-cytototoxic against normal JB6 Cl41 and cancer HT-29 and MDA-MB-231 cell lines at concentrations up to 100 µM, however, they are able to suppress cell proliferation and colony formation of cancer HT-29 and MDA-MB-231 cells.

3. Experimental Section

3.1. General Procedures

The ^{1}H- and ^{13}C- NMR spectra were recorded on a Bruker Avance III 500 HD (Bruker, Germany) spectrometer at 500.13 and 125.76 MHz and a Bruker Avance III 700 spectrometer (Bruker, Germany) at 700.13 and 176.04 MHz, respectively. Chemical shifts (ppm) were internally referenced to the corresponding residual solvent signals δ_H 3.30/δ_C 49.0 for CD_3OD. HRESIMS mass spectra were recorded on a Bruker Impact II Q-TOF mass spectrometer (Bruker, Bremen, Germany); the samples were dissolved in MeOH (c 0.001 mg/mL). Optical rotation was measured using the Perkin Elmer 343 polarimeter (Waltham, MA, USA). IR spectra were recorded on a Bruker OPUS Vector-22 infrared spectrophotometer. HPLC separations were carried out on an Agilent 1100 Series chromatograph (Agilent Technologies, Santa Clara,.CA, USA) equipped with a differential refractometer; Discovery C18 (5 µm, 250 × 10 mm, Supelco, Bellefonte, PA, USA) and YMC-Pack Pro C18 (5 µm, 250 × 4.6 mm, YMC CO., LTD, Kyoto, Japan) columns were used. Low pressure column liquid chromatography was performed using Amberlite XAD-2 (20 to 60 mesh, Supelco, Bellefonte, PA., USA), and silica gel KSK (50 to 160 µm, Sorbpolimer, Krasnodar, Russia). Sorbfil silica gel plates (4.5 × 6.0 cm, 5 to 17 µm, Sorbpolimer, Krasnodar, Russia) were used for thin-layer chromatography.

3.2. Animal Material

Specimens of *Anthenoides laevigatus* Liao & A.M. Clark, 1989 (order Valvatida, family Goniasteridae) were collected in January 2018 from the coastal waters of the Qui Nhon Province (Vietnam), at a depth of 20 to 30 m, and were identified by Dr. Do Cong Thung, the Institute of Marine Resources and Environment, VAST, Hai Phong, Vietnam. A voucher specimen (no. SBAL 01-2018 (was deposited at the Institute of Natural Products Chemistry, VAST, Vietnam.

3.3. Extraction and Isolation

The fresh animals (2.2 kg) were chopped into small pieces and extracted four times by MeOH with heating. The MeOH extract was evaporated *in vacuo*, and the residue (37 g) was dissolved in H_2O (1.3 L). The H_2O-soluble fraction was passed in two portions through an Amberlite XAD-2 column (7.5 × 28 cm) and eluted with distilled H_2O until a negative chloride ion reaction was obtained, followed by elution with EtOH. The combined EtOH eluate was evaporated to give a brownish material (4.4 g). The resulting total fraction was chromatographed on a Si gel column (6.5× 15 cm) using CH_3Cl-EtOH (stepwise gradient, 3:1→1:2, *v*/*v*), EtOH, and EtOH-H_2O (stepwise gradient, 20:1→9:1, *v*/*v*) to give ten main fractions (1–10). Fractions 4 and 5 mainly contained the mixtures of polyhydroxylated steroids based on TLC data on Si gel plates in the eluent system toluene-EtOH (9:5, *v*/*v*). HPLC separation of fraction 4 (110 mg) on a Discovery C18 column with 55% aq. EtOH (1.5 mL/min) as an eluent system yielded pure **5** (4.1 mg, t_R 48.8 min) and subfraction 4.1 and 4.2 that were further purified on a YMC-Pack Pro C18 column with 78% aq. MeOH (0.9 mL/min) as an eluent system to give pure **2** (0.9 mg, t_R 15.3 min), **3** (0.9 mg, t_R 14.1 min), and **4** (1.1 mg, t_R 10.4 min). HPLC separation of fraction 5 (182 mg) on a Discovery C18 column with 55% aq. EtOH (1.5 mL/min) as an eluent system gave pure **1** (49.6 mg, t_R 21.0 min) and **6** (23.6 mg, t_R 26.1 min).

3.4. Compound Characterization Data

(25S)-5α-cholestane-3β,5,6β,14,15α,16β,26-heptaol (**1**): Colorless amorphous powder; $[\alpha]_D^{25}$ + 9.5 (*c* 0.8, MeOH); IR (KBr) ν_{max} 3401, 2943, 2870, 1385, 1048, 1017, 964 cm^{-1}; (+)HRESIMS *m*/*z* 507.3292 $[M + Na]^+$ (calcd for $C_{27}H_{48}O_7Na$, 507.3292); (−)HRESIMS *m*/*z* 483.3321 $[M - H]^-$ (calcd for $C_{27}H_{47}O_7$, 483.3327); ^{1}H-NMR data, see Table 1; ^{13}C-NMR data, see Table 2.

(25S)-5α-cholestane-3β,6β,14,15α,16β,26-hexaol (**2**): Colorless amorphous powder; $[\alpha]_D^{25}$ + 14.4 (*c* 0.1, MeOH); (+)HRESIMS *m*/*z* 491.3337 $[M + Na]^+$ (calcd for $C_{27}H_{48}O_5Na$, 491.3343); (−)HRESIMS *m*/*z* 467.3379 $[M - H]^-$ (calcd for $C_{27}H_{47}O_6$, 467.3378); ^{1}H-NMR data, see Table 1; ^{13}C-NMR data, see Table 2.

(25S)-5β-cholestane-3β,6β,14,15α,16β,26-hexaol (**3**): Colorless amorphous powder; $[\alpha]_D^{25}$ + 13.3 (*c* 0.1, MeOH); (+)HRESIMS *m*/*z* 491.3341 $[M + Na]^+$ (calcd for $C_{27}H_{48}O_6Na$ 491.3343); (−)HRESIMS *m*/*z* 467.3380 $[M - H]^-$ (calcd for $C_{27}H_{47}O_6$, 467.3378); ^{1}H-NMR data, see Table 1; ^{13}C-NMR data, see Table 2.

(25S)-5β-cholestane-3β,4α,6β,14,15α,16β,26-heptaol (**4**): Colorless amorphous powder; $[\alpha]_D^{25}$ + 15.5 (*c* 0.1, MeOH); (+)HRESIMS *m*/*z* 507.3290 $[M + Na]^+$ (calcd for $C_{27}H_{48}O_7Na$, 507.3292); (−)HRESIMS *m*/*z* 483.3327 $[M - H]^-$ (calcd for $C_{27}H_{47}O_7$, 483.3327); ^{1}H-NMR data, see Table 1; ^{13}C-NMR data, see Table 2.

3.5. Preparation of the MTPA Esters of Compound 1

Aliquots (0.7 mg each) of compound **1** were treated with *S*-(+)- and *R*-(−)-α-methoxy-α-(trifluoromethyl)-phenylacetyl (MTPA) chloride (0.7 µL) in dry pyridine (180 µL) for 3 h at room temperature. After removal of the solvent, the products were purified on a Si gel column (0.8 × 5 cm) using $CHCl_3$-EtOH (50:1, *v*/*v*) to obtain the corresponding (*R*)- and (*S*)-MTPA esters of **1**.

3,15,26-tri-(*R*)-MTPA ester of **1**: Selected ^{1}H-NMR (500.13 MHz, CD_3OD): δ_H 0.75 (1H, m, H-7), 0.87 (3H, d, *J* = 6.7 Hz, H_3-21), 0.91 (3H, d, *J* = 6.8 Hz, H_3-27), 1.07 (3H, s, H_3-18), 1.12 (3H, s, H_3-19), 1.49 (1H, m, H′-7), 1.83 (1H, m, H-25), 2.10 (1H, m, H-17), 3.05 (1H, dd, *J* = 3.6, 2.5 Hz, H-6), 4.15 (1H, m, H-16), 4.15 (1H, dd, *J* = 10.7, 5.7 Hz, H-26), 4.17 (1H, dd, *J* = 10.7, 6.0 Hz, H′-26), 5.03 (1H, d, *J* = 3.1 Hz, H-15), 5.39 (1H, m, H-3).

3,6,15,26-tetra-(*S*)-MTPA ester of **1**: Selected ^{1}H-NMR (500.13 MHz, CD_3OD): δ_H 0.62 (3H, s, H_3-19), 0.87 (3H, d, *J* = 6.7 Hz, H_3-21), 0.90 (3H, d, *J* = 6.8 Hz, H_3-27), 1.02 (3H, s, H_3-18), 1.33 (1H, m, H-7), 1.82 (1H, m, H-25), 2.02 (1H, dd, *J* = 11.2, 8.2 Hz, H-17), 2.09 (1H, m, H′-7), 4.08 (2H, m, H-16,

H-26), 4.09 (1H, m, H-16), 4.24 (1H, dd, J = 10.7, 5.5 Hz, H′-26), 4.83 (1H, t, J = 3.0 Hz, H-6), 5.19 (1H, d, J = 3.4 Hz, H-15), 5.32 (1H, m, H-3).

3.6. Bioactivity Assay

3.6.1. Reagents

Phosphate buffered saline (PBS), L-glutamine, penicillin-streptomycin solution (10,000 U/mL, 10 μg/mL) were from Sigma-Aldrich company (St. Louis, MO, USA). MTS reagent (3-(4,5-dimethylthiazol-2-yl)-5-(3-carboxymethoxyphenyl)-2-(4-sulfophenyl)-2H-tetrazolium) was purchased from Promega (Madison, Wisconsin, USA). The Basal Medium Eagle (BME), Dulbecco's modified Eagle's medium (DMEM), and McCoy's 5A modified medium (McCoy's 5A), trypsin, fetal bovine serum (FBS), and agar were purchased from ThermoFisher Scientific (Waltham, MA, USA).

3.6.2. Cell Lines and Culture Conditions

Mouse epidermal cells JB6 Cl41 (ATCC® no. CRL-2010™), human colorectal carcinoma HT-29 (ATCC® no. HTB-38™), and human breast adenocarcinoma MDA-MB-231 (ATCC® HTB-26™) cells were obtained from the American Type Culture Collection (Manassas, VA, USA).

JB6 Cl41 cells grown in MEM supplemented with 5% fetal bovine serum (FBS), HT-29 cells were cultured in McCoy's 5A with 10% FBS, and MDA-MB-231 cells were grown in DMEM with 10% FBS. The cell cultures were maintained at 37 °C in humidified atmosphere containing 5% CO_2.

3.6.3. MTS Assay

To determine the cytotoxic activity of compounds **1**, **5**, and **6**, JB6 Cl41, HT-29, and MDA-MB-231 cells were seeded at a density of 1.0×10^4cells/200 μL of complete MEM/5% FBS, McCoy's 5A/10% FBS, and DMEM/10% FBS media, respectively, in 96-well plates. After incubation for 24 h, the cells were treated with tested compounds in the range of concentration 5–100 μM, while the control was treated with the complete medium only. Cells were cultured for additional 24 h at 37 °C in 5% CO_2 atmosphere. Subsequently, MTS reagent (20 μL) was added to each well, and the cells were incubated for an additional 3 h at 37 °C in 5% CO_2. Absorbance was measured at 490/630 nm by a Power Wave XS microplate reader (BioTek, Winooski, VT, USA). All tested samples were carried out in triplicates. Compound's concentration causing 50% of cell viability inhibition (IC_{50}) were calculated.

To analyze the anti-proliferative activity of compounds **1**, **5**, and **6**, the cells (1.0×10^4cells/200 μL) were treated with tested compounds at concentration of 20 μM and incubated for an additional 24, 48, and 72 h at 37 °C in 5% CO_2. MTS reagent (20 μL) was added to each well, and the cells were incubated for an additional 3 h at 37 °C in 5% CO_2. Absorbance was measured at 490/630 nm using a microplate reader. All tested samples were analyzed in triplicates.

3.6.4. Soft Agar Assay

Cells (2.4×10^4/mL) were grown in 1 mL of 0.3% Basal Medium Eagle's agar containing 10% FBS. The cells were treated by compounds **1**, **5**, and **6** at non-toxic concentration of 5, 10, and 20 μM. The cultures were maintained at 37 °C in 5% CO_2 incubator for 2 weeks and the number and size of the colonies were determined using a Motic microscope AE 20 (XiangAn, Xiamen, China) and ImageJ software bundled with 64-bit Java 1.8.0_112 (NIH, Bethesda, Maryland, USA).

3.6.5. Statistical Analysis

Results are expressed as the mean ± standard deviation (SD). Student's T test was used to evaluate the data with the following significance levels: *$p < 0.05$, **$p < 0.01$, ***$p < 0.001$. All assays were performed in at least three independent experiments.

4. Conclusions

Four new polyhydroxylated steroids were isolated along with two previously known related steroids from the Vietnamese starfish *A. laevigatus* and their chemical structures were elucidated. Two new compounds have the common 5α-cholestane skeleton, while the other two new compounds have the 5β-cholestane skeleton, which are rare among starfish steroids. Previously, only two steroid compounds with the *cis*-A/B ring junction were known from two species of the starfish *L. clathrata* and *T. novaecaledoniae*. Three of the substances that were isolated from *A. laevigatus* did not show cytotoxic effects against normal JB6 Cl41 and human colorectal cancer HT-29 and breast cancer MDA-MB-231 cells, however they suppressed cell proliferation and colony formation of cancer HT-29 and MDA-MB-231 cells.

Supplementary Materials: The following are available online. Copies HRESIMS (Figures S1, S9, S16, and S23), ^{1}H-NMR (Figures S2, S10, S17, and S24), ^{13}C-NMR (Figures S3, S11, S18, and S25), COSY (Figures S4, S12, S19, and S26), HSQC (Figures S5, S13, 20, and S27), HMBC (Figures S6, S14, S21, and S28), and ROESY (Figures S7, S15, S22, and S29) spectra of compounds **1**, **2**, **3**, and **4**, respectively. Copies ^{1}H-NMR (Figure S8) spectra of (*R*)- and (*S*)-MTPA esters of **1**, 1D TOCSY (Figure S30) spectrum of **4**.

Author Contributions: A.A.K. isolated the metabolites, elucidated their structures, and prepared manuscript; D.T.H. and T.T.T.T. isolated the metabolites and elucidated their structures; N.V.I. and T.V.M. analyzed the compounds and edited the manuscript; O.S.M. carried out biological evaluation of compounds and analysis of the results; A.I.K. performed the acquisition and interpretation of NMR spectra; R.S.P. did the acquisition and interpretation of mass spectra; N.T.T.H. and P.Q.L. helped with collection of animals and manuscript editing. All authors have read and agreed to the published version of the manuscript.

Funding: This research was partly funded by the RFBR (Russian Foundation for Basic Research), grant number 20-03-00014 and the VAST (Vietnam Academy of Science and Technology), grant numbers QTRU02.03/18-19 and QTRU04.05/18-19.

Acknowledgments: The study was carried out on the equipment of the Collective Facilities Center "The Far Eastern Center for Structural Molecular Research (NMR/MS) of PIBOC FEB RAS". The authors are grateful to Do Cong Thung (Institute of Environment and Marine Resources of the VAST, Hai Phong, Vietnam) for species identification of the starfish.

Conflicts of Interest: The authors declare no conflict of interest.

References

1. Minale, L.; Riccio, R.; Zollo, F. Steroidal oligoglycosides and polyhydroxysteroids from Echinoderms. *Chem. Org. Nat.* **1993**, *62*, 75–308.
2. Stonik, V.A. Marine polar steroids. *Russ. Chem. Rev.* **2001**, *70*, 673–715. [CrossRef]
3. Iorizzi, M.; De Marino, S.; Zollo, F. Steroidal oligoglycosides from the Asteroidea. *Curr. Org. Chem.* **2001**, *5*, 951–973. [CrossRef]
4. Stonik, V.A.; Ivanchina, N.V.; Kicha, A.A. New polar steroids from starfish. *Nat. Prod. Commun.* **2008**, *3*, 1587–1610. [CrossRef]
5. Dong, G.; Xu, T.H.; Yang, B.; Lin, X.P.; Zhou, X.F.; Yang, X.W.; Liu, Y.H. Chemical constituents and bioactivities of starfish. *Chem. Biodivers.* **2011**, *8*, 740–791. [CrossRef] [PubMed]
6. Ivanchina, N.V.; Kicha, A.A.; Stonik, V.A. Steroid glycosides from marine organisms. *Steroids* **2011**, *76*, 425–454. [CrossRef] [PubMed]
7. Ivanchina, N.V.; Kicha, A.A.; Malyarenko, T.V.; Stonik, V.A. *Advances in Natural Products Discovery*; Gomes, A.R., Rocha-Santos, T., Duarte, A., Eds.; Nova Science Publishers: New York, NY, USA, 2017; Volume 6, pp. 191–224.
8. Gomes, A.R.; Freitas, A.C.; Rocha-Santos, T.A.P.; Duarte, A.C. Bioactive compounds derived from echinoderms. *Rsc Adv.* **2014**, *4*, 29365–29382. [CrossRef]
9. Lazzara, V.; Arizza, V.; Luparello, C.; Mauro, M.; Vazzana, M. Bright spots in the darkness of cancer: A review of starfishes-derived compounds and their anti-tumor action. *Mar. Drugs* **2019**, *17*, 617. [CrossRef] [PubMed]
10. Malyarenko, O.S.; Malyarenko, T.V.; Kicha, A.A.; Ivanchina, N.V.; Ermakova, S.P. Effects of polar steroids from the starfish *Patiria (=Asterina) pectinifera* in combination with X-ray radiation on colony formation and apoptosis induction of human colorectal carcinoma cells. *Molecules* **2019**, *24*, 3154. [CrossRef] [PubMed]

11. Ha, D.T.; Kicha, A.A.; Kalinovsky, A.I.; Malyarenko, T.V.; Popov, R.S.; Malyarenko, O.S.; Ermakova, S.P.; Thuy, T.T.T.; Long, P.Q.; Ivanchina, N.V. Asterosaponins from the tropical starfish Acanthaster planci and their cytotoxic and anticancer activities in vitro. *Nat. Prod. Res.* **2019**, *19*, 1–8. [CrossRef]
12. Kicha, A.A.; Ha, D.T.; Ivanchina, N.V.; Malyarenko, T.V.; Kalinovsky, A.I.; Dmitrenok, P.S.; Ermakova, S.P.; Malyarenko, O.S.; Hung, N.A.; Thuy, T.T.T.; et al. Six new polyhydroxysteroidal glycosides, anthenosides S1–S6, from the starfish *Anthenea Sibogae*. *Chem. Biodivers.* **2018**, *15*, e1700553. [CrossRef] [PubMed]
13. Malyarenko, T.V.; Ivanchina, N.V.; Malyarenko, O.S.; Kalinovsky, A.I.; Dmitrenok, P.S.; Evtushenko, E.V.; Minh, C.V.; Kicha, A.A. Two new steroidal monoglycosides, anthenosides A_1 and A_2, and revision of the structure of known anthenoside A with unusual monosaccharide residue from the starfish. *Anthenea Aspera Mol.* **2018**, *23*, 1077. [CrossRef] [PubMed]
14. Malyarenko, T.V.; Kharchenko, S.D.; Kicha, A.A.; Ivanchina, N.V.; Dmitrenok, P.S.; Chingizova, E.A.; Pislyagin, E.A.; Evtushenko, E.V.; Antokhina, T.I.; Minh, C.V.; et al. Anthenosides L-U, steroidal glycosides with unusual structural features from the starfish *Anthenea aspera*. *J. Nat. Prod.* **2016**, *79*, 3047–3056. [CrossRef] [PubMed]
15. Kicha, A.A.; Kalinovsky, A.I.; Malyarenko, T.V.; Ivanchina, N.V.; Dmitrenok, P.S.; Menchinskaya, E.S.; Yurchenko, E.A.; Pislyagin, E.A.; Aminin, D.L.; Huong, T.T.T.; et al. Cyclic steroid glycosides from the starfish *Echinaster luzonicus*: Structures and immunomodulatory activities. *J. Nat. Prod* **2015**, *78*, 1397–1405. [CrossRef] [PubMed]
16. Iorizzi, M.; Bryan, P.; McClintock, J.; Minale, L.; Palagiano, E.; Maurelli, S.; Riccio, R.; Zollo, F. Chemical and biological investigation of the polar constituents of the starfish *Luidia clathrata*, collected in the gulf of Mexico. *J. Nat. Prod* **1995**, *58*, 653–671. [CrossRef] [PubMed]
17. Ivanchina, N.V.; Kicha, A.A.; Kalinovsky, A.I.; Dmitrenok, P.S.; Dmitrenok, A.S.; Chaikina, E.L.; Stonik, V.A.; Gavagnin, M.; Cimino, G. Polar steroidal compounds from the Far Eastern starfish *Henricia leviuscula*. *J. Nat. Prod.* **2006**, *69*, 224–228. [CrossRef] [PubMed]
18. Minale, L.; Pizza, C.; Riccio, R.; Squillace-Greco, O.; Zollo, F.; Pusset, J.; Menou, J.L. New polyhydroxylated sterols from the starfish Luidia Maculate. *J. Nat. Prod.* **1984**, *47*, 784–789. [CrossRef]
19. Vanderach, D.J.; Djerassi, C. Marine natural products. Synthesis of four naturally occurring 20β-H cholanic acid derivatives. *J. Org. Chem.* **1978**, *43*, 1442–1448. [CrossRef]
20. De Riccardis, F.; Minale, L.; Riccio, R.; Giovannitti, B.; Iorizzi, M.; Debitus, C. Phosphated and sulfated marine polyhydroxylated steroids from the starfish *Tremaster Novaecaledoniae*. *Gazz. Chim. Ital.* **1993**, *123*, 79–86.
21. Kicha, A.A.; Ivanchina, N.V.; Malyarenko, T.V.; Kalinovsky, A.I.; Popov, R.S.; Stonik, V.A. Six new polyhydroxylated steroids conjugated with taurine, microdiscusols A-F, from the Arctic starfish *Asterias Microdiscus*. *Steroids* **2019**, *150*, 108458. [CrossRef] [PubMed]

Sample Availability: Samples of the compounds are available from the authors.

molecules
MDPI

Article

New Isomalabaricane-Derived Metabolites from a *Stelletta* sp. Marine Sponge

Sophia A. Kolesnikova [1,*], Ekaterina G. Lyakhova [1], Anastasia B. Kozhushnaya [2], Anatoly I. Kalinovsky [1], Dmitrii V. Berdyshev [1], Roman S. Popov [1] and Valentin A. Stonik [1,2]

[1] G.B. Elyakov Pacific Institute of Bioorganic Chemistry, Far Eastern Branch of Russian Academy of Sciences, Pr. 100-let Vladivostoku 159, 690022 Vladivostok, Russia; elyakhova@inbox.ru (E.G.L.); kaaniw@piboc.dvo.ru (A.I.K.); berdyshev@piboc.dvo.ru (D.V.B.); prs_90@mail.ru (R.S.P.); stonik@piboc.dvo.ru (V.A.S.)
[2] School of Natural Sciences, Far Eastern Federal University, Sukhanova Str. 8, 690000 Vladivostok, Russia; kozhushnaia.ab@mail.ru
* Correspondence: sovin81@inbox.ru; Tel.: +7-423-231-1168

Abstract: In continuation of our studies on a Vietnamese collection of a *Stelletta* sp., sponge we have isolated two new isomalabaricane triterpenoids, stellettins Q and R (**1** and **2**), and four new isomalabaricane-derived *nor*-terpenoids, stellettins S-V **3–6**, along with previously known globostelletin N. Among them, compound **3** contains an acetylenic fragment, unprecedented in the isomalabaricane family and extremely rare in other marine sponge terpenoids. The structures and absolute configurations of all new compounds were established by extensive NMR, MS, and ECD analyses together with quantum-chemical modeling. Additionally, according to obtained new data we report the correction in stereochemistry of two asymmetric centers in the structures of two known isomalabaricanes, 15*R*,23*S* for globostelletin M and 15*S*,23*R* for globostelletin N.

Keywords: isomalabaricanes; *Stelletta* sp.; marine sponge; terpenoid; structure elucidation

Citation: Kolesnikova, S.A.; Lyakhova, E.G.; Kozhushnaya, A.B.; Kalinovsky, A.I.; Berdyshev, D.V.; Popov, R.S.; Stonik, V.A. New Isomalabaricane-Derived Metabolites from a *Stelletta* sp. Marine Sponge. *Molecules* **2021**, *26*, 678. https://doi.org/10.3390/molecules26030678

Academic Editor: Akihito Yokosuka
Received: 29 December 2020
Accepted: 25 January 2021
Published: 28 January 2021

1. Introduction

Malabaricanes are rather a small group of tricarbocyclic triterpenoids found in different tropical flowering terrestrial plants [1–3]. Isomalabaricanes, which differ from malabaricanes in the configuration of C-8 asymmetric center and have an α-oriented CH_3-30, are known as metabolites of four genera of marine sponges—*Stelletta*, *Jaspis*, *Geodia* and *Rhabdastrella*—belonging to the class Demospongiae. Some of them are highly cytotoxic against tumor cells [4]. Since the first isolation of three yellow highly conjugated isomalabaricane-type triterpenoids from the marine sponge *Jaspis stellifera* in 1981 [5] more than 130 isomalabaricanes and related natural products have been reported from the abovementioned sponge genera. It was noticed that *Stelletta* metabolites are quite different depending on the collection. Indeed, isomalabaricane triterpenoids were mainly found as very complex mixtures in tropical sponge samples, while boreal and cold-water sponges contain mostly alkaloids and lipids. From a chemo-ecological point of view, this indicates that studied sponges are able to produce different types of secondary metabolites in order to adapt to the various living conditions [6]. In confirmation, our attempt to find isomalabaricanes in a cold-water *Stelletta* spp., collected in 2019 in the Sea of Okhotsk, was unsuccessful, as the characteristic yellow pigments were not detected by thin layer chromatography in the extracts of these sponges.

Additionally, in result of the chemical investigation of the sponge *Stelletta tenuis*, Li et al. identified two naturally occurring α-pyrones, namely gibepyrones C and F, along with three isomalabaricane-type triterpenoids [7]. These α-pyrones were supposed to be the oxidation products of the co-occurring stellettins [6]. Gibepyrone F had previously been isolated from the fungal plant pathogen *Gibberella fujikuroi* [8], as well as from the sponge *Jaspis stellifera* [9]. These findings allow to presume that symbiotic microorganisms in the corresponding sponges are involved in the generation of some metabolites.

Diverse isomalabaricane-type *nor*-terpenoids, containing less than 30 carbon atoms in their skeleton systems, have been found together with isomalabaricanes several times [10–12]. Their presence could be explained either by oxidative degradation of C_{30} metabolites or by precursor role of *nor*-terpenoids in the biosynthesis of these compounds [12,13]. However, the biogenesis of isomalabaricane compounds in sponges remains to be mysterious so far.

Recently, we have reported the isolation of two isomalabaricane-type *nor*-terpenoids, cyclobutastellettolides A and B, and series of known isomalabaricanes from a *Stelletta* sp. [14] We suppose that new data on structural variety of isomalabaricane derivatives supported with strong evidence on stereochemistry could someday shed light on their origin.

In the present work, an investigation of the chemical components of a *Stelletta* sp. from Vietnamese waters was continued. Herein, we report the isolation and structural elucidation of six new compounds **1–6** and known globostelletin N [15].

2. Results and Discussion

The frozen sample of a marine sponge *Stelletta* sp. was finely chopped and extracted with EtOH, then the extract was concentrated under reduced pressure and subjected to Sephadex LH-20 and silica gel column chromatography followed by normal- and reversed-phase HPLC procedures (Figure S73) to afford new stellettins Q-V **1–6** together with known globostelletin N [15] (Figure 1).

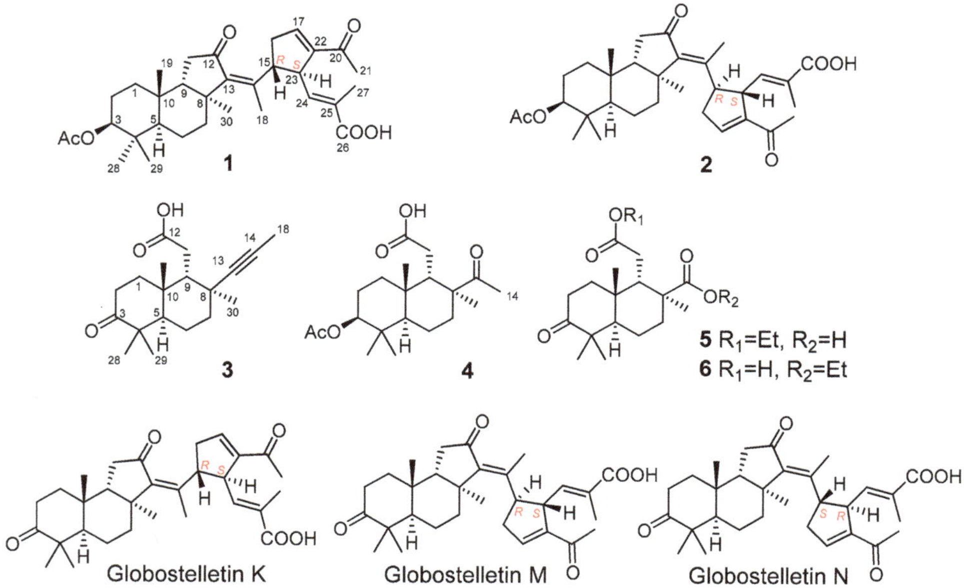

Figure 1. Structures of compounds **1–6** and globostelletins K, M, and N.

Stellettin Q (**1**) was isolated as a yellow oil with molecular formula $C_{32}H_{44}O_6$ deduced by HRESIMS (Figure S3). The NMR data of **1** (Table 1; Figures S4 and S5) were closely related to the spectral characteristics of isomalabaricane globostelletin K (Figure 1, Figures S55 and S56) initially found in the marine sponge *Rabdastrella globostellata* [15] and also co-isolated from the studied *Stelletta* sp. [14].

Table 1. ^{1}H and ^{13}C NMR data of **1** (700 and 176 MHz) and **2** (500 and 126 MHz) in $CDCl_3$.

No. [1]	1		2	
	δ_H mult (*J* in Hz)	δ_C	δ_H mult (*J* in Hz)	δ_C
1α	1.57, td (13.0, 3.9)	33.1	1.57, m	32.9
1β	1.38, m		1.36, m	
2α	1.82, m	25.1	1.81, m	25.1
2β	1.71, m		1.68, m	
3α	4.54, dd (11.7, 5.2)	80.8	4.53, dd (11.6, 5 2)	80.7
4		38.2		38.2
5α	1.73, m	46.5	1.75, m	46.4
6α	1.70, m	18.2	1.68, m	18.2
6β	1.48, m		1.48, m	
7α	1.99, m	37.7	2.06, m	38.5
7β	2.00, m		1.93, m	
8		44.2		43.8
9β	1.75, m	50.3	1.77, m	50.3
10		35.4		35.4
11α	2.13, m	36.4	2.15, m	36.4
11β	2.13, m		2.15, m	
12		206.6		206.9
13		146.8		146.0
14		147.9		147.5
15	4.75, m	45.0 [2]	3.22, dt (9.2, 6.0)	48.0
16α	2.27, m	37.9	2.52, m	38.4
16β	3.02, ddt (19.4, 9.3, 2.5)		2,89, ddt (19.4, 9.2, 2.7)	
17	6.80, br s	144.6	6.78, dd (4.3, 2 5)	143.6
18	1.79, s	16.2 [2]	2.06, s	15.7
19	1.02, s	22.4	1.01, s	22.4
20		195.6		195.1
21	2.29, s	26.9	2.30, s	27.0
22		146.6		146.7
23	3.86, br t (8.3)	47.5	3.95, m	48.0
24	6.57, br d (10.2)	145.9	6.59, dd (10.6, 1.5)	144.6
25		126.3		127.3
26		171.0		170.7
27	1.88, br s	12.6	1.86, d (1.3)	12.5
28	0.91, s	29.0	0.90, s	29.0
29	0.89, s	16.9	0.88, s	17.0
30	1.29, s	24.1	1.23, s	26.4
OAc	2.06, s	171.021.2	2.05, s	170.921.2

[1] Assignments were made with the aid of HSQC, HMBC and ROESY data. [2] The values were found from HSQC experiment.

The detailed analysis of 2D spectra (COSY, HSQC, HMBC etc.) of **1** supported the main structure (Figure 2 and Figures S6–S9). The signals of methyl group at δ_H 2.06, s; δ_C 21.2 and acetate carbonyl at δ_C 171.0, together with the HMBC correlation of axial proton H-3 at δ_H 4.54, dd (11.7, 5.2) to that carbonyl, revealed the O-acetyl substitution at C-3 in the ring A. Moreover, the signal of C-3 at δ_C 80.8 instead of ketone signal at δ_C 219.2 in ^{13}C NMR spectrum of globostelletin K also demonstrated the 3-acetoxy-tricyclic core in **1**, while the 3β-orientation of acetoxy group was confirmed by strong correlations of H-3/H-5 and CH_3-28 observed in the ROESY spectrum. The 13*Z* geometry in **1** was in agreement with the signal of CH_3-18 at δ_H 1.79, s and its ROESY correlation with CH_3-30. As well as *E* configuration of 24(25)-double bond was found from the *W*-path COSY correlation of protons H-24/CH_3-27.

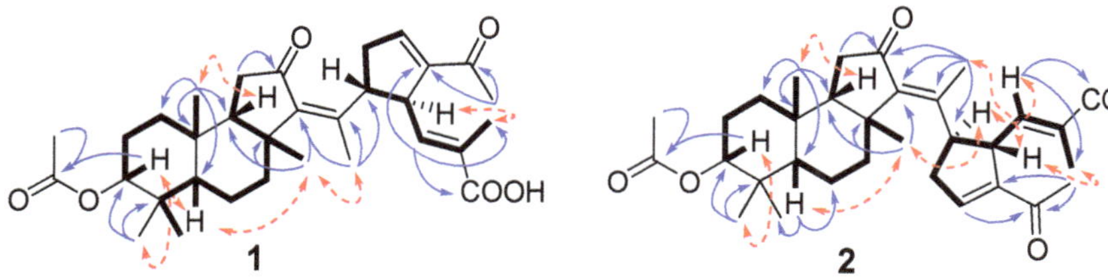

Figure 2. Selected COSY (▬), HMBC (⌒) and ROESY (⌒) correlations of **1** and **2**.

The ^{1}H- and ^{13}C-NMR signals of the side chain of **1** as well as the form of ECD curve (Figure S10) were analogous to those of globostelletin K [15] (Figure S57) suggesting the same stereochemistry of the side chain. This assignment was in a good agreement with the computational ECD results performed using density functional theory (DFT) with the nonlocal exchange-correlation functional B3LYP [16], the polarization continuum model (PCM) [17] and split-valence basis sets 6-31G(d), implemented in the Gaussian 16 package of programs [18] (Figure S66). The 15*R*,23*S* absolute configuration, providing 13*Z*,24*E* geometry and *trans*−*syn*−*trans*-fused tricyclic moiety with 3*β*-oriented acetoxy group fully satisfies the similarity of the experimental and theoretical ECD spectra of **1** (Figure 3). In detailes, statistically avereged curve (Figure 3) follows the shape of the experimental one, although even more close coincidence was indicated for theoretically less probable conformer (Figure S67). In addition, we could conclude that the presence of 3-acetoxy or 3-oxo functions in the structures of the corresponding compounds insignificantly affects the shape of their ECD curves. According to obtained new data we pose the same 15*R*,23*S* stereochemistry for globostelletin K (Figure S69).

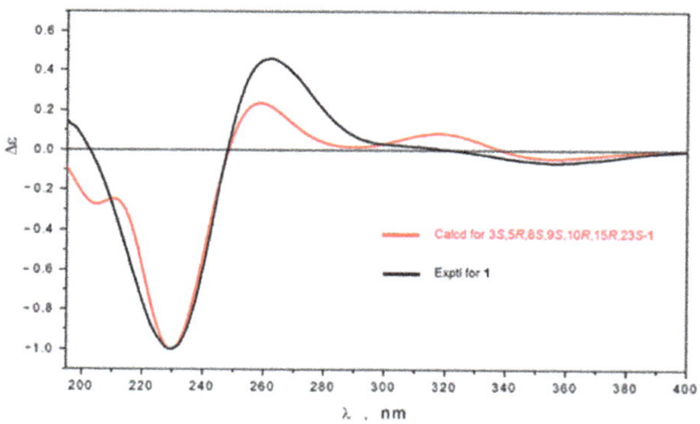

Figure 3. Comparison of experimental and theoretical ECD spectra of stellettin Q (**1**).

Stellettin R (**2**) has a molecular formula of $C_{32}H_{44}O_6$ as it was established on the basis of HRESIMS (Figure S11). Spectral data (Table 1, Figure 2, Figures S12 and S13) were consistent with known globostelletin M [15] (Figure 1, Figures S59 and S60) possessing an isomalabaricane core connected with 13*E* double bond (CH_3-18: δ_H 2.06, s). However, like stellettin Q, it contains 3*β*–acetoxy group (δ_H 4.53, dd (11.6, 5.2); δ_C 80.7; δ_H 2.05, s; δ_C 170.9; 21.2). Concerning to the relative configuration of the cyclopentene unit in the side chain of **2**, the ROESY cross-peaks between H-15/H-24 and H-23/CH_3-18 ascertained a *trans*-relationship of the vicinal protons H-15 and H-23. Careful examination of the chemical shifts for CH-15 (δ_H 3.22, dt (9.2, 6.0); δ_C 48.0) and CH-23 (δ_H 3.95, m; δ_C 48.0) showed the values similar to those of globostelletin M and differed from globostelletin N (Figure 1, Figures S63 and S64) isolated by Li et al. [15] and co-isolated by us. Moreover, the ECD spectrum of **2** (Figure S18) displayed the same curve and peaks as those published for globostelletin M (Figure S61).

However, structure modeling as well as calculation of ECD spectra for possible stereoisomers of **2** demonstrated a good agreement between experimental and theoretical spectra for 15*R*,23*S* absolute configuration (Figure 4) quite differ from 15*S*,23*S* reported for globostelletin M [15]. This inconsistence encouraged us to re-investigate the stereochemistry of co-isolated globostelletins M and N. We have obtained NMR and ECD spectra of the both compounds (Figures S59–S65) and they were identical to those provided as supplementary data by Li et al. [15]. At the same time, our computational results suggested globostelletin M to possess the same 15*R*,23*S* absolute configuration (Figure S70) of cyclopentene unit as **2**, while globostelletin N has 15*S*,23*R* stereochemistry (Figure S71). Based on the data we believe that previously published research comprises some inaccuracies and the stereochemistry of these centres in corresponding isomalabaricanes should be revised. It was noted that the isomalabaricane-type terpenoids undergo a photoisomerization of the side chain 13-double bond during the isolation and storage [19,20]. We consider compounds **1** and **2** to be the 13*Z*/*E* pair of the same 15*R*,23*S* isomer.

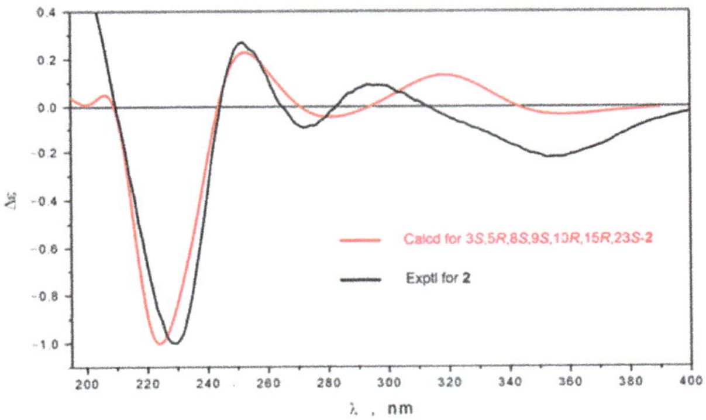

Figure 4. Comparison of experimental and theoretical ECD spectra of stellettin R (**2**).

The molecular formula $C_{19}H_{28}O_3$ of stellettin S (**3**) calculated from HRESIMS data (Figure S19) showed **3** to be a rather smaller molecule then classical C_{30}-isomalabaricanes, intriguing due to the lack of a significant part in the molecule, when compared with the majority of known isomalabaricanes and their derivatives. The ^{13}C- and DEPT NMR spectra (Table 2; Figures S21 and S22) exhibited 19 resonances, including those of carbonyl carbon at δ_C 216.5 (C-3) and carboxyl carbon at δ_C 178.8 (C-12) as well as two down-shifted quaternary carbons at δ_C 88.1 (C-13), 77.8 (C-14). ^{1}H- and ^{13}C-NMR spectra (Figures S20 and S21) revealed five methyls, two methylene, two methine groups and seven quaternary carbons, suggesting an isoprenoid nature. In the HSQC spectrum (Figure S23) four methyl singlets (δ_H 1.06, 1.08, 1.25, and 1.62) correlated with carbon signals at δ_C 21.6 (CH_3-29), 25.9 (CH_3-28), 30.8 (CH_3-30) and 23.3 (CH_3-19), respectively, while singlet of one more methyl group at δ_H 1.80 gave a cross-peak with high field signal at δ_C 3.7 (CH_3-18). The further inspection of 2D spectra (Figure 5 and Figures S23–S26) revealed the bicyclic framework resembling the core of globostelletin A (Figure 5), isolated from the sponge *Rhabdastrella globostellata* [13].

This was confirmed by the key long-range HMBC correlations from gem-dimethyl group (CH_3-28 and 29) to C-3, C-4 and C-5; from H-5 to C-1, C-4, C-6, C-9 and C-10; from methyl CH_3-19 to C-1, C-9 and C-10; from methyl CH_3-30 to C-7, C-8 and C-9 as well as from the methylene of carboxymethyl group (CH_2-11) to C-8, C-9, C-10 and carboxyl carbon C-12 (Figure 5 and Figure S24). The empirical formula, besides bicyclic system and two carbonyls, required two additional degrees of unsaturation which were accounted for an acetylenic bond in a short side chain. The NMR signals of two quaternary carbons at δ_C 88.1 (C-13), 77.8 (C-14) and methyl (CH_3-18) at δ_H 1.80, δ_C 3.7 were finally attributed to the methylacetylenic substituent at C-8, that was confirmed by HMBC correlations from CH_3-30 to C-8 and C-13, along with that from CH_3-18 to C-7, C-8, C-9, C-13, C-14 and CH_3-30. Analogous methylacetylenic substituent was characterized previously with similar chemical shifts in a series of synthetic alkynes [21].

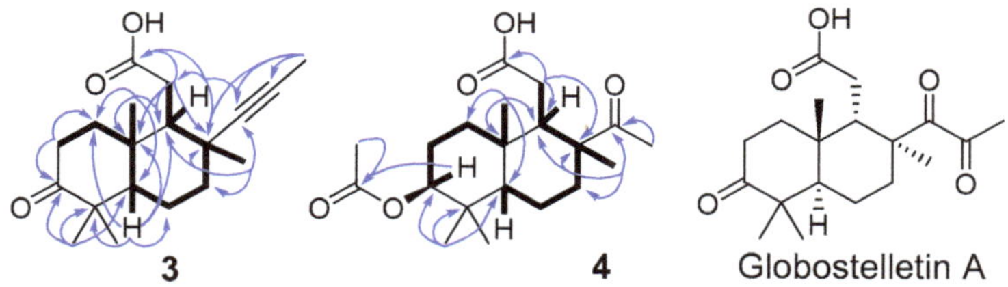

Figure 5. Selected COSY (▬) and HMBC (↶) correlations of **3** and **4** and structure of known globostelletin A.

Interestingly, the NMR signal of CH_3-19 (δ_H 1.62, s) was notably downfield shifted in comparison with that in a number of isomalabaricanes and their derivatives spectra. We explained it by the joint influence of the methylacetylene and carboxymethyl groups. The quantum chemical calculations (Figure S66) of the chemical shifts for structure **3** confirmed the down-shifted position of the proton signal of CH_3-19 and afforded its theoretical chemical shift value of δ_H 1.69 ppm.

The relative stereochemistry of **3** was determined by ROESY experiment (Figure 6 and Figure S26). A *trans*-fusion of the bicyclic system was shown by key NOE interactions. The correlations between CH_3-19/CH_3-29, CH_3-28/H-5, H-5/H_a-11, CH_3-19/H-9, and CH_3-30/H_b-11 showed the β-orientations of CH_3-19 and H-9, whereas H-5, CH_2-11, and CH_3-30 were α-oriented. The chair conformation of the ring B with equatorial positions of H-9 and CH_3-30 corresponded to the long-range COSY correlation between H-9 and H_β-7 (Figure 5 and Figure S25) together with ROESY correlations H-5/H_a-11 and H_α-7/H_b-11. Taking into consideration the relative stereochemistry of the compound **3** along with above mentioned absolute stereochemistry of the C_{30} congeners **1** and **2** as well as the fact of co-isolation of cyclobutastellettolides A and B [14] with the same absolute configurations we suggested the 5*S*, 8*R*, 9*R*, 10*R* absolute stereochemistry of stelletin S (**3**).

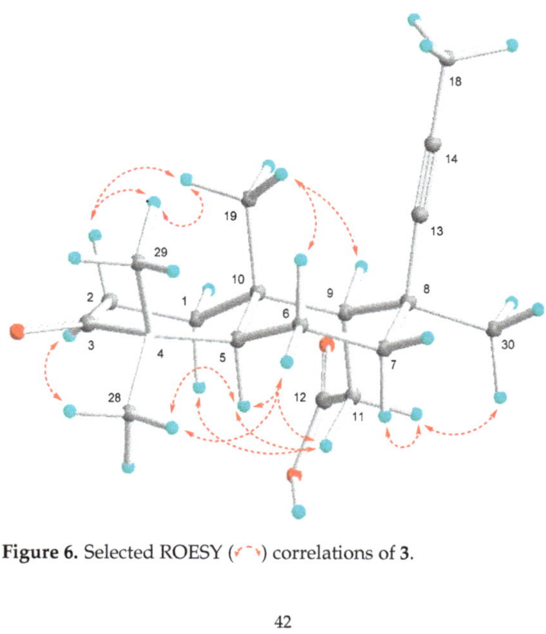

Figure 6. Selected ROESY (↶) correlations of **3**.

Table 2. ^{1}H and ^{13}C NMR data (700 and 176 MHz) of **3–6** in $CDCl_3$.

No. [1]	3		4		5		6	
	δ_H mult (*J* in Hz)	δ_C	δ_H mult (*J* in Hz)	δ_C	δ_H mult (*J* in Hz)	δ_C	δ_H mult (*J* in Hz)	δ_C
1α	1.76, m	35.9	1.33, m	34.3	1.78, m	35.5	1.79, m	35.6
1β	1.57, ddd (13.3, 6.3, 3.7)		1.28, m		1.57, ddd (13.5, 6.0, 4.0)		1.60, m	
2α	2.33, m	34.8	1.72, m	23.3	2.31, m	34,8	2.32, dt (15.4, 4.5)	34.8
2β	2.70, ddd (15.6, 12.3, 6.4)		1.68, m		2.65, ddd (15.4, 12.6, 6.1)		2.66, ddd (15.4, 12.6, 6.1)	
3		216.5	α: 4.44, dd (11.2, 5.1)	80.3		216.1		216.0
4		47.7		37.6		47.5		47.5
5α	1.33, dd (12.4, 2.7)	47.5	1.01, dd (11.9; 2.7)	46.2	1.40, dd (12.6; 3.0)	46.7	1.39, dd (12.7; 3.0)	46.8
6α	1.47, dq (13.6, 3.2)	21.1	1.65, m	17.9	1.46, m	20.4	1.47, m	20.5
6β	1.83, qd (13.2, 3.4)		1.47, m		1.79, m		1.78, m	
7α	1.25, td (13.2, 3.7)	36.5	1.51, m	30.0	1.11, m	31.2	1.08, m	31.5
7β	1.74, m		1.59, m		2.20, br d (15.1)		2.23, br d (14.2)	
8		34.1		52.5		44.5		44.5
9β	2.22, t (5.1)	51.2	2.31 br d (8.0)	49.4	2.78 br t (5.0)	48.3	2.74 br t (4.6)	48.3
10		38.5		38.6		38.5		38.4
11a	2.40, dd (18.1, 5.8)	31.1	2.58, dd (18.9, 7.9)	34.1	2.41, dd (17.7, 5.3)	31.1	2.48, dd (18.1, 5.4)	30.5
11b	2.33, dd (17.7, 4.8)		1.70, br d (18.9)		2.30, m		2.38, dd (18.1, 4.8)	
12		178.8		175.2 [2]		173.7		177.6 [3]
13		88.1		213.0		183.8		178.0 [3]
14		77.8	2.12, s	24.7				
15–17								
18	1.80, s	3.7						
19	1.62, s	23.3	1.27, s	24.6	1.24, s	20.8	1.14, s	20.7
20–27								
28	1.08, s	25.9	0.89, s	27.9	1.07, s	25.7	1.07, s	25.7
29	1.06, s	21.6	0.87, s	16.2	1.00, s	21.5	0.99, s	21.5
30	1.25, s	30.8	1.31,s	24.6	1.17, s	27.8	1.12, s	27.7
OAc			2.05, s	170.9 21.2				
OEt					4.15, q (7.1), 2H, 1.26, t (7.1), 3H	60.8 14.1	4.23, dq (10.9, 7.1), H 4.13, dq (10.9, 7.1), H 1.31, t (7.1), 3H	60.6 14.0

[1] Assignments were made with the aid of HSQC, HMBC and ROESY data. [2] The values were found from HMBC experiment. [3] These signals could be interchanged.

Stellettin T (**4**) with the molecular formula $C_{20}H_{32}O_5$ seemed to be another isomalabaricane-type derivative. The ispection of NMR data (Table 2; Figure 5 and Figures S29–S33) revealed the same type of 9-carboxymethyl substituted bicyclic core as was deduced for compound **3**. It contains 3β-acetoxy group, confirmed with the signals of CH-3 (δ_H 4.44, dd (11.2, 5.1); δ_C 80.3), methyl of acetate group (δ_H 2.05, s; δ_C 21.2) and acetate carbon (δ_C 170.9). According to the ^{13}C NMR spectrum and molecular formula, compound **4** has one carbonyl less side chain then known globostelletin A [13]. Based on this data and HMBC correlations from CH_3-14 (δ_H 2.12, s) and CH_3-30 (δ_H 1.31, s) to C-13 (δ_C 213.0), the acetyl was connected with C-8 (δ_C 52.5). The key ROESY correlations H-3/H-5, H-5/H_a-11, H-14/CH_3-19, H-9/CH_3-19 and H-9/CH_3-29 (Figure S34) suggested configurations at C-5, C-8, C-9 and C-10 identical to those of co-isolated isomalabaricanes.

The structures of stellettins U (**5**) and V (**6**) corresponded to the same $C_{19}H_{30}O_5$ molecular formula deduced from HRESIMS (Figures S36 and S45). In comparison with co-isolated metabolites, the spectral data of compounds **5** and **6** revealed bicyclic core with keto group at C-3, gem-dimethyl group at C-4 and two angular methyls at C-8 and C-10 (Table 2). Additionally, ^{1}H- and ^{13}C-NMR spectra of compound **5** (Figures S37 and S38) demonstrated signals of two carbonyls (δ_C 173.7 and 183.8) and one ethoxy group (δ_H 4.15, q (7.1); δ_C 60.8 and δ_H 1.26, t (7.1); δ_C 14.1). HMBC experiment (Figure 7 and Figure S41) allowed to place the carboxy group at C-8 and ethyl ester at C-11 on the basis of congruous correlations from methylenes –<u>CH_2</u>-CH_3 (δ_H 4.15, q (7.1), 2H) and CH_2-11 (δ_H 2.41, dd (17.7, 5.3) and 2.30, m) to <u>carboxyl C</u>-12 (δ_C. 173.7) and also from methyl CH_3-30 (δ_H 1.17, s) to carboxyl C-13 (δ_C 183.8). The relative stereochemistry of **5** was determined by ROESY spectral analysis (Figure S43). Correlation between H-9 (δ_H 2.78, br t (5.0)) and CH_3-19 (δ_H 1.24, s) indicated their β-orientation. Meanwhile, a ROESY correlation between H-5 (δ_H 1.40, dd (12.6, 3.0))/H_a-11 (δ_H 2.41, dd (17.7, 5.3)) and H_b-11 (δ_H 2.30, m)/CH_3-30 (δ_H 1.17, s) confirmed the α-orientation of H-5, -CH_2-COOEt and CH_3-30. The above-mentioned results were in agreement with the spatial structure of isomalabaricane derivatives.

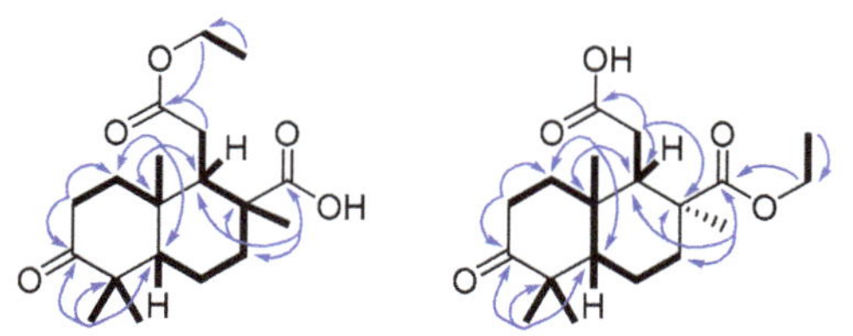

Figure 7. Selected COSY (▬) and HMBC (↶) correlations of **5** and **6**.

Compound **6** was an isomer of compound **5**, differed by the NMR signals (Figures S46 and S47) of carboxylic carbons (δ_C 177.6 and 178.0), methyl C-19 (δ_H 1.14, s), methylene CH_2-11 (δ_H 2.48, dd (18.1, 5.4) and 2.38, dd (18.1, 4.8)) and ethoxy group (δ_H 4.23, dq (10.9, 7.1); 4.13, dq (10.9, 7.1) and 1.31, t (7.1)). The key HMBC correlations (Figure 7) satisfied the proposed structure of **6**. However, since the values of carboxyl carbons shifts for **6** are close, distinguishing their correlations and direct ester positioning without data for isomer **5** brought some uncertainty. To avoid future difficulties with structurally related esters we calculated carbon chemical shift values for two isomers **5** and **6** (Figure S66). It was shown, that theoretical δ_C C-13 (**5**) = 192.7 and δ_C C-12 (**5**) = 182.4 gave the $\Delta\delta_{C(13\text{-}12)}$ = 10.3 ppm close to experimental value $\Delta\delta_{C(13\text{-}12)}$ = 10.1 ppm, while theoretical and experimental $\Delta\delta_{C(13\text{-}12)}$ for compound **6** were of 0.4 ppm (clcd δ_C C-13 (**6**) = 183.2, δ_C C-12 (**6**) = 182.8).

ROESY correlations of **6** supported the relative stereochemistry similarly to that of compound **5**. In fact, we detected expected NOE interactions H-9 (δ_H 2.74, br t (4.6))/CH_3-19 (δ_H 1.14, s); H-5 (δ_H 1.39, dd (12.7; 3.0))/H_a-11 (δ_H 2.48, dd (18.1, 5.4)) and H_b-11 (δ_H 2.38, dd (18.1, 4.8))/CH_3-30 (δ_H 1.12, s). Therefore, derivatives **5** and **6** possess the same stereochemistry as other co-isolated isomalabaricanes. Although compounds **5** and **6** are rather artificial products derived during EtOH extraction, the isolated pair of esters allowed to reliably establish the position of the ether group based on the chemical shifts of C-12 and C-13.

Both compounds were supposed to be the half-ester derivatives of the hypothetical dicarboxylic acid. The anhydrous form of the acid was reported by Ravi et al. [5] as a product of ozonolysis of isomalabaricane precursor [22,23]. Moreover, Ravi et al. obtained dimethyl and monomethyl esters of the acid and did not point the place of esterification in the case of the latter.

Among isolated new compounds **1–6**, we find stellettin S (**3**) the most intriguing, since occurrences of acetylene-containing isoprenoids are rare and not so far reported in the isomalabaricane series. To date, several biosynthetic pathways leading to the alkyne formation in natural products has been supported with identified and characterized gene clusters. In the first case, acetylenases, a special family of desaturases, catalyze the dehydrogenation of olefinic bonds in unsaturated fatty acids to afford acetylenic functionalities [24,25]. Next, acetylenases are also used to form the terminal alkyne in polyketides [26]. One more biosynthetic route results in a terminal alkyne formation in acetylenic amino acids and involves consequent transformations by halogenase BesD, oxidase BesC and lyase BesB [27]. Finally, two recent papers describe the molecular basis for the formation of alkyne moiety in acetylenic prenyl chains occurring in a number of meroterpenoids [28,29]. The abovementioned reports highlight hot trends in a scientific search for enzymatic machineries leading to the biologically significant and synthetically applicable acetylene bond in natural compounds. We believe that isolation of the new terpenoidal alkyne **3** could inspire further investigations of the *Stelletta* spp. sponges and associated microorganisms through genome mining.

According to obtained new data we also report the correction in stereochemistry of two asymmetric centers in globostelletins M (Figure S70) and N (Figure S71). Really, their ECD and NMR spectra in comparison with those of globostelletin K and stellettins Q and R (Figures S59–S71) clearly show rather 15*R*,23*S* configuration for globostelletin M instead of previously reported 15*S*,23*S* [15] as well as 15*S*,23*R* stereochemistry for globostelletin N instead of 15*R*,23*R* [15].

3. Materials and Methods

3.1. General Experimental Procedures

Optical rotations were measured on Perkin-Elmer 343 digital polarimeter (Perkin Elmer, Waltham, MA, USA). UV-spectra were registered on a Shimadzu UV-1601PC spectrophotometer (Shimadzu Corporation, Kyoto, Japan). ECD spectra were obtained on a Chirascan plus instrument (Applied Photophysics Ltd., Leatherhead, UK). ^{1}H-NMR (500.13 MHz, 700.13 MHz) and ^{13}C-NMR (125.75 MHz, 176.04 MHz) spectra were recorded in $CDCl_3$ on Bruker Avance III HD 500 and Bruker Avance III 700 spectrometers (Bruker BioSpin, Bremen, Germany). The ^{1}H- and ^{13}C-NMR chemical shifts were referenced to the solvent peaks at δ_H 7.26 and δ_C 77.0 for $CDCl_3$. HRESIMS analyses were performed using a Bruker Impact II Q-TOF mass spectrometer (Bruker). The operating parameters for ESI were as follows: a capillary voltage of 3.5 kV, nebulization witH-Nitrogen at 0.8 bar, dry gas flow of 7 L/min at a temperature of 200 °C. The mass spectra were recorded within *m/z* mass range of 100–1500. The instrument was operated using the otofControl (ver. 4.1, Bruker Daltonics) and data were analyzed using the DataAnalysis Software (ver. 4.4, Bruker Daltonics). Column chromatography was performed on Sephadex LH-20 (25–100 µm, Pharmacia Fine Chemicals AB, Uppsala, Sweden), silica gel (KSK, 50–160 mesh, Sorbfil, Krasnodar, Russia) and YMC ODS-A (12 nm, S-75 um, YMC Co., Ishikawa, Japan). HPLC were carried out using an Agilent 1100 Series chromatograph equipped with a differential refractometer (Agilent Technologies, Santa Clara, CA, USA). The reversed-phase columns YMC-Pack ODS-A (YMC Co., Ishikawa, Japan, 10 mm × 250 mm, 5 µm and 4.6 mm × 250 mm, 5 µm) and Discovery HS F5-5 (SUPELCO Analytical, Bellfonte, PA, USA, 10 mm

$\times$ 250 mm, 5 μm) were used for HPLC. Yields are based on dry weight (212.1 g) of the sponge sample.

3.2. Animal Material

The *Stelletta* sp. sponge sample (wet weight 1.3 kg) was collected by SCUBA diving at the depth of 7–12 m near Cham Island (15°54.3′ N, 108°31.9′ E) in the Vietnamese waters of the South China Sea during the 038-th cruise of R/V "Academik Oparin" in May 2010. The species was identified and described [14] by Dr. Boris B. Grebnev from G. B. Elyakov Pacific Institute of Bioorganic Chemistry, FEB RAS (PIBOC, Vladivostok, Russia). A voucher specimen (PIBOC O38-301) has been deposited at the collection of marine invertebrates in PIBOC.

3.3. Extraction and Isolation

The frozen sponge was chopped and extracted with EtOH (1.7 L $\times$ 3) (Figure S73). The EtOH soluble materials (52.5 g) were concentrated, dissolved in distilled H_2O (100 mL) and partitioned in turn with EtOAc (100 mL $\times$ 3). The EtOAc extracts were concentrated to a dark brown gum (15.7 g) that was further separated on a Sephadex LH-20 column (2 $\times$ 95 cm, $CHCl_3$/EtOH, 1:1) to yield three fractions. Fraction 2 (10.2 g) was separated into nine subfractions using step-wise gradient silica gel column chromatography (4 $\times$ 15 cm, $CHCl_3 \rightarrow$EtOH). Subfraction 2.4 (3.4 g) eluted with $CHCl_3$/EtOH (80:1–10:1) was subjected to a silica gel column (4 $\times$ 15 cm, $CHCl_3$/EtOH, 100:1$\rightarrow$10:1) to obtain four subfractions. The fourth subfraction 2.4.4 (255.5 mg) was subjected to reversed-phase HPLC (YMC-Pack ODS-A, 70% EtOH) to give four subsubfractions (2.4.4.1-4) that were subjected to rechromatography. The HPLC fractionation of 2.4.4.1 (YMC-Pack ODS-A, 60% EtOH) gave cyclobutastellettolide A (7.7 mg, 0.004%), mixtures of globostelletins E+F (~2:1; 4.0 mg, 0.002%), K (3.4 mg, 0.002%), and M (1.7 mg, 0.002%), that were purified by HPLC procedures (Discovery HS F5-5, 60% EtOH), as it was reported previously [14]. One more component of this subsubfraction was purified (Discovery HS F5-5, 60% EtOH) to yield stellettin V (**6**, 2.6 mg, 0.001%). The subsubfraction 2.4.4.2, subjected to HPLC (Discovery HS F5-5, 70% EtOH) gave cyclobutastellettolide B [14] (1.6 mg, 0.004%) and stellettin T (**4**, 1.2 mg, 0.0006%), purified by HPLC (Discovery HS F5-5, 70% EtOH). The subsubfraction 2.4.4.3 contained cyclobutastellettolide B (1.4 mg) and stellettin Q (**1**, 0.9 mg, 0.0008%) isolated using reversed-phase HPLC (Discovery HS F5-5, 70% EtOH). The third subfraction 2.4.3 (641.5 mg) was divided four times (~160 mg $\times$ 4) using reversed-phase column chromatography (1 $\times$ 5 cm, YMC-Pack ODS-A, 50% EtOH and 100% EtOH) to yeild two subsubfractions. The subsubfraction eluted with 50% EtOH was separated by HPLC (YMC-Pack ODS-A, 70% EtOH) to afford a number of compounds and mixes for further purification. Then, stellettin U (**5**, 10.3 mg, 0.005%) as well as globostelletins N (9.2 mg, 0.004%) and M (1.5 mg) were isolated by HPLC (Discovery HS F5-5) in 80% MeOH. The HPLC procedures (Discovery HS F5-5) in 80% EtOH were used to obtain individual stellettins R (**2**, 2.1 mg, 0.001%), S (**3**, 6.9 mg, 0.003%) and portion of stellettin Q (**1**, 0.9 mg) that was purified by HPLC (Discovery HS F5-5) rechromatography in 65% CH_3CN. Finally, one more portion of cyclobutastellettolide B (5.0 mg) was obtained from the subfraction by HPLC (Discovery HS F5-5) in 85% MeOH.

3.4. Compound Characteristics

Stellettin Q (**1**): Yellow oil; $[\alpha]_D{}^{25}$–65.0 (*c* 0.1, $CHCl_3$); ECD (*c* 8.6 $\times$ 10^{-4} M, EtOH) λ_{max} ($\Delta\varepsilon$) 195 (4.15), 229 (−27.92), 262 (12.78), 355 (−1.75) nm; ^{1}H- and ^{13}C-NMR data ($CDCl_3$), Table 1; HRESIMS *m/z* 523.3065 $[M–H]^-$ (calcd for $C_{32}H_{43}O_6$ 523.3065).

Stellettin R (**2**): Yellow oil; $[\alpha]_D{}^{25}$–24.0 (*c* 0.2, $CHCl_3$); ECD (*c* 1.3 $\times$ 10^{-3} M, EtOH) λ_{max} ($\Delta\varepsilon$) 195 (4.06), 229 (−5.33), 252 (1.46), 273 (−0.48), 295 (0.47), 353 (−1.20) nm; ^{1}H- and ^{13}C- NMR data ($CDCl_3$), Table 1; HRESIMS *m/z* 523.3068 $[M–H]^-$ (calcd for $C_{32}H_{43}O_6$ 523.3065).

Stellettin S (**3**): Slightly yellow oil; $[\alpha]_D^{25}$ + 31.5 (*c* 0.2, $CHCl_3$); ECD (*c* 5.3 × 10^{-3} M, EtOH) λ_{max} ($\Delta\varepsilon$) 289 (0.34), 321 (−0.14) nm; ^{1}H- and ^{13}C-NMR data ($CDCl_3$), Table 2; HRESIMS *m/z* 303.1966 [M–H]$^-$ (calcd for $C_{19}H_{27}O_3$ 303.1966).

Stellettin T (**4**): Slightly yellow oil; $[\alpha]_D^{25}$–22.0 (*c* 0.1, $CHCl_3$); ECD (*c* 3.4 × 10^{-3} M, EtOH) λ_{max} ($\Delta\varepsilon$) 208 (−0.98), 249 (0.58), 280 (−0.27), 321 (−0.29) nm; ^{1}H- and ^{13}C-NMR data ($CDCl_3$), Table 2; HRESIMS *m/z* 351.2176 [M–H]$^-$ (calcd for $C_{20}H_{31}O_5$ 351.2177).

Stellettin U (**5**): Slightly yellow oil; $[\alpha]_D^{25}$ 0.0 (*c* 0.2, $CHCl_3$); ECD (*c* 5.7 × 10^{-3} M, EtOH) λ_{max} ($\Delta\varepsilon$) 197 (−0.73), 232 (0.27), 261 (0.45), 295 (0.01), 321 (−0.34) nm; ^{1}H- and ^{13}C-NMR data ($CDCl_3$), Table 2; HRESIMS *m/z* 337.2022 [M–H]$^-$ (calcd for $C_{19}H_{29}O_5$ 337.2020).

Stellettin V (**6**): Slightly yellow oil; $[\alpha]_D^{25}$ + 32.9 (*c* 0.17, $CHCl_3$); ECD (*c* 7.7 × 10^{-3} M, EtOH) λ_{max} ($\Delta\varepsilon$) 210 (−0.04), 220 (0.08), 238 (−0.23), 269 (0.38), 291 (0.26), 330 (0.13), 357 (−0.09) nm; ^{1}H- and ^{13}C-NMR data ($CDCl_3$), Table 2; HRESIMS *m/z* 337.2025 [M–H]$^-$ (calcd for $C_{19}H_{29}O_5$ 337.2020).

Globostelletin N (Figure 1): Slightly yellow oil; $[\alpha]_D^{25}$ + 17.8 (*c* 0.23, MeOH); ECD (*c* 1.2 × 10^{-3} M, EtOH) λ_{max} ($\Delta\varepsilon$) 196 (−4.18), 228 (9.44), 256 (−1.72), 289 (0.97), 340 (−1.97) nm; ^{1}H- and ^{13}C-NMR spectra ($CDCl_3$) corresponded to previously reported data [15] (Figures S62–S64); HRESIMS *m/z* 479.2808 [M–H]$^-$ (calcd for $C_{30}H_{39}O_5$ 479.2803).

4. Conclusions

To summarize, the present report describes the isolation and structural elucidation of six metabolites **1–6** from a tropical marine sponge belonging the genus *Stelletta*. A combination of NMR methods, supported with computational quantum-chemical modeling allowed us to establish the structures and absolute stereochemistry of two isomalabaricanes **1** and **2**, while the structures and configurations of four isomalabaricane-derived terpenoids **3–6** were suggested on the basis of spectral data and biogenetic considerations. Stellettin S (**3**) represents the first acetylene-containing isomalabaricane-related compound. Additionally, according to new data the absolute stereochemistry of the C-15 and C-23 asymmetric centers of known globostelletins M and N were corrected.

Supplementary Materials: The following are available online. Figure S1: article title, authors affiliations and contact information, Figure S2: contents of Supplementary Materials, Figures S3–S10: HRESIMS, ^{1}H- and ^{13}C-NMR, HSQC, HMBC, COSY, ROESY spectra ($CDCl_3$, 700 MHz) and ECD spectrum (EtOH) of stellettin Q (**1**), respectively, Figures S11–S18: HRESIMS, ^{1}H- and ^{13}C-NMR, HSQC, HMBC, COSY, ROESY spectra ($CDCl_3$, 500 MHz) and ECD spectrum (EtOH) of stellettin R (**2**), respectively, Figures S19–S27: HRESIMS, ^{1}H- and ^{13}C-NMR, DEPT, HSQC, HMBC, COSY, ROESY spectra ($CDCl_3$, 700 MHz) and ECD spectrum (EtOH) of stellettin S (**3**), respectively, Figures S28–S35: HRESIMS, ^{1}H- and ^{13}C-NMR, HSQC, HMBC, COSY, ROESY spectra ($CDCl_3$, 700 MHz) and ECD spectrum (EtOH) of stellettin T (**4**), respectively, Figures S36–S44: HRESIMS, ^{1}H- and ^{13}C-NMR, DEPT, HSQC, HMBC, COSY, ROESY spectra ($CDCl_3$, 700 MHz) and ECD spectrum (EtOH) of stellettin U (**5**), respectively, Figures S45–S53: HRESIMS, ^{1}H- and ^{13}C-NMR, DEPT, HSQC, HMBC, COSY, ROESY spectra ($CDCl_3$, 700 MHz) and ECD spectrum (EtOH) of stellettin V (**6**), respectively, Figures S54–S57: HRESIMS, ^{1}H- and ^{13}C-NMR spectra ($CDCl_3$) and ECD spectrum (EtOH) of globostelletin K, respectively, Figures S58–S61: HRESIMS, ^{1}H- and ^{13}C-NMR spectra ($CDCl_3$) and ECD spectrum (EtOH) of globostelletin M, Figures S62–S65: HRESIMS, ^{1}H and ^{13}C NMR spectra ($CDCl_3$) and ECD spectrum (EtOH) of globostelletin N, respectively, Figure S66: theoretical modeling details, Figures S67 and S68: optimized geometries and statistical weights of main and minor conformations of stellettin Q (**1**) and stellettin R (**2**), respectively, Figures S69–S71: the computational ECD results for globostelletins K, M and N, respectively. Figure S72: the computational ECD results calculated for all possible 15,23-stereoisomers of studied globostelletins K, M, N, and new stellettins Q (**1**) and R (**2**), Figure S73: the isolation scheme.

Author Contributions: Conceptualization, S.A.K., E.G.L.; methodology, S.A.K. and E.G.L.; formal analysis, S.A.K., E.G.L. and D.V.B.; investigation, S.A.K., E.G.L. and A.B.K.; data curation S.A.K. and E.G.L.; writing—original draft preparation, S.A.K. and E.G.L.; writing—review and editing, S.A.K., E.G.L. and V.A.S.; visualization, S.A.K. and E.G.L.; NMR data providing, A.I.K.; HRESI MS spectra acquisition and interpretation, R.S.P.; supervision, V.A.S.; project administration, V.A.S.; funding acquisition, V.A.S. All authors have read and agreed to the published version of the manuscript.

Funding: This research was funded by the RSF (Russian Science Foundation), grant number 20-14-00040.

Acknowledgments: The study was carried out on the equipment of the Collective Facilities Center "The Far Eastern Center for Structural Molecular Research (NMR/ MS) of PIBOC FEB RAS". Authors thank the Vietnamese Academy of Science and Technology for collaboration and participation in previous joint publication [14].

Conflicts of Interest: The authors declare no conflict of interest.

Sample Availability: Samples of the compounds are available from the authors.

References

1. Chawla, A.; Dev, S. A new class of triterpenoids from *Ailanthus malabarica* DC derivatives of malabaricane. *Tetrahedron Lett.* **1967**, *8*, 4837–4843. [CrossRef]
2. Ziegler, H.L.; Stærk, D.; Christensen, J.; Olsen, C.E.; Sittie, A.A.; Jaroszewsky, J.W. New dammarane and malabaricane triterpenes from *Caloncoba echinata*. *J. Nat. Prod.* **2002**, *65*, 1764–1768. [CrossRef] [PubMed]
3. Achanta, P.S.; Gattu, R.K.; Belvotagi, A.R.V.; Akkinepally, R.R.; Bobbala, R.K.; Achanta, A.R.V.N. New malabaricane triterpenes from the oleoresin of *Ailanthus malabarica*. *Fitoterapia* **2015**, *100*, 166–173. [CrossRef]
4. Ebada, S.S.; Lin, W.; Proksch, P. Bioactive sesterterpenes and triterpenes from marine sponges: Occurrence and pharmacological significance. *Mar. Drugs* **2010**, *8*, 313–346. [CrossRef]
5. Ravi, B.N.; Wells, R.J.; Croft, K.D. Malabaricane triterpenes from a Fijian collection of the sponge *Jaspis stellifera*. *J. Org. Chem.* **1981**, *46*, 1998–2001. [CrossRef]
6. Wu, Q.; Nay, B.; Yang, M.; Ni, Y.; Wang, H.; Yao, L.; Li, X. Marine sponges of the genus *Stelletta* as promising drug sources: Chemical and biological aspects. *Acta Pharm. Sin. B.* **2019**, *9*, 237–257. [CrossRef] [PubMed]
7. Li, Y.; Tang, H.; Tian, X.; Lin, H.; Wang, M.; Yao, M. Three new cytotoxic isomalabaricane triterpenes from the marine sponge *Stelletta tenuis*. *Fitoterapia* **2015**, *106*, 226–230. [CrossRef] [PubMed]
8. Barrero, A.F.; Oltra, J.E.; Herrador, M.M.; Cabrera, E.; Sanchez, J.F.; Quílez, J.F.; Rojas, F.J.; Reyes, J.F. Gibepyrones: α-pyrones from *Gibberella fujikuroi*. *Tetrahedron* **1993**, *49*, 141–150. [CrossRef]
9. Tang, S.; Xu, R.; Lin, W.; Duan, H. Jaspiferin A and B: Two new secondary metabolites from the South China sea sponge *Jaspis stellifera*. *Rec. Nat. Prod.* **2012**, *6*, 398–401.
10. Dung, D.T.; Hang, D.T.T.; Nhiem, N.X.; Quang, T.H.; Tai, B.H.; Yen, P.H.; Hoai, N.T.; Thung, D.C.; Minh, C.V.; Kiem, P.V. Rhabdaprovidines D–G, four new 6,6,5-tricyclic terpenoids from the Vietnamese sponge *Rhabdastrella providentiae*. *Nat. Prod. Commun.* **2018**, *13*, 1251–1254. [CrossRef]
11. Kiem, P.V.; Dung, D.T.; Yen, P.H.; Nhiem, N.X.; Quang, T.H.; Tai, B.H.; Minh, C.V. New isomalabaricane analogues from the sponge *Rhabdastrella providentiae* and their cytotoxic activities. *Phytochem. Lett.* **2018**, *26*, 199–204. [CrossRef]
12. Tang, S.; Pei, Y.; Fu, H.; Deng, Z.; Li, J.; Proksch, P.; Lin, W. Jaspolides A–F, six new isomalabricane-type terpenoids from the sponge *Jaspis* sp. *Chem. Pharm. Bull.* **2006**, *54*, 4–8. [CrossRef] [PubMed]
13. Li, J.; Xu, B.; Cui, J.; Deng, Z.; de Voogd, N.J.; Proksch, P.; Lin, W. Globostelletins A–I, cytotoxic isomalabaricane derivatives from the marine sponge *Rhabdastrella globostellata*. *Bioorg. Med. Chem.* **2010**, *18*, 4639–4647. [CrossRef] [PubMed]
14. Kolesnikova, S.A.; Lyakhova, E.G.; Kalinovsky, A.I.; Berdyshev, D.V.; Pislyagin, E.A.; Popov, R.S.; Grebnev, B.B.; Makarieva, T.N.; Minh, C.V.; Stonik, V.A. Cyclobutastellettolides A and B, C_{19} norterpenoids from a *Stelletta* sp. marine sponge. *J. Nat. Prod.* **2019**, *82*, 3196–3200. [CrossRef] [PubMed]
15. Li, J.; Zhu, H.; Ren, J.; Deng, Z.; de Voogd, N.J.; Proksch, P.; Lin, W. Globostelletins J–S, isomalabaricanes with unusual cyclopentane sidechains from the marine sponge *Rhabdastrella globostellata*. *Tetrahedron* **2012**, *68*, 559–565. [CrossRef]
16. Stephens, P.J.; Devlin, F.J.; Chabalowski, C.F.; Frisch, M.J. Ab initio calculation of vibrational absorption and circular dichroism spectra using density functional force fields. *J. Phys. Chem.* **1994**, *98*, 11623–11627. [CrossRef]
17. Miertuš, S.; Scrocco, E.; Tomasi, J. Electrostatic interaction of a solute with a continuum. A direct utilizaion of ab initio molecular potentials for the prevision of solvent effects. *Chem. Phys.* **1981**, *55*, 117–129. [CrossRef]
18. Frisch, M.J.; Trucks, G.W.; Schlegel, H.B.; Scuseria, G.E.; Robb, M.A.; Cheeseman, J.R.; Scalmani, G.; Barone, V.; Mennucci, B.; Petersson, G.A.; et al. *Gaussian 16*; Revision B.01.; Gaussian, Inc.: Wallingford, CT, USA, 2016.
19. Ryu, G.; Matsunaga, S.; Fusetani, N. Globostellatic Acids A−D, new cytotoxic isomalabaricane triterpenes from the marine sponge *Stelletta globostellata*. *J. Nat. Prod.* **1996**, *59*, 512–514. [CrossRef]
20. Zampella, A.; D'Auria, M.V.; Debitus, C.; Menou, J.-L. New isomalabaricane derivatives from a new species of *Jaspis* sponge collected at the Vanuatu islands. *J. Nat. Prod.* **2000**, *63*, 943–946. [CrossRef]
21. Deng, J.; Wu, J.; Tian, H.; Bao, J.; Shi, Y.; Tian, W.; Gui, J. Alkynes from furans: A general fragmentation method applied to the synthesis of the proposed structure of aglatomin B. *Angew. Chem.* **2018**, *57*, 3617–3621. [CrossRef]

22. McCabe, T.; Clardy, J.; Minale, L.; Pizza, C.; Zollo, F.; Riccio, R. A triterpenoid pigment with the isomalabaricane skeleton from the marine sponge *Stelletta* sp. *Tetrahedron Lett.* **1982**, *23*, 3307–3310. [CrossRef]
23. McCormick, J.L.; McKee, T.C.; Cardellina, J.H.; Leid, M.; Boyd, M.R. Cytotoxic triterpenes from a marine sponge, *Stelletta* sp. *J. Nat. Prod.* **1996**, *59*, 1047–1050. [CrossRef]
24. Lee, M. Identification of non-heme diiron proteins that catalyze triple bond and epoxy group formation. *Science* **1998**, *280*, 915–918. [CrossRef] [PubMed]
25. Jeon, J.E.; Kim, J.-G.; Fischer, C.R.; Mehta, N.; Dufour-Schroif, C.; Wemmer, K.; Mudgett, M.B.; Sattely, E. A pathogen-responsive gene cluster for highly modified fatty acids in tomato. *Cell* **2020**, *180*, 176–187. [CrossRef] [PubMed]
26. Zhu, X.; Liu, J.; Zhang, W. *De novo* biosynthesis of terminal alkyne-labeled natural products. *Nat. Chem. Biol.* **2015**, *11*, 115–120. [CrossRef] [PubMed]
27. Marchand, J.A.; Neugebauer, M.E.; Ing, M.C.; Lin, C.-I.; Pelton, J.G.; Chang, M.C.Y. Discovery of a pathway for terminal-alkyne amino acid biosynthesis. *Nature* **2019**, *567*, 420–424. [CrossRef]
28. Chen, Y.; Naresh, A.; Liang, S.; Lin, C.; Chein, R.; Lin, H. Discovery of a dual function cytochrome P450 that catalyzes enyne formation in cyclohexanoid terpenoid biosynthesis. *Angew. Chem.* **2020**, *59*, 13537–13541. [CrossRef]
29. Lv, J.; Gao, Y.; Zhao, H.; Awakawa, T.; Liu, L.; Chen, G.; Yao, X.; Hu, D.; Abe, I.; Gao, H. Biosynthesis of biscognienyne B involving a cytochrome P450-dependent alkynylation. *Angew. Chem.* **2020**, *59*, 13531–13536. [CrossRef]

molecules

MDPI

Article

Structural Analysis of Oxidized Cerebrosides from the Extract of Deep-Sea Sponge *Aulosaccus* sp.: Occurrence of Amide-Linked Allylically Oxygenated Fatty Acids

Elena A. Santalova *, Vladimir A. Denisenko and Pavel S. Dmitrenok

G.B. Elyakov Pacific Institute of Bioorganic Chemistry, Far Eastern Branch of the Russian Academy of Sciences, Pr. 100 let Vladivostoku 159, 690022 Vladivostok, Russia; vladenis@piboc.dvo.ru (V.A.D.); paveldmt@piboc.dvo.ru (P.S.D.)

* Correspondence: santalova@piboc.dvo.ru or santalova.e.a@yandex.ru; Tel. +7-(4232)-31-14-30; Fax: +7-(4232)-31-40-50

Academic Editors: Daniela Rigano and RuAngelie Edrada-Ebel

Received: 1 November 2020; Accepted: 19 December 2020; Published: 21 December 2020

Abstract: The structural elucidation of primary and secondary peroxidation products, formed from complex lipids, is a challenge in lipid analysis. In the present study, rare minor oxidized cerebrosides, isolated from the extract of a far eastern deep-sea glass sponge, *Aulosaccus* sp., were analyzed as constituents of a multi-component RP-HPLC (high-performance liquid chromatography on reversed-phase column) fraction using NMR (nuclear magnetic resonance) spectroscopy, mass spectrometry, GC (gas chromatography), and chemical transformations (including hydrogenation or derivatization with dimethyl disulfide before hydrolysis). Eighteen previously unknown β-D-glucopyranosyl-(1→1)-ceramides (**1a–a″**, **1b–b″**, **2a–a″**, **2b–b″**, **3c–c″**, **3d–d″**) were shown to contain phytosphingosine-type backbones (2*S*,3*S*,4*R*,11*Z*)-2-aminoeicos-11-ene-1,3,4-triol (in **1**), (2*S*,3*S*,4*R*,13*Z*)-2-aminoeicos-13-ene-1,3,4-triol (in **2**), and (13*S**,14*R**)-2-amino-13,14-methylene-eicosane-1,3,4-triol (in **3**). These backbones were *N*-acylated with straight-chain monoenoic (2*R*)-2-hydroxy acids that had allylic hydroperoxy/hydroxy/keto groups on C-17′ in the 15′*E*-23:1 chain (**a–a″**), C-16′ in the 17′*E*-23:1 (**b–b″**) and 14′*E*-22:1 (**c–c″**) chains, and C-15′ in the 16′*E*-22:1 chain (**d–d″**). Utilizing complementary instrumental and chemical methods allowed for the first detailed structural analysis of a complex mixture of glycosphingolipids, containing allylically oxygenated monoenoic acyl chains.

Keywords: glycosphingolipids; cerebrosides; peroxidation products; structure elucidation; allylic thioether; NMR; ESI-MS; GC-MS; mass spectra; glass sponge

1. Introduction

Lipid hydroperoxides are labile compounds derived from lipids containing carbon-carbon double bonds. The formation of these primary peroxidation products occurs in enzymatic and non-enzymatic (autooxidation, photo-oxidation) reactions [1]. Lipid hydroperoxides, formed in biological systems, have not only multiple damaging effects on cellular macromolecules, but are also important regulators of many cellular processes [2]. In some pathological situations, lipid hydroperoxides are generated at higher than normal rates. Such overproduction is implicated in several human diseases and exposures including atherosclerosis, cancer, diabetes, acute lung injury, chronic alcohol exposure, and neurodegenerative disorders. The complex nature of lipid peroxidation and its potential biological significance have attracted the attention of scientists across many disciplinary fields, ranging from chemistry and biochemistry to biology and clinical science (see review [3] and references cited herein).

Recent progress in the characterization and quantification of oxygenated fatty acids (FAs) has facilitated a better understanding of lipid oxidation, but the methods currently available still suffer from unresolved sensitivity and selectivity problems [4]. The main reason for these problems is the complexity of the product profile formed by oxidation from a single lipid molecular species. In particular, peroxidation of any unsaturated fatty acyl group may generate a mixture of allylic hydroperoxides with different double bond positions and/or configurations. These hydroperoxides are easily decomposed or go through further reactions to form different secondary peroxidation products including epoxides, allylic alcohols, α,β-unsaturated ketones (enones), and chain-cleavage products [1]. The constituents of these mixtures are difficult to separate, isolate, and identify. Additionally, natural extracts are characterized, as a rule, by extremely low levels of oxidized lipids, causing further difficulties in their isolation and analysis [3,4]. Undoubtedly, structural elucidation of primary and secondary peroxidation products, formed from complex lipids, remains a challenge for lipid analysts.

Monoenes are much less prone to undergo oxidation than polyenes. Respectively, direct addition of peroxyl radicals to monounsaturated lipids is not generally observed with the exceptions of cholesterol [3] and related sterols. In particular, sphingolipids, containing mainly saturated or monounsaturated hydrocarbon chains, are poor substrates for peroxidation. We found only one report on the isolation of peroxidized sphingolipids from a natural source, but these compounds formed due to oxidation of unique sphingoid base moiety with conjugated double bonds. Namely, some stereoisomeric glycosphingolipids, containing endoperoxide and allylic hydroperoxide functions in their dienoic sphingoid base moieties, were isolated from the extract of the sponge *Axinella corrugate* [5]. The locations of the double bonds and peroxide-containing groups were determined by ^{1}H,^{1}H-COSY (proton-proton correlation spectroscopy), HMBC (heteronuclear multiple-bond correlation spectroscopy), and ROESY (rotating-frame nuclear Overhauser effect spectroscopy) experiments due to the proximity of the previously mentioned functionalities to the polar portion of the sphingolipids. Methanolysis (MeOH/HCl) of the peroxidized glycosphingolipids was used to release methyl esters of saturated FAs, but oxidized dienoic sphingoid bases were not obtained, apparently, due to rapid decomposition under acidic conditions.

In mass spectrometric (MS) studies of oxidative stress markers (disease biomarkers), oxidation of some standard glycosphingolipids, containing monoenoic FAs, was induced by a Fenton reaction (H_2O_2/$FeCl_2$) or UVA (ultraviolet A) irradiation [6,7]. In particular, Couto et al. obtained galactosylceramides with allylic hydroperoxy, hydroxy, or keto groups in FA chains. These cerebrosides were characterized using ESI-MS (electrospray ionization mass spectrometry) and HPLC-MS (high-performance liquid chromatography-mass spectrometry) methods (including tandem mass spectrometry (MS/MS)) [6]. In our continued study on lipids from far eastern marine sponges [8], similar oxidized cerebrosides (Figure 1), presumably derived from glucosylceramides with monounsaturated fatty acyl groups, were found in the extract of a deep-sea glass sponge, *Aulosaccus* sp. As seen in Figure 1, the depicted compounds contain an allylic hydroperoxy, hydroxy, or keto group in each acyl chain, but their monoenoic sphingoid base moieties are not oxidized. A fraction of these minor cerebrosides, along with fractions of non-oxidized cerebrosides [8], was isolated from the sponge extract using high-performance liquid chromatography on reversed-phase column (RP-HPLC). The present study presents structural elucidation of rare oxidized glycosphingolipids using mass spectrometry (ESI-MS and gas-chromatography-mass spectrometry (GC-MS)), ^{1}H-, ^{13}C-NMR (nuclear magnetic resonance) spectroscopy, GC, chemical transformations, and optical rotation measurement.

A classical approach to the analysis of any complex lipid includes methanolysis or hydrolysis, which is followed by analyses of liberated derivatized simple lipids and sugar. However, prolonged high temperatures and treatments with acidic or alkaline solutions, required for solvolysis of *N*-acyl lipids, presents significant problems with respect to the potentially labile allylic oxygen-containing groups of the acyl chains. To solve these problems, we used catalytic hydrogenation to fix the starting positions of the oxygenated groups prior to chemical degradation of the oxidized glycosphingolipids. In addition, attempts were made to fix the double bonds of allylic substructures by reacting with dimethyl disulfide (DMDS) because, in our preliminary research, the DMDS adduct of methyl oleate did not lose *S*-methyl groups during hydrolysis with HCl in MeCN-H_2O. To minimize possible allylic rearrangements (1,3-isomerizations) and other alterations, we avoided elevated temperatures and strong acid/base conditions in the derivatization reactions before hydrolysis. Thus, our attention was mainly focused on procedures suitable for an initial detailed structural analysis of a complex mixture of glycosphingolipids, containing an allylic hydroperoxy, hydroxy, or keto group in the monoenoic acyl chain.

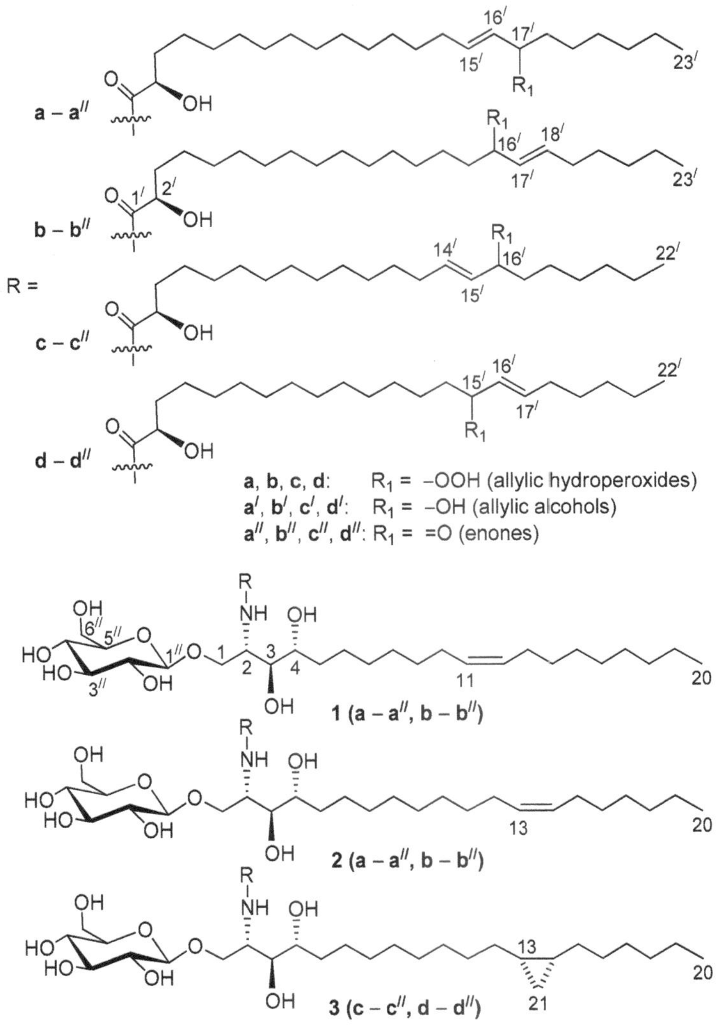

Figure 1. Oxidized cerebrosides from the extract of a sponge *Aulosaccus* sp.

2. Results and Discussion

Isomeric allylic hydroperoxides **1a**, **1b**, **2a**, **2b**, **3c**, **3d**, related isomeric allylic alcohols **1a′**, **1b′**, **2a′**, **2b′**, **3c′**, **3d′**, and isomeric enones **1a″**, **1b″**, **2a″**, **2b″**, **3c″**, **3d″** (Figure 1) were constituents of a single RP-HPLC fraction. In the beginning of this study, the fraction contained 46% allylic hydroperoxides, 43% allylic alcohols, and 11% enones, calculated using relative intensities of the characteristic, isolated

^{1}H-NMR signals of *trans*-olefinic protons (see below). The final results of this study yielded 25%, 54%, and 21% of these allylically oxygenated compounds, respectively. Therefore, further transformation of allylic hydroperoxides to allylic alcohols and enones occurred upon analysis (registrations of NMR and MS spectra, preparation of samples, storage, etc.).

In UPLC-MS (ultra-performance liquid chromatography–mass spectrometry) analysis of oxidized cerebrosides, base peak chromatogram, and extracted-ion chromatograms provided limited information because many components eluted simultaneously. In particular, peaks for major isomeric allylic hydroperoxides **1a** and **1b** were not well-resolved (Supplementary Materials, Figure S1a,b).

2.1. Positive and Negative Ion Mode ESI-MS/MS Analyses of Oxidized Cerebrosides

The molecular formulae, $C_{49}H_{93}NO_{12}$ for allylic hydroperoxides (**1a, 1b, 2a, 2b, 3c, 3d**), $C_{49}H_{93}NO_{11}$ for allylic alcohols (**1a$^/$, 1b$^/$, 2a$^/$, 2b$^/$, 3c$^/$, 3d$^/$**), and $C_{49}H_{91}NO_{11}$ for enones (**1a$^{//}$, 1b$^{//}$, 2a$^{//}$, 2b$^{//}$, 3c$^{//}$, 3d$^{//}$**), were determined by HR-ESI-MS (high resolution ESI-MS) analyses in positive-((+)ESI-MS) and negative-((−)ESI-MS) ion modes. Complementary (+)ESI- and (−)ESI-MS/MS analyses of these glucosylceramides, containing 2-hydroxy acyl chains and phytosphingosine-type backbones, resulted in a series of fragment ions, as shown in Scheme 1.

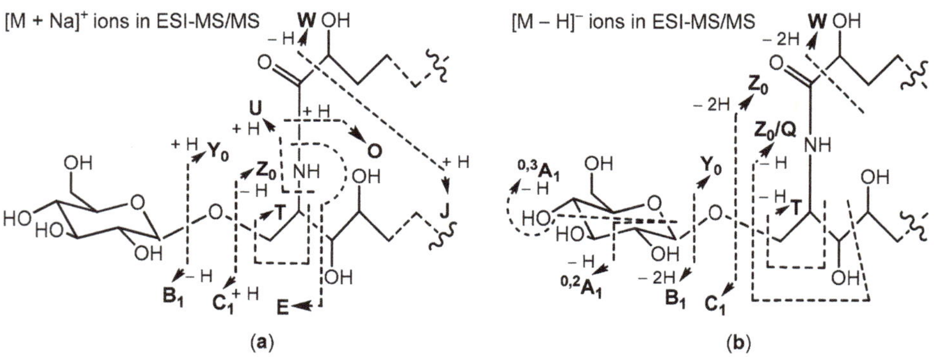

Scheme 1. Designations of the fragment ions observed in the ESI-MS/MS spectra of (**a**) $[M + Na]^+$ and (**b**) $[M - H]^-$ ions of the cerebrosides reported here. These designations are consistent with the ones proposed by Ann and Adams [9]. In addition, $[Z_0/Q - C_3H_5N$ (55 Da)] ions (not shown) were observed in the MS/MS spectra of the $[M - H]^-$ ions. The fragment Z_0/Q was also referred to as a Z_0/K ion [10].

Many of the fragment ions, shown in Scheme 1, have also been detected in our ESI-MS/MS studies of non-oxidized cerebrosides isolated from *Aulosaccus* sp. Namely, (+)ESI-MS/MS spectra of sodium adducts from non-oxidized glycosphingolipids have been characterized by prominent peaks, corresponding to $[M + Na]^+$ (base peak), Y_0, Z_0, O, and C_1 ions, and by small peaks, representing $[M + Na - H_2O]^+$, E, and B_1 ions. However, the (+)ESI-MS/MS spectrum of the $[M + Na]^+$ ion of isomeric hydroperoxy cerebrosides (Figure 2) showed other relative abundances for a variety of these ions. In particular, the O ions of m/z 528.35, representing isomeric monoglucosylated monounsaturated C_{20} sphingoid base backbones **1** and **2**, constituted the base peak of this spectrum. The spectrum also exhibited a homologous, less abundant $O^/$ ion (m/z 542.36), containing a monoglucosylated cyclopropane C_{21} sphingoid base backbone **3**. A relatively low intensity pseudo-molecular ion peak (m/z 910.65, $[M + Na]^+$) and very small peaks, corresponding to Y_0 (m/z 748.60) and Z_0 (m/z 730.58) ions (not shown), were observed. The presence of the $[M + Na - H]^+$ peak, comparable with the pseudo-molecular ion peak of the hydroperoxides, was explained by hydrogen atom abstraction, followed by electronic delocalization in the resulting radical, which might yield rearranged products. At the same time, the (+)ESI-MS/MS spectrum revealed more abundant $[M + Na - 102]^+$ (m/z 808.55), $[M + Na - 88]^+$ (m/z 822.56), $[Y_0 - 102]$ (m/z 646.505), $[Y_0 - 88]$ (m/z 660.52), $[Z_0 - 102]$ (m/z 628.49), and [$Z_0 - 88$] (*m*/*z* 642.50) ions. These fragments and fragment [$W - 102$] (*m*/*z* 275.20), each with a terminal α,β-unsaturated aldehyde, were thought to arise from specific α-cleavages of lipid hydroperoxides (Figure 2). Related fragment ions, formed by $C_nH_{2n+2}O$ losses from $[M + Na]^+$ ions, were also found in the MS/MS studies of some monoenoic [6] and polyenoic [11–13] FA moieties or free FAs, in which an allylic hydroperoxy group was between double bond(s) and a terminal methyl group. In general, the $[M + Na - C_nH_{2n+2}O]^+$ ions were also more abundant in MS/MS spectra of those compounds compared to their $[M + Na]^+$ ions.

Ions, possibly formed by Hock cleavage [14,15], were insignificant in our (+)ESI-MS/MS analysis of allylic hydroperoxides. These fragments included *m*/*z* 796.55 (for **1b**, **2b**, and **3d**) and 782.54 (for the allylic isomers of **1a**, **2a**, and **3c**) ions, as illustrated in Scheme S1 (Supplementary Materials) and Figure 2 (allylic rearrangements for acyl chains **a** and **c** are not shown). In contrast, the relatively abundant ions, presumably formed by analogous cleavage from hemiacetal derivatives, were reported for MS/MS fragmentations of $[M + Na]^+$ precursors of some free polyenoic FAs in which an allylic hydroperoxy group was between the double bond system and C-1 [12,13].

Favorable cleavage of the isomers, formed by allylic rearrangements of acyl chains **b** and **d** (Figure 2), occurred, losing an 88 Da fragment. At the same time, compounds with parent acyl chains **b** and **d** underwent other favorable fragmentations, yielding a distinct group of homologous ions (from *m*/*z* 668.44 to 738.50), containing the most significant fragment of *m*/*z* 724.49. We suggest that the occurrence of these ions may be connected with homolysis of a weak RO-/-OH bond, formation of an alkoxyl radical, and subsequent formation of a radical centered on a remote and non-activated carbon atom of a saturated hydrocarbon chain. This may lead to fast cyclization that, in turn, leads to MS fragmentations of the resulting cyclic ethers, shown in Scheme 2. The proposed cyclization reaction is reminiscent of the formation of cyclic (mainly, five-membered) ethers from acyclic saturated monohydroxy alcohols, occurring via radical (primarily alkoxyl) intermediates under appropriate chemical and thermal conditions (for reviews, see References [16,17]). In this process, secondary aliphatic alcohols yielded 2,5-dialkyltetrahydrofurans [18]. A possible mechanism for the formation of such products includes 1,5-transposition of the radical center from the oxygen atom of the alkoxyl radical to δ-carbon atom, involving a 1,5-hydrogen transfer through a chair-like six-membered transition state. For a linear hydrocarbon chain with an initial alkoxyl radical, the 1,5-hydrogen atom transfer may be considered the most common reaction, even though intramolecular abstractions of the hydrogen atom from other positions (1,4-migrations, 1,6-migrations, and 1,7-migrations, etc.) may be observed [17]. Similar processes may have occurred in our MS/MS experiment because alkyl hydroperoxides are known to form alkoxyl radicals by thermal or photolytic decomposition [17,19]. Apparently, the ability to undergo favorable cyclic transition states affected the fragmentation process, leading to the formation of major homologous ions, as illustrated in Scheme 2.

Ion intensity profiles, obtained for cerebrosides with allylic hydroxy or keto groups, were, in general, similar to those of non-oxidized cerebrosides found in *Aulosaccus* sp. In particular, the MS/MS spectra of sodiated molecular ions of allylic alcohols (Figure 3a: *m*/*z* 894.66 $[M + Na]^+$) and enones (Figure 3b: *m*/*z* 892.64 $[M + Na]^+$) showed predominant peaks of pseudo-molecular ions, along with peaks of lower intensities, which represented $[M + Na - H_2O]^+$, Y_0, Z_0, and *O* ions. However, unlike $[M + Na]^+$ ions of non-oxidized cerebrosides, the sodium adducts of allylic alcohols and enones fragmented to give discernible ions of *W*-type and minor *U* and *T* ions. Additionally, the acyl-containing ions of allylic alcohols had a tendency to lose one, or even two, hydrogen atoms. In this case, a trend was observed toward the increased loss of hydrogen atoms with decreasing ion masses. For example, a significant difference was noted between the relative intensities of $[M + Na]^+$ and $[M + Na - H]^+$ peaks ($[M + Na - 2H]^+$ ions were not even detected), but the intensities of *W*, [W – H], and [W – 2H] peaks were comparable (Figure 3a).

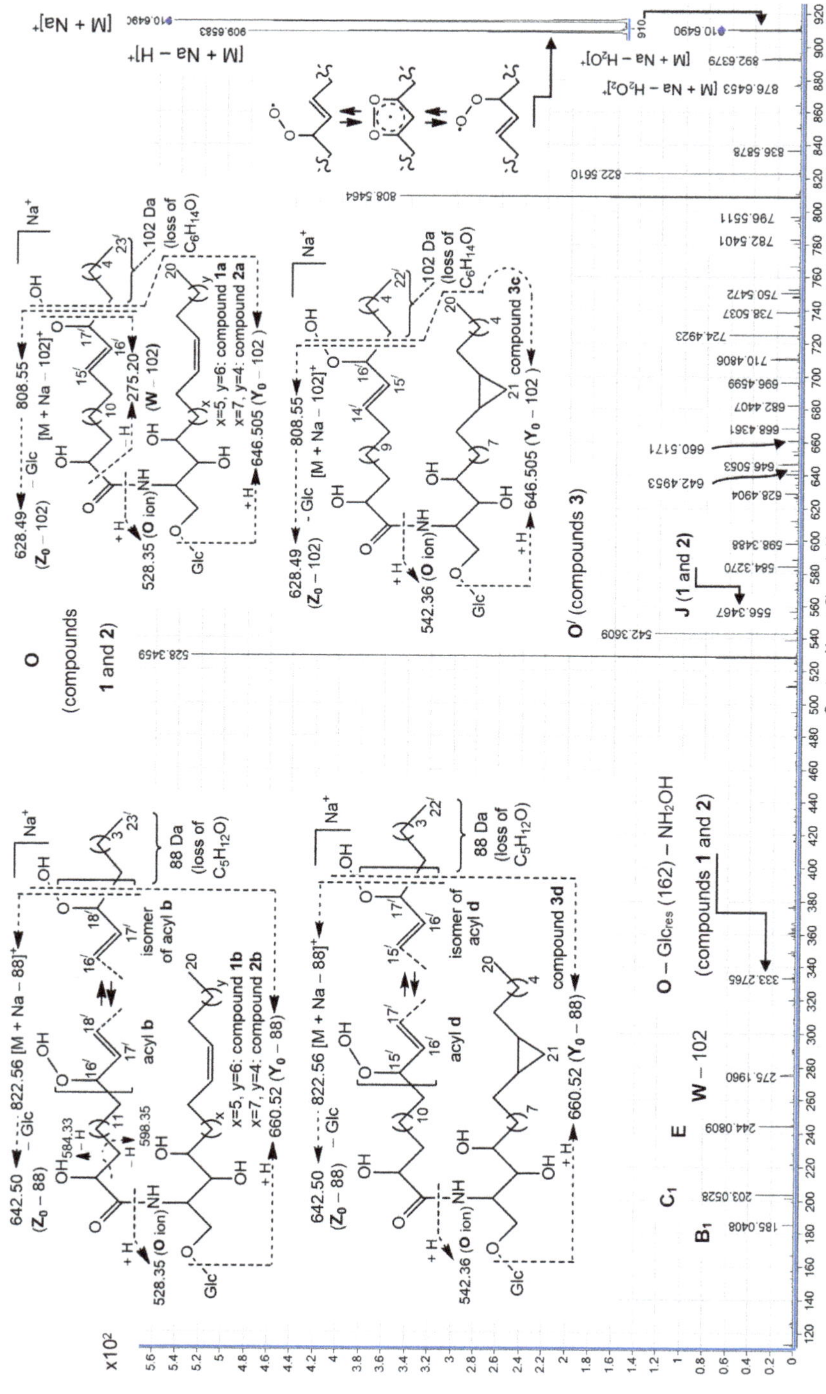

Figure 2. (+)ESI-MS/MS spectrum of $[M + Na]^+$ ion of isomeric allylic hydroperoxides **1a**, **1b**, **2a**, **2b**, **3c**, and **3d**.

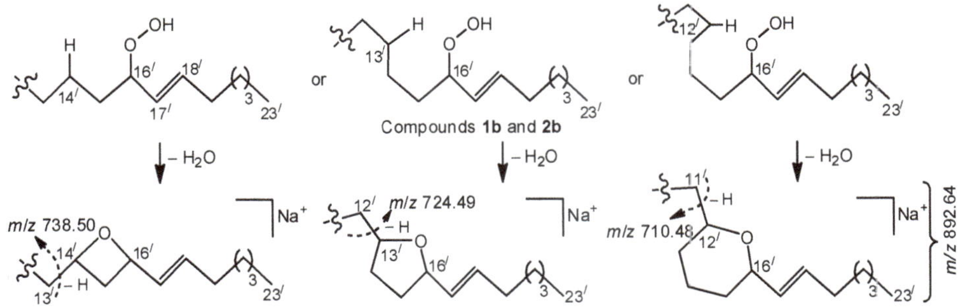

Scheme 2. Possible formation of the homologous ions of *m*/*z* 738.50, 724.49, and 710.48.

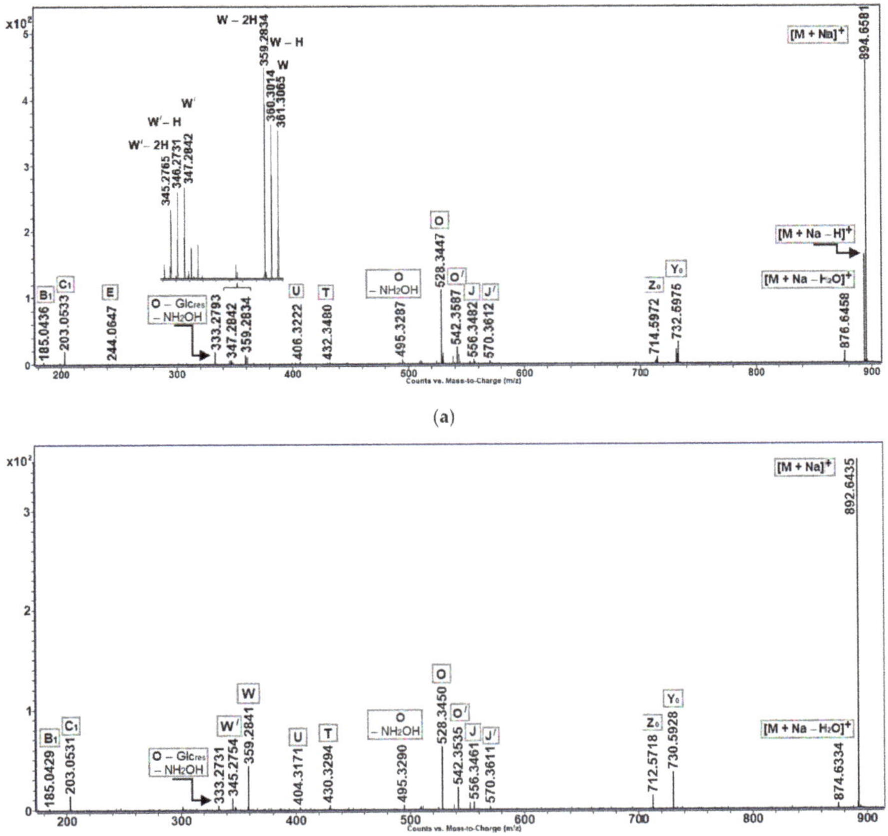

(a)

(b)

Figure 3. (+)ESI-MS/MS spectra of $[M + Na]^+$ ions of isomeric (**a**) allylic alcohols **1a′**, **1b′**, **2a′**, **2b′**, **3c′**, **3d′** and (**b**) enones **1a″**, **1b″**, **2a″**, **2b″**, **3c″**, **3d″**.

In (−)ESI-MS/MS experiments with $[M - H]^-$ and $[M + Cl]^-$ ions (Figure 4a–c), allylic hydroperoxides again produced more fragments than allylic alcohols and enones. The main feature of (−)ESI-MS/MS fragmentation of $[M - H]^-$, Y_0, Z_0, Z_0/Q, *T*, *W*, and other precursor ions, containing an allylic hydroperoxy group, was the loss of water to produce fragments with enone functionality in acyl chains, as described for $[M - H]^-$ ions of hydroperoxy-eicosatetraenoic acids [20]. In particular, the homologous ions of *m/z* 436.4 ([$Z_0/Q - H_2O$]) and 422.4 ([$Z_0'/Q - H_2O$]), containing acyl C_{23} and C_{22} chains, respectively (Figure 4a), were also observed in (−)ESI-MS/MS spectrum of enones (Figure 4c). Then, relatively low-intensity pseudo-molecular ion peaks ($[M + Cl]^-$ and $[M - H]^-$) and very small peaks, corresponding to Y_0 and Z_0 ions (not shown), were seen in the (−)ESI-MS/MS spectrum of allylic hydroperoxides. A discernible [M − 2H] peak was comparable with a pseudo-molecular [M − H] peak. This spectrum also exhibited $[M + Cl - 102]^-$, $[M + Cl - 88]^-$, $[M - H - 102]^-$, $[M - H - 88]^-$, and [Z_0/Q − 102] ions, interpreted as α-cleavage ions with a terminal α,β-unsaturated aldehyde. Additionally, the two minor ions of *m/z* 743.6 and 729.55 could be formed by α-cleavages of compounds in which an allylic hydroperoxy group was between a double bond and C-1′. In particular, the *m/z* 743.6 ions could be fragments of isomeric compounds **1b**, **2b** (C-15′–C-16′ bond fission), and **3d** (C-14′–C-15′ bond fission), while the less abundant *m/z* 729.55 ions could be fragments of the allylic isomers of compounds **1a**, **2a** (C-14′–C-15′ bond fission), and **3c** (C-13′–C-14′ bond fission).

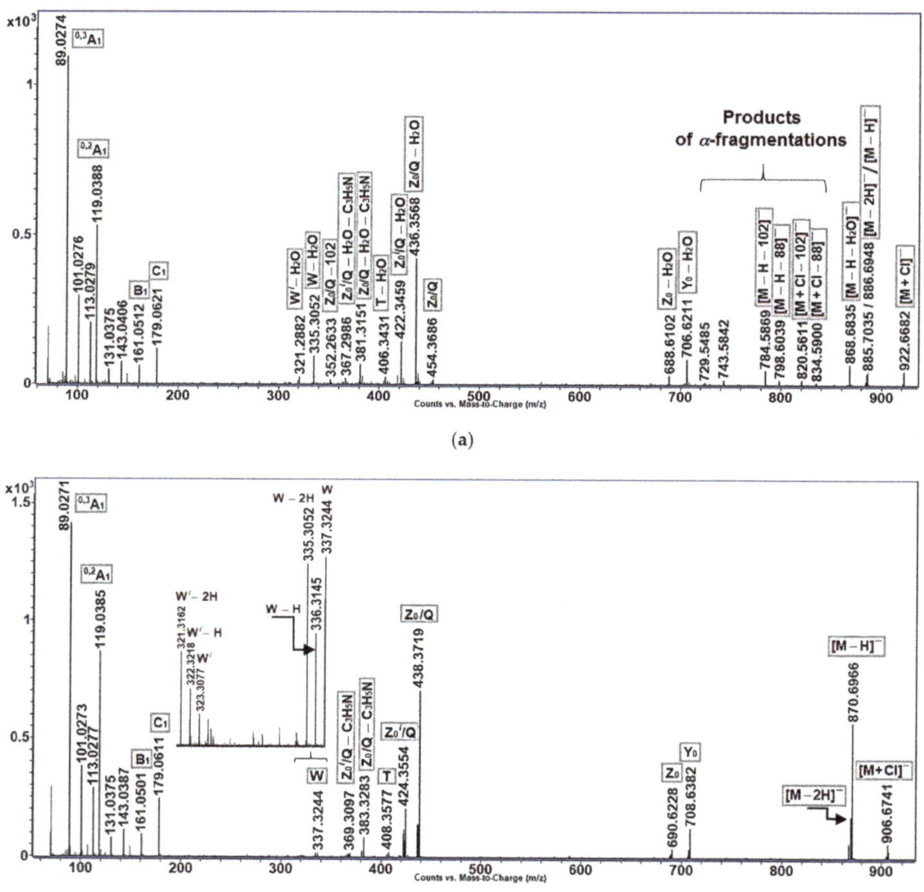

(a)

(b)

Figure 4. *Cont.*

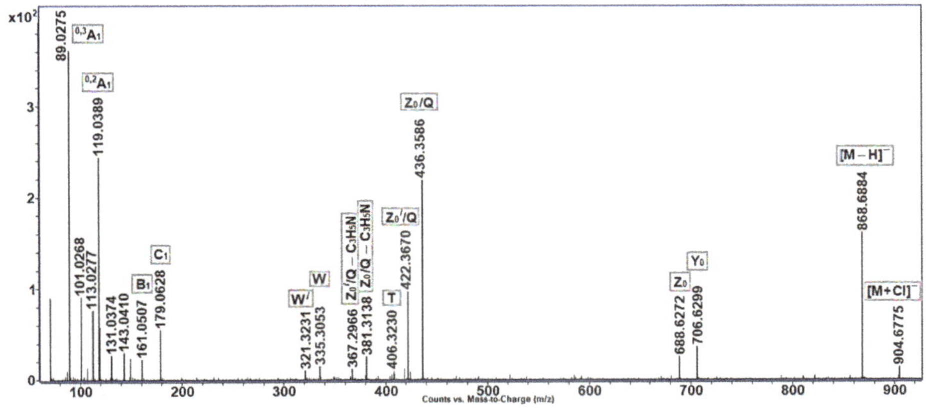

(c)

Figure 4. (−)ESI-MS/MS spectra of $[M + Cl]^-$ and $[M - H]^-$ ions of isomeric (**a**) allylic hydroperoxides **1a**, **1b**, **2a**, **2b**, **3c**, **3d**, (**b**) allylic alcohols **1a/**, **1b/**, **2a/**, **2b/**, **3c/**, **3d/**, and (**c**) enones **1a//**, **1b//**, **2a//**, **2b//**, **3c//**, **3d//**.

Like the (−)ESI-MS/MS spectra of the non-oxidized cerebrosides of *Aulosaccus* sp., those of allylic alcohols (Figure 4b) and enones (Figure 4c) exhibited significant peaks corresponding to $[M - H]^-$ and Z_0/Q ions, with lower intensity peaks representing $[M + Cl]^-$, Y_0, Z_0, $[Z_0/Q - C_3H_5N]$, and W ions. The negatively charged acyl-containing ions of allylic alcohols (Figure 4b), like their positively charged acyl-containing counterparts (Figure 3a), tended to lose hydrogen atoms in the MS/MS experiment.

2.2. NMR Characterization of Oxidized Cerebrosides

The ^{1}H-NMR and ^{13}C-NMR spectra (CD_3OD) of oxidized cerebrosides (Table 1, Figures S2 and S3), as well as the corresponding spectra of the non-oxidized cerebrosides of *Aulosaccus* sp. [8], showed signals of β-glucopyranosyl-(1→1)-ceramides that had monoenoic or cyclopropane-containing phytosphingosine-type backbones, *N*-acylated with 2-hydroxy FAs. In particular, H-2 (δ_H 4.25, m) of the sphingoid base moieties displayed a characteristic cross signal with *N*-acyl C-1/ (δ_C 177.7), as demonstrated by an HMBC experiment with sphingolipids **1**–**3**. ^{1}H,^{1}H-COSY diagram indicated that several protons, starting from $-O-CH_2-$ (δ_H 3.80 and 4.045, dd, CH_2-1) and ending with alkyl $-CH_2-$ (δ_H 1.31 and 1.55, m, CH_2-6), formed a linear spin system of phytosphingosine-type moieties of **1**–**3**. Another spin system consisted of CH-2/ (δ_H 4.01, dd), CH_2-3/ (δ_H 1.60 and 1.74, m), and CH_2-4/ (δ_H 1.42, m) protons of 2-hydroxy acyl chains. The signals of a β-glucopyranoside moiety were sequentially assigned by ^{1}H,^{1}H-COSY, HSQC (heteronuclear single-quantum correlation spectroscopy), and HMBC experiments, starting from the signal of anomeric CH-1// (δ_H 4.28, d, J = 7.8 Hz; δ_C 105.3). Accordingly, cross signals CH_2-1/**C**-1// and CH-1///**C**-1 were observed in the HMBC diagram of glycosides **1**–**3**. Then, the ^{1}H-and ^{13}C-NMR spectra of these sphingolipids showed signals of long hydrocarbon chains ($-(CH_2)_n-$, δ_H 1.22–1.42, m, δ_C 30.6–32.0) and terminal methyl groups (δ_H 0.89–0.90, several overlapping triplets, δ_C 15.0–15.05, broad signal). The δ values of allylic CH_2 (δ_C 28.7–28.85, δ_H 2.02, m) and olefinic CH (δ_H 5.34, m) were used to characterize *cis*-double bonds of backbones **1** and **2**. These data were in good agreement with ^{1}H-NMR and ^{13}C-NMR (CD_3OD) data on some cerebrosides [21] and FA standards [22] with isolated *cis*-double bonds (*cis*-isomers: δ_C 28.7–28.9, δ_H 2.02–2.03 m, allylic CH_2, and δ_H 5.335–5.34 m, olefinic CH, *trans*-isomers: δ_C 34.15–34.2, δ_H 1.97–1.975 m, allylic CH_2, and δ_H 5.37–5.38 m, olefinic CH). A *cis*-cyclopropane ring in backbone **3** caused three upfield shifted signals at δ_H −0.33 (dt, J = 4.1, 5.3 Hz, H-21a), 0.58 (ddd, J = 4.1, 8.3, 8.3 Hz, H-21b), and 0.67 (m, H-13, H-14).

Table 1. ^{1}H-(500 MHz) and ^{13}C-(125 MHz) NMR data (CD_3OD) [1] for the RP-HPLC fraction, containing compounds **1a–a//**, **1b–b//**, **2a–a//**, **2b–b//**, **3c–c//**, and **3d–d//**.

Position	δ_H (mult., J Hz)	δ_C
Sphingoid base moieties		
1	3.80 (dd, 3.9, 10.5), 4.045 (dd, 6.2, 10.5)	70.6
2	4.25 (m)	52.3
3	3.60 (t, 6.0)	76.1
4	3.51 (m)	73.6
5	1.39 (m), 1.64 (m)	33.7
6	1.31 (m), 1.55 (m)	27.6
CH_2-pool	1.22–1.42 (m)	30.6–32.0
Allylic CH_2 of **1** and **2**	2.02 (m)	28.7–28.85
Olefinic CHof **1** and **2**	5.34 (m)	131.4–131.5
12, 15 of **3**	1.17 (m), 1.37 (m)	30.5
13, 14 of **3**	0.67 (m)	17.5
21 of **3**	−0.33 (dt, 4.1, 5.3), 0.58 (ddd, 4.1, 8.3, 8.3)	12.2
18	1.28 (m)	33.5
19	1.31 (m)	24.3
20 of **1** and **2**	0.895 (t, 6.9)	15.0
20 of **3**	0.899 (t, 7.0) [2]	15.0
N-Acyl moieties (some general features [3])		
1/	–	177.7
2/	4.01 (dd, 4.0, 7.7)	73.6
3/	1.60 (m), 1.74 (m)	36.4
4/	1.42 (m)	26.7
CH_2-pool	1.22−1.42 (m)	30.6–32.0
(n−2) CH_2	1.28 (m)	33.7
(n−1) CH_2	1.32 (m)	24.2
CH_3	0.893 and 0.897 (t, 7.0) [2]	15.05
β-Glucopyranoside moiety		
1//	4.28 (d, 7.8)	105.3
2//	3.17 (dd, 7.8, 9.1)	75.6
3//	3.35 (m)	78.5
4//	3.27 (m)	72.2
5//	3.27 (m)	78.6
6//	3.66 (m), 3.865 (dd, 1.1, 11.9)	63.3

[1] All signals were assigned based on data of ^{1}H–^{1}HCOSY, HSQC, and HMBC experiments. [2] The two resolved triplets were observed in 700-MHz spectra. [3] δ_H and δ_C values for acyl fragments with allylic –OOH/–OH/=O are given separately (Figure 5).

Apart from the previously mentioned NMR resonances, the signals of *trans*-monoenoic acyl moieties with an allylic hydroperoxy, hydroxy, or keto group (Figure 5a–c) were observed in the NMR spectra of the RP-HPLC fraction, containing oxidized cerebrosides. The δ_H values (CD_3OD) of the signals for –**HC**(OOH)– (4.16, dt), –**HC**(OH)– (3.94, q), and –CH=**CH**–CO– (6.105, d) were close to δ_H values ($CDCl_3$) of signals of corresponding protons in the *trans*-monoenoic allylic hydroperoxides (4.2, q), allylic alcohols (4.0, q), and enones (6.1, d), respectively, prepared from oleic acid [23]. Deviations of δ_H values (this study) from those in other experiments [23] arose from solvent effects. According to ^{1}H,^{1}H-COSY correlations, a spin system, assigned to acyl chains **a–d** (Figure 5a), included protons of allylic hydroperoxide, especially two olefinic CH (δ_H 5.35 and 5.665, m) and allylic CH (δ_H 4.16, dt, J = 6.6, 7.0 Hz), bearing –OOH. Allylic CH (δ_H 3.94, q, J = 6.8 Hz), bearing –OH, and two olefinic CH (δ_H 5.395 and 5.59, m) groups, belonging to allylic alcohol, were part of a spin system within acyl chains **a/–d/** (Figure 5b). In experiments, involving selective irradiation of allylic protons, the coupling constants J = 15.6 Hz and J = 15.4 Hz for the *trans*-olefinic protons of the allylic hydroperoxides and allylic alcohols, respectively, were detected. The characteristic NMR resonances of two *trans*-alkenyl CH (δ_H 6.105, d, J = 15.9 Hz, and 6.91, dt, J = 7.0, 15.9 Hz), conjugated C=O (δ_C 204.3), and α-CH_2 (δ_H 2.565, t, J = 7.4 Hz) groups were used for determining substructures of enones in acyl chains **a**$^{//}$–**d**$^{//}$ (Figure 5c). The structures of the allylic hydroperoxides, allylic alcohols, and enones were confirmed by HMBC correlations, as depicted in Figure 5.

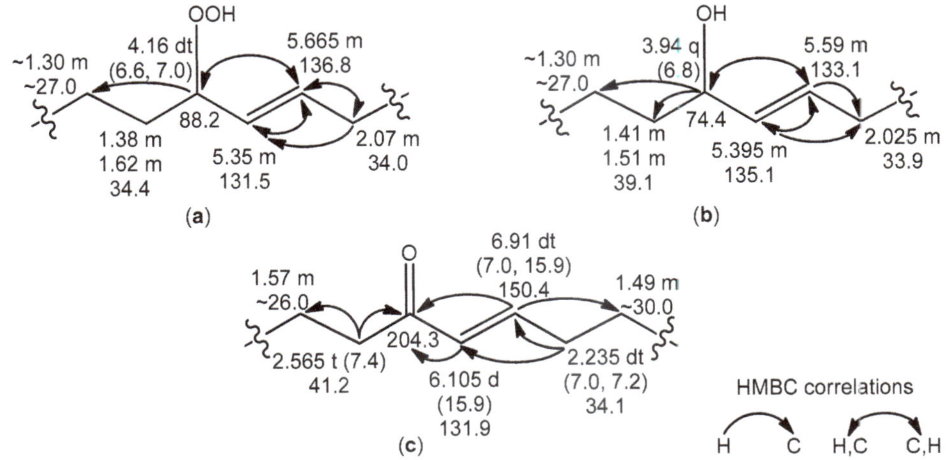

Figure 5. The δ_H (500 MHz) and δ_C (125 MHz) values and key HMBC correlations for acyl fragments with allylic (**a**) hydroperoxy (acyl chains **a–d**), (**b**) hydroxy (acyl chains **a**$^{/}$–**d**$^{/}$), and (**c**) keto (acyl chains **a**$^{//}$–**d**$^{//}$) groups (solvent: CD_3OD).

Some very weak signals in the ^{1}H-NMR, ^{1}H,^{1}H-COSY, and HSQC spectra of oxidized cerebrosides found in the present study were attributed to *cis*-double bonds of allylic hydroperoxides and allylic alcohols (Appendix A, Figure A1). The complete structures of these compounds could not be elucidated due to their trace amounts.

2.3. Analyses of FAs, Sphingoid Bases, and Sugar Obtained from Oxidized Cerebrosides

The RP-HPLC oxidized cerebroside fraction was divided into two parts (parts 1 and 2), which were treated using different chemical procedures before hydrolysis. Then, we applied MeCN/HCl hydrolysis [24] for chemical degradation of cerebrosides. In our experience [8], this procedure causes less disruption of spingoid bases than methanolysis (MeOH/HCl), which is most widely used in studies of complex lipids.

Analysis of FAs from Part 1. Part 1 of the oxidized cerebrosides was subjected to hydrogenation (with Adams' catalyst) to fix the positions of allylic oxygen-containing groups before hydrolysis. Upon hydrolysis, liberated FAs were acetylated and methylated. The ^{1}H-NMR spectrum ($CDCl_3$) of FA derivatives showed proton signals of mid-chain substructures, including –$\mathbf{H_2}$C–CO–$\mathbf{CH_2}$– at δ_H 2.38 (t, J = 7.4 Hz) and –CH(OAc)– at δ_H 4.855 (m). GC-MS analysis (electron impact ionization) of these derivatives revealed methyl esters of 2-acetyloxy C_{23} and C_{22} acids, containing an isolated keto or acetyloxy group or no additional oxygenated group. Similar products, namely keto, hydroxy, and non-oxygenated acid derivatives, have been previously reported for the hydrogenation (in EtOH over Adams' catalyst) of allylic 9-hydroperoxides and 10-hydroperoxides, obtained from methyl oleate [25].

In the present report, mass spectra exhibited base peaks at m/z 339 and 325 ([M − MeOCO − $CH_2CO]^+$) for methyl esters of 2-acetyloxy keto C_{23} (440 $[M]^+$) and C_{22} (426 $[M]^+$) acids, respectively (Figures S4–S7). Ions produced by cleavage β to a keto group and ions formed from the methyl end of the molecules by cleavage α to the keto group were prominent in the mass spectra, as described for MS fragmentations of several methyl esters of oxo (keto) FAs [26]. In particular, mass spectra of methyl esters of C_{23} and C_{22} acids displayed homologous pairs of m/z 127/142 and 113/128 fragments, containing methyl ends of the molecules with keto groups on the (*n*–8) or (*n*–7) carbon, respectively. These keto group positions were confirmed by a number of ions containing the polar end of the FA esters that also produced daughter ions due to loss of AcOH, CH_2CO, or MeOH (Figures S4–S7).

Hydrogenation of part 1, followed by hydrolysis, acetylation, and methylation, also yielded the methyl esters of 2-acetyloxy C_{23} and C_{22} acids, containing an additional isolated acetyloxy group. Expectedly, $[M]^+$ peaks were absent in the mass spectra of the methyl esters of these diacetylated acids, but $[M - CH_3CO]^+$ and $[M - AcOH]^+$ ions were observed in high mass regions (Figures S8–S11). Additionally, these compounds, as derivatives of 2-acetyloxy FAs, fragmented to give abundant $[M - MeOCO - AcOH]^+$ ions of *m*/*z* 365 and 351 for methyl esters of C_{23} and C_{22} acids, respectively. Isomers, containing an isolated acetyloxy group in different positions, were discerned based on the presence of diagnostic α-cleavage ions and more abundant product ions, formed by elimination of CH_2CO or AcOH from α-fragments, as described for acetates of secondary alcohols [27]. In particular, the α-fragments included the ions of *m*/*z* 385 and 399 (Figures S8 and S9) for the derivatives of 2,16-, and 2,17-diacetyloxy C_{23} acids, respectively, and the ions of *m*/*z* 371 and 385 (Figures S10 and S11) for the 2,15- and 2,16-diacetyloxy C_{22} acid derivatives, respectively. Moreover, some homologous ions, which arose from C-1-/-C-2 bond fission and cleavage α to an isolated acetyloxy group, were specific for the different positions of this group in acyl chains. For example, the abundant ions of *m*/*z* 283 and 297 (Figures S8 and S9), which were detected in the mass spectra of the methyl esters of isomeric 2-acetyloxy C_{23} acids, confirmed the presence of second acetyloxy groups on C-16 and C-17, respectively. Similarly, *m*/*z* 269 and 283 ions in the mass spectra of the methyl esters of 2-acetyloxy C_{22} acids (Figures S10 and S11) indicated a second acetoxy group on C-15 and C-16, respectively. Additionally, α-fragmentation of isomers with an acetyloxy group in the (*n*–8) or (*n*–7) positions and subsequent loss of AcOH gave rise to *m*/*z* 111 and 97 ions, respectively, containing the methyl end of acyl chains.

As a result of the FA analyses, hydrogenated minor derivatives of allylically oxygenated FAs were also detected. While the major derivatives formed from fatty acyl groups containing hydroperoxy/hydroxy/keto groups in the (*n*–8) or (*n*–7) positions, the minor derivatives were rearranged products of these FA moieties (Appendix B, Figures S12–S20).

Analysis of FAs from Part 2. In this investigation, we tried to use *S*-methyl groups as markers for oxidized cerebroside double bonds. However, enones with polarized double bonds that did not react with DMDS under mild conditions, and labile allylic hydroperoxides were not suitable for this purpose. Therefore, enones and allylic hydroperoxides were converted into allylic alcohols that were expected to add to DMDS.

The oxidized cerebrosides from part 2 were acetylated to increase solubility of these relatively polar compounds in DMDS and other low-polarity or non-polar organic solvents. In this process, allylic hydroperoxides were transformed into enones (Scheme 3), as reported by Porter and Wujek [23]. The mixture obtained after acetylation was treated with $NaBH_4/CeCl_3$ [28] to convert enones into allylic alcohols. The resulting derivatives, containing an allylic hydroxy or acetyloxy group, were treated with DMDS, and the products of this reaction were hydrolyzed in MeCN/HCl. Liberated 2-hydroxy FAs were acetylated, methylated, and analyzed by a GC-MS method that revealed major mono(methylthio) and minor tris(methylthio) derivatives (Schemes 3 and 4). The mono(methylthio) compounds, containing an allylic methylthio group, were characterized by ^{1}H-NMR resonances ($CDCl_3$) of two *trans*-olefinic CH (δ_H 5.175 dd, *J* = 9.0, 15.2 Hz, and 5.40 dt, *J* = 6.9, 15.2 Hz) and one CH (δ_H 2.99 m), bearing a –SMe group (δ_H 1.97 s). The presence of the mid-chain allylic substructure was confirmed by a TOCSY (total correlation spectroscopy) experiment.

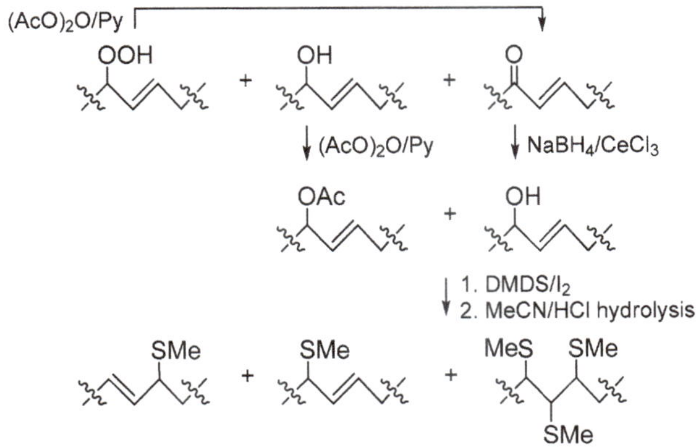

Scheme 3. Transformations of acyl substructures, containing allylic hydroperoxy/hydroxy/keto groups, into methylthio derivatives.

Cleavage patterns for methyl esters of 2-acetyloxy C_{23} (470 $[M]^+$) and C_{22} (456 $[M]^+$) acids, containing an allylic methylthio group, are depicted in Scheme 4. The positional isomers of these allylic thioethers were only partially GC-separated. The mass spectra of isomeric allylic thioethers (Figures S21–S28) were characterized by diagnostic peaks, corresponding to ions produced by α-cleavage. The fragments, formed by cleavage α to the carbon carrying an allylic methylthio group on the side remote from the carboxyl group included *m*/*z* 385 and 399 ions for the methyl esters of 2-acetyloxy C_{23} acids with –SMe group in the (*n*–7) and (*n*–6) positions, respectively (Figures S22 and S24). The peaks at *m*/*z* 371 and 385, observed in the mass spectra of methyl esters of homologous 2-acetyloxy C_{22} acids, represented fragments of the same origin (Figures S26 and S28). α-fragmentation of other mono(methylthio) derivatives gave rise to the ions at *m*/*z* 171 and 157, containing the methyl end of isomers with an allylic –SMe group in the (*n*–9) and (*n*–8) positions, respectively (Figures S21, S23, S25, and S27). Relative abundances of α-fragments in average mass spectrum were used to quantify isomer distribution, and approximately equal amounts of the four isomeric allylic thioethers were found. This finding may reflect the fact that these *S*-methyl derivatives were possibly products of allylic rearrangements, occurring prior to GC-MS analysis.

The minor tris(methylthio) derivatives of the methyl esters of 2-acetyloxy C_{23} (564 $[M]^+$) or C_{22} (550 $[M]^+$) acids produced more diagnostic fragments than the previously mentioned major mono(methylthio) derivatives. Expectedly, the cleavages of minor *S*-methylated compounds occurred between methylthio-carrying carbons to yield substantial fragment ions, as illustrated in Scheme 4 and Figures S29–S32. A cluster of four GC peaks for isomeric tris(methylthio) derivatives was observed. According to fragmentation patterns, there were two peaks representing positional isomers and two peaks that represented stereoisomers of these regioisomers on the chromatogram.

The results of the transformations of allylic alcohols and their acetates into methylthio derivatives (Scheme 3) were confirmed by experiments with model compounds, methyl esters of 11-hydroxy and 8-acetyloxy elaidic acids (prepared from methyl oleate, Appendix C, Scheme A1, Figures S33–S37). Under the conditions used here, the allylic alcohol acetates reacted with DMDS to give major allylic thioethers and minor tris(methylthio) derivatives. Allylic alcohols reacted with DMDS to give major DMDS adducts and minor allylic thioethers and tris(methylthio) derivatives. However, in contrast to the DMDS adducts of monoenes with an isolated double bond, the bis(methilthio) derivatives of allylic alcohols were destroyed during MeCN/HCl hydrolysis.

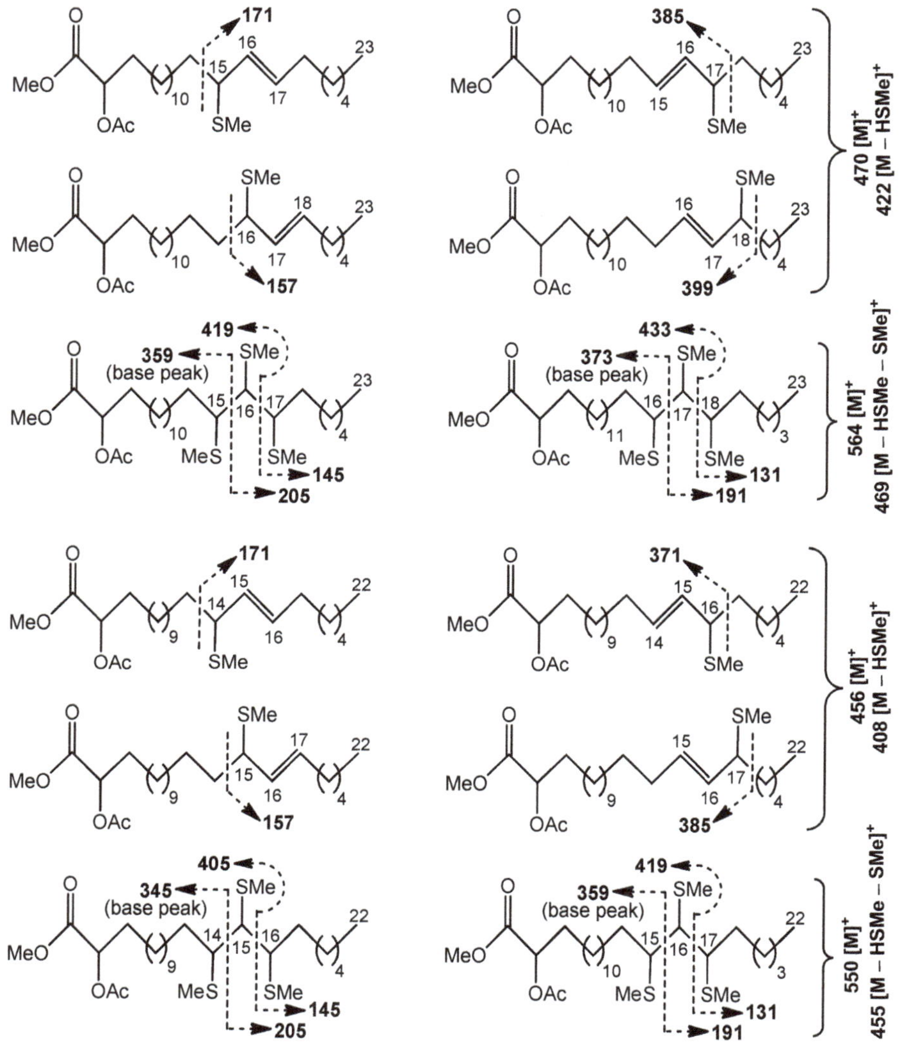

Scheme 4. GC-MS cleavage patterns for the methyl esters of 2-acetyloxy C_{23} and C_{22} acids, containing an allylic methylthio group or three methylthio groups.

Previously, the synthesis of allylic thioethers from allylic alcohols and thiols was reported by Zhang et al. ([29]: iodine-catalyzed process) and Tabarelli et al. ([30]: catalyst-free approach). Regio-isomeric mixtures of allylic thioethers were produced when the allylic alcohol contained two different substituents. To explain the presence of regio-isomer products of 1,3-isomerization, the allylic cation, formed by water loss from the allylic alcohol, was proposed to be an intermediate in the reaction pathway [30]. The existence of a similar mechanism explains the formation of regio-isomeric allylic thioethers and their tris(methilthio) derivatives in the iodine-catalyzed reaction of allylic alcohols and their acetates with DMDS reported here. Additionally, allylic thioethers, which could undergo 1,3-isomerization under acidic conditions, were possibly formed from the bis-DMDS adducts of allylic alcohols during MeCN/HCl hydrolysis. As a result of these rearrangements, the FA methylthio derivatives, obtained from oxidized cerebrosides, gave characteristic mass spectra, permitting locations of three-carbon allylically oxygenated substructures, rather than double bonds in the starting acyl chains.

We clarified the acyl structures (Figure 1) using a logical approach. The GC-MS analyses of the methyl esters of 2-acetyloxy methylthio FAs (Scheme 4) revealed that the three-carbon allylically functionalized substructures of oxidized cerebrosides included C-15′–C-16′–C-17′ and C-16′–C-17′–C-18′ fragments for 2-hydroxy C_{23} acyl chains and C-14′–C-15′–C-16′ and C-15′–C-16′–C-17′ fragments for 2-hydroxy C_{22} acyl chains. According to the GC-MS analyses of the hydrogenated FA derivatives, the amide-linked FAs of oxidized cerebrosides contained hydroperoxy, hydroxy, or keto groups in the (*n*–8) or (*n*–7) positions, more specifically in the 16′ and 17′ positions of 2-hydroxy C_{23} acyl chains and the 15′ and 16′ positions of 2-hydroxy C_{22} acyl chains. A priori, an allylic hydroperoxy, hydroxy, or keto group should be located in the terminal points of the previously mentioned three-carbon substructures. Consequently, the C-15′–C-16′–C-17′ fragments of the C_{23} acyl chains (**a–a″**) contained such groups in position C-17′ and the double bond between C-15′ and C-16′, while the C-16′–C-17′–C-18′ fragments of the other C_{23} acyl chains (**b–b″**) had 16′-hydroperoxy/hydroxy/oxo groups and 17′,18′-double bonds. Similarly, the C-14′–C-15′–C-16′ fragments of the C_{22} acyl chains (**c–c″**) contained 16′-hydroperoxy/hydroxy/oxo groups and 14′,15′-double bonds, while the C-15′–C-16′–C-17′ fragments of the other C_{22} acyl chains (**d–d″**) had 15′-hydroperoxy/hydroxy/oxo groups and double bonds between C-16′ and C-17′.

The mono(methylthio) and tris(methylthio) derivatives of allylically oxygenated FAs were also used to explain other structural peculiarities of the starting material. Upon hydrogenation, all the methylthio derivatives lost *S*-methyl groups, giving saturated hydrocarbon chains. In particular, the transformations, shown in Scheme 3, with subsequent hydrodesulfurization allowed us to convert –CH=CH–CH(OOH/OH)– and –CH=CH–CO– substructures to $–CH_2–CH_2–CH_2–$ chains without reducing other oxygen-containing groups in the molecules. As a result, the unbranched structures of the parent amide-linked FAs were clarified using retention times of the 2-acetyloxy tricosanoic and docosanoic acid methyl esters, obtained from the methylthio derivatives. Then, these methyl esters were converted into (2*S*)-oct-2-yl esters of 2-hydroxy acids. The resulting (2*S*)-oct-2-yl esters of 2-hydroxy tricosanoic and docosanoic acids coeluted in GC analyses with the reference (2*S*)-oct-2-yl ester of (2*R*)-2-hydroxy tricosanoic or docosanoic acids, respectively, indicating (2*R*)-configurations of 2-hydroxy acids liberated from oxidized cerebrosides.

Thus, we used two complementary approaches for determining oxidized acyl chain structures in glycosphingolipids. Analysis of the FA derivatives from part 1 indicated the allylic oxygen-containing group positions, but hydrogenation resulted in the loss of information regarding the positions of double bonds in the starting material (approach 1). In the analysis of part 2 (approach 2), the data on the locations of three-carbon allylically oxygenated substructures in the FA derivatives and the information on their straight-chain structures and (2*R*)-configurations were obtained. Although there was no direct information about the position of the double bond in the FA esters, the combined data of approaches 1 and 2 allowed for determination of the double bond and allylic hydroperoxy, hydroxy, or keto group locations in each acyl chain.

Analyses of Sphingoid Bases and Sugar from Parts 1 and 2. The sphingoid bases, liberated by hydrolysis of oxidized cerebrosides, were obtained as acetylated derivatives. The ^{1}H-NMR data (δ_H values of CH_2-1–CH-4; $CDCl_3$) and optical rotation value ($[\alpha]^{25}_D = +27.9$, $CHCl_3$) of the hydrogenated sphingoid base acetates, isolated from the hydrolysate of part 1, indicated their (2*S*,3*S*,4*R*)-configuration [31]. The ^{1}H-NMR spectrum of these compounds also showed signals of the terminal methyl groups of dominant normal-chain (δ_H 0.88, t, J = 6.9 Hz) and minor cyclopropane-containing (δ_H 0.89, t, J = 6.9 Hz) constituents. Under our conditions of hydrogenation, ring opening was not the dominant process for the minor constituent, so the signals of *cis*-cyclopropane protons at δ_H −0.33 (dt, J = 4.2, 5.5 Hz), 0.56 (ddd, J = 4.2, 8.3, 8.3 Hz), and 0.645 (m) [32] were observed. The hydrolysis of derivatized oxidized cerebrosides from part 2 with subsequent acetylation of products gave three sphingoid base derivatives. Two spingoid base derivatives were the DMDS adducts of acetylated isomeric monoenoic

C_{20} compounds. The mass spectrum of the DMDS adduct of major acetylated C_{20} monoene exhibited significant peaks at *m/z* 432 $[M - H_3CSC_9H_{18}]^+$, 372 $[M - H_3CSC_9H_{18} - AcOH]^+$ (base peak), and 173 $[H_3CSC_9H_{18}]^+$, indicating 11,12-double bond in backbone **1**. Key peaks in the mass spectrum of the DMDS adduct of isomeric minor acetylated C_{20} monoene, which have a longer retention time in GC-MS, were observed at *m/z* 460 $[M - H_3CSC_7H_{14}]^+$, 400 $[M - H_3CSC_7H_{14} - AcOH]^+$ (base peak), and 145 $[H_3CSC_7H_{14}]^+$, indicating Δ13 unsaturation in backbone **2**. A third acetylated sphingoid base, derived from cyclopropane-containing backbone **3**, did not give a DMDS adduct. In the mass spectrum of this compound, an $[M - AcOH]^+$ ion fragmented to give discernible peaks at *m/z* 380 $[M - AcOH - (CH_2)_6CH_3]^+$ (α cleavage to a cyclopropane ring, at the C-14–C-15 position) and 394 $[M - AcOH - (CH_2)_5CH_3]^+$ (β-cleavage, at the C-15–C-16 position) that was characteristic of the peracetate of the C_{21} sphingoid base containing a ring between C-13 and C-14.

Methylthiolation of *cis*-monoenes and *trans*-monoenes with DMDS, as an anti-addition, leads to the threo-adducts and erythro-adducts, respectively [33,34]. The threo-diastereomers and erythro-diastereomers can be easily distinguished by NMR shifts of protons and carbons in and close to the 1,2-bis(alkylthio) moiety [35]. In addition to the data presented in the NMR study of Knothe and Steidley [35], we used δ_H values of the DMDS adducts of two standards, methyl palmitoleate (Figure 6a) and its *trans*-isomer (Figure 6b), to confirm the configurations of double bonds in the monoenoic sphingoid base moieties of the starting oxidized cerebrosides. The δ_H values for CH_2, α to the –CH(SMe)–CH(SMe)– moiety, were obtained through correlations in $^1H,^1H$-COSY diagrams. The ^1H-NMR and $^1H,^1H$-COSY spectra ($CDCl_3$) of the DMDS derivatives of acetylated sphingoid bases, derived from backbones **1** and **2**, showed superimposed resonances of two vicinal CH groups (δ_H 2.685, m), linked to –SMe (δ_H 2.10, s), and two α-CH_2 groups (δ_H 1.84, m; 1.32, m), characteristic of threo-diastereomers. Consequently, *cis*-monoenes (**1** and **2**) were precursors of these compounds.

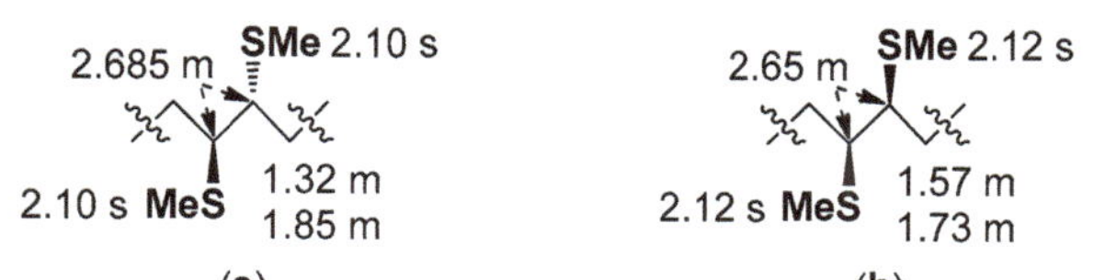

Figure 6. Some δ_H values (700 MHz, $CDCl_3$) for the *S*-methylated substructures of (**a**) methyl threo-9,10-bis(methylthio)hexadecanoate and (**b**) methyl erythro-9,10-bis(methylthio)hexadecanoate. Unlike the signals of the methylthio groups of the threo-isomer, the signals of –SMe groups of the erythro-isomer were two partially resolved singlets at δ_H 2.121 and 2.119.

GC analyses of peracetylated (2*R*)- and (2*S*)-oct-2-yl glucosides showed a D-configuration of glucose, released from parts 1 and 2 [36].

2.4. Oxidized Cerebrosides from the Extract of Aulosaccus sp.: Structures and Possible Origins

As a result of our study, structures of 18 previously unknown compounds, found in the complex mixture of the oxidized cerebrosides from the extract of *Aulosaccus* sp., were elucidated. These β-D-glucopyranosyl-(1→1)-ceramides (**1a–a//**, **1b–b//**, **2a–a//**, **2b–b//**, **3c–c//**, **3d–d//**) were shown to contain phytosphingosine-type backbones, (2*S*,3*S*,4*R*,11*Z*)-2-aminoeicos-11-ene-1,3,4-triol (in **1**), (2*S*,3*S*,4*R*,13*Z*)-2-aminoeicos-13-ene-1,3,4-triol (in **2**), and (13*S**,14*R**)-2-amino-13,14-methylene-eicosane-1,3,4-triol (in **3**). These backbones were *N*-acylated with straight-chain monoenoic (2*R*)-2-hydroxy acids that had allylic hydroperoxy/hydroxy/keto groups on C-17/ in the 15/*E*-23:1 chain (**a–a//**), C-16/ in the 17/*E*-23:1 (**b–b//**) and 14/*E*-22:1 (**c–c//**) chains, and C-15/ in the 16/*E*-22:1 chain (**d–d//**). Cerebrosides, having backbones **1**, **2**, and **3**, comprised, respectively, 60%, 20%, and 20% of the mixture. The percentages were calculated from the integration of signals of a cyclopropane ring in the ^1H-NMR spectra of this mixture (backbone **3**) and from relative intensities of GC peaks, represented by DMDS derivatives of two acetylated isomeric monoenoic sphingoid bases (backbones **1** and **2**, Δ^{11}:Δ^{13}

≈ 3:1). The GC-MS analysis of the hydrogenation products of amide-linked FAs indicated that the **a**:**b**, **c**:**d**, **a**/:**b**/, **c**/:**d**/, **a**//:**b**//, and **c**//:**d**// isomer ratios were approximately 1:1.Therefore, the employed complementary instrumental and chemical methods clarified structures of oxidized cerebrosides in a complex mixture, without requiring isolation or complete separation.

Previously, the possible precursors of the oxidized cerebrosides were found in the major RP-HPLC fraction of glycosphingolipids, isolated from the extract of *Aulosaccus* sp. These potential precursors were β-D-glucopyranosyl-(1→1)-ceramides that contained backbones **1** (60% in the fraction) and **2** (20%), *N*-acylated with (2*R*,16*Z*)-2-hydroxytricos-16-enoic acid, and backbone **3** (20%), *N*-acylated with (2*R*,15*Z*)-2-hydroxydocos-15-enoic acid [8]. Perhaps, peroxidation of the amide-linked FAs occurred symmetrically about the *cis*-(*n*–7) double bond, so a hydroperoxy, hydroxy, or keto group, found in the major *trans*-monoenoic peroxidation products, was located at each of the carbon atoms, which originally formed the double bond (namely, in the (*n*–8) and (*n*–7) positions, ≈1:1). In particular, structures **a**–**a**// and **b**–**b**// could be formed from $C_{23}\Delta^{16Z}$ acyl chain while structures **c**–**c**// and **d**–**d**// could be formed from the $C_{22}\Delta^{15Z}$ acyl chain.

According to the product composition, photo-oxidation and autooxidation [1] are possible mechanisms involved in the formation of the oxidized cerebrosides from the extract of *Aulosaccus* sp. However, we would like to point to another possible origin of the oxidized cerebrosides in *Aulosaccus* sp., taking into account the relationship between these oxidation products and other compounds isolated from the same sponge sample. In particular, some bacterial branched-chain, cyclopropane-containing FAs, and their monoenoic precursors were present in significant amounts in *Aulosaccus* sp. [32], and an overwhelming number of the sterols (stanols, Δ^5-, Δ^7-, and $\Delta^{8(14)}$-sterols) of this sponge were oxidized to the corresponding 3-ketosteroids [37] The occurrence of these FAs and steroids in *Aulosaccus* sp. suggested this sponge was associated with actinobacteria, known as sponge-specific microorganisms [38] and sterol degraders [39]. Cholesterol oxidase, produced by a variety of actinobacteria [40], could catalyze the transformations of the previously mentioned sterols into 3-ketosteroids [41] with the generation of H_2O_2. We suggest that H_2O_2 production in the enzymatic oxidation of *Aulosaccus* sp. sterols led to oxidative transformations of a certain part of cerebrosides, located in the membranes of eukaryotic cells together with sterols.

3. Materials and Methods

3.1. General Procedures

^{1}H-, ^{13}C-NMR, ^{1}H,^{1}H-COSY, HSQC, HMBC, NOESY (Nuclear Overhauser Effect Spectroscopy), and TOCSY spectra (in CD_3OD or $CDCl_3$) were recorded on Bruker Avance III HD 500 and Bruker Avance III 700 spectrometers (Bruker BioSpin, Bremen, Germany) at 125 MHz (^{13}C), 500 (^{1}H), and 700 (^{1}H) MHz. Chemical shifts (ppm) were internally referenced to the corresponding residual solvent signals δ_H 3.306/ δ_C 49.6 for CD_3OD. For the spectra of compounds in $CDCl_3$, TMS was used as an internal standard. ESI mass spectra were recorded on a Q-TOF 6510 mass spectrometer (Agilent Technologies, Santa Clara, CA, USA). GC-MS analyses were carried out on a Hewlett Packard HP6890 GC System (Hewlett-Packard Company, Palo Alto, CA, USA) with an HP-5MS (J&W Scientific, Folsom, CA, USA) capillary column (30.0 m × 0.25 mm), helium as the carrier gas, and 70 eV ionizing potential. GC analyses were done on an Agilent 6850 Series GC System chromatograph (Agilent Technologies, Santa Clara, CA, USA) equipped with an HP-1 (Agilent technology, Santa Clara, CA, USA) capillary column (30 m × 0.32 mm), the carrier gas was helium (flow rate 1.7 mL/min), and the detector temperature was 300 °C (or 280 °C for per-acetylated (2*R*)-oct-2-yl glucosides). The GC-MS analyses of FA esters and per-acetylated sphingoid bases were done using the injector temperature of 270 °C and the temperature program 100 °C (1 min)–10 °C/min–280 °C (30 min). The GC analyses of per-acetylated (2*R*)-oct-2-yl derivatives of D-Glc and L-Glc were carried out using the injector temperature of 150 °C and the temperature program 100 °C (0.5 min)–5 °C/min–250 °C (10 min). Optical rotation was measured on a Perkin–Elmer polarimeter, model 343 (Perkin-Elmer

GmbH, Überlingen, Germany). Column chromatography was performed using silica gel (50/100 μm, Sorbpolimer, Krasnodar, Russia). RP-HPLC separations were performed using a Du Pont Series 8800 Instrument (DuPont, Wilmington, DE, USA) with an RIDK-102 refractometer (Laboratorni pristroje, Praha, Czechoslovakia). Two columns (Agilent Technologies, Santa Clara, CA, USA), Agilent ZORBAX Eclipse XDB-C8 (4 × 150 mm), and Agilent ZORBAX SB-C18 (4.6 × 250 mm), were used for the HPLC.

3.2. Animal Material

The sample of the genus *Aulosaccus* (phylum Porifera, class Hexactinellida, order Lyssacinosida, family Rossellidae), 930 g (wet weight), was collected in July 2011 by dredging (with small Sigsbi trail) from a 505-m depth near Iturup Island (Kuril Islands, 45°01.05′N, 147°00.03′E) during a cruise onboard the r/v "Academik Oparin". The species was identified by Dr A.L. Drozdov (A.V. Zhirmunskii Institute of Marine Biology, Far Eastern Branch of RAS, Vladivostok, Russia). The collected sponge was stored at −15 °C.

3.3. Extraction and Isolation

The isolation of total cerebroside sum (238 mg) from the EtOH extract of *Aulosaccus* sp. was described in our previous article [8]. This sum was separated by RP-HPLC (Agilent ZORBAX Eclipse XDB-C8 column, EtOH-H_2O, 85:15, *v/v*). A mixture of the most polar minor cerebrosides (22.4 mg) was re-chromatographed (EtOH-H_2O-$CHCl_3$, 75:25:5, *v/v*) to give a major fraction (4.4 mg) containing oxidized cerebrosides: (+)HR-ESI-MS: 910.6606 ($[M + Na]^+$ of isomeric allylic hydroperoxides, $C_{49}H_{93}NO_{12}Na^+$; calc. 910.6590), 894.6649 ($[M + Na]^+$ of isomeric allylic alcohols, $C_{49}H_{93}NO_{11}Na^+$; calc. 894.6641), and 892.6534 ($[M + Na]^+$ of isomeric enones, $C_{49}H_{91}NO_{11}Na^+$; calc. 892.6484); (−)HR-ESI-MS: 922.6387 ($[M + Cl]^-$ of isomeric allylic hydroperoxides, $C_{49}H_{93}NO_{12}Cl^-$; calc. 922.6386), 906.6425 ($[M + Cl]^-$ of isomeric allylic alcohols, $C_{49}H_{93}NO_{11}Cl^-$; calc. 906.6437), and 904.6353 ($[M + Cl]^-$ of isomeric enones, $C_{49}H_{91}NO_{11}Cl^-$; calc. 904.6281); ^{1}H- and ^{13}C-NMR (CD_3OD): see Table 1 and Figures S2 and S3. This fraction was divided into two parts. The constituents of parts 1 and 2 were analyzed using different chemical procedures.

3.4. Analyses of FAs, Sphingoid Bases, and Sugar from Oxidized Cerebrosides (Parts 1 and 2)

The constituents of part 1 (3.0 mg) were hydrogenated in EtOH over Adams' catalyst at room temperature (10 h). The resulting hydrogenation product was hydrolyzed in MeCN (0.45 mL) and 5 N HCl (0.05 mL) for 4 h at 75 °C in a capped vial. Hydrolysate was concentrated in vacuo, partitioned between $CHCl_3$ (1.0 mL) and H_2O (3 × 1.0 mL). The H_2O extract was evaporated to dryness to give sugar-containing subfraction. The $CHCl_3$ layer was concentrated in vacuo, and a dry residue was acetylated with Ac_2O (0.2 mL) in pyridine (0.2 mL), overnight. The acetylated material was subjected to chromatography over silica gel column (3.0 cm × 1.2 cm), using hexane-ethylacetate (5:1→2:1→1:1, *v/v*) system, and then $CHCl_3$-EtOH (1:1, *v/v*) system. Elution with 60 mL of hexane-ethylacetate (2:1, *v/v*) gave a subfraction of per-acetylated sphingoid bases (0.8 mg). 2-Acetyloxy FA (0.7 mg) was eluted with 20 mL of $CHCl_3$-EtOH (1:1, *v/v*) system. The subfraction of the acetates of sphingoid bases was re-purified by column chromatography (SiO_2, 3.5 cm × 1.2 cm) eluting with 60 mL of hexane-ethyl acetate (2:1, *v/v*). The eluate was evaporated in vacuo yielding a mixture (0.5 mg) of (2*S*,3*S*,4*R*)-2-aminoeicosane-1,3,4-triol (*n*-t20:0, using shorthand nomenclature [42]) and (2*S*,3*S*,4*R*,13*S**,14*R**)-2-amino-13,14-methylene-eicosane-1,3,4-triol (*cis*-13,14-methylene-t21:0): $[\alpha]^{25}_D$ = +27.9 (*c* = 0.03, $CHCl_3$), ^{1}H-NMR (500 MHz, $CDCl_3$): 5.94 (d, *J* = 9.4 Hz, NH), 5.10 (dd, *J* = 3.0, 8.1 Hz, H-3), 4.94 (dt, *J* = 3.0, 9.9 Hz, H-4), 4.475 (m, H-2), 4.29 (dd, *J* = 4.8, 11.6 Hz, H-1b), 4.01 (dd, *J* = 3.0, 11.6 Hz, H-1a), 2.08 (s, $-OCOCH_3$ at C-3), 2.05 (two s, $-OCOCH_3$ at C-4 and at C-1), 2.025 (s, $-NHCOC\mathbf{H_3}$), 1.64 (m, H_2C-5), 1.45–1.05 (m, $-(CH_2)_n-$), 0.88 (t, *J* = 6.9 Hz, H_3C-20 of *n*-t20:0), 0.89 (t, *J* = 6.9 Hz, H_3C-20 of *cis*-13,14-methylene-t21:0), 0.65 (m, H-13, H-14 of *cis*-13,14-methylene-t21:0), 0.565 (ddd, *J* = 4.1, 8.3, 8.3 Hz, H-21b of *cis*-13,14-methylene-t21:0), −0.33 (dt, *J* = 4.1, 5.5 Hz, H-21a of *cis*-13,14-methylene-t21:0); (2*S*,3*S*,4*R*)-2-aminoeicosane-1,3,4-triol MS, *m/z* (relative intensity, %):

452/453/454 [M − AcOH − H/M − AcOH/M − AcNH$_2$]$^+$ (1.0/0.3/0.9), 380 [M − AcOCH$_2$ − AcOH]$^+$ (15), 338 [M − AcOCH$_2$ − AcOH − CH$_2$CO]$^+$ (14), 333 [M − 3AcOH]$^+$ (19), 320 [M − AcOCH$_2$ − 2AcOH]$^+$ (23),144 [AcOCH$_2$CH(NHAc)]$^+$ (90), 102 [AcOCH$_2$CH(NHAc) − CH$_2$CO]$^+$ (45), 84 [AcOCH$_2$CH(NHAc) − AcOH]$^+$ (100); (13*S**,14*R**)-2-amino-13,14-methylene-eicosane-1,3,4-triol MS, *m*/*z* (relative intensity, %): 525 [M]$^+$ (2), 465 [M − AcOH]$^+$ (21), 406 [M − AcOH − AcNH$_2$]$^+$ (9), 394 (6), 380 (5), 332 [M − AcOCH$_2$ − 2AcOH]$^+$ (12), 144 [AcO-CH$_2$CH(NHAc)]$^+$ (52), 102 [AcOCH$_2$CH(NHAc) − CH$_2$CO]$^+$ (52), 84 [AcOCH$_2$CH(NHAc) − AcOH]$^+$ (100). The components of the subfraction of 2-acetyloxy FAs were methylated by a standard method with *N*-nitroso-*N*-methylurea. The resulting compounds were purified by column chromatography (SiO$_2$, 3.5 cm × 1.2 cm) eluting with 44 mL of hexane-ethyl acetate (10:1, *v*/*v*). The elution gave a mixture of the methyl esters of 2-acetyloxy FAs (0.5 mg): ^{1}H-NMR (CDCl$_3$, 700 MHz): 4.99 (dd, *J* = 6.6, 6.9 Hz, H-2), 4.855 (m, mid-chain –CH(OAc)– of diacetyloxy acids), 3.74 (s, H$_3$CO– at C-1), 2.38 (t, *J* = 7.4 Hz, mid-chain –**H$_2$**C–CO–**CH$_2$**– of keto acids), 2.135 (s, −OCOCH$_3$ at C-2), 1.82 (m, H$_2$C-3), 1.45–1.15 (m, –(CH$_2$)$_n$–), 0.88 (a series of triplets about this point, *J* ≈ 7.0 Hz), the methyl esters of keto C$_{23}$ and C$_{22}$ acids MS: Figures S4–S7, S12–S16; the methyl esters of diacetyloxy C$_{23}$ and C$_{22}$ acids MS: Figures S8–S11, S17–S20, the methyl ester of saturated 2-acetyloxy C$_{23}$ acid MS: 426 [M]$^+$ (0.04%), 384 [M − CH$_2$CO]$^+$ (100%), 352 [M − CH$_2$CO − MeOH]$^+$ (34%), and the methyl ester of saturated 2-acetyloxy C$_{22}$ acid MS: 412 [M]$^+$ (0.04%), 370 [M − CH$_2$CO]$^+$ (100%), 338 [M − CH$_2$CO − MeOH]$^+$ (35%).

Components of part 2 (1.4 mg) were acetylated with Ac$_2$O (0.2 mL) in pyridine (0.2 mL), overnight. After vacuum drying, CHCl$_3$ (two drops) and CeCl$_3$·7H$_2$O (18.6 mg) were added to the acetylated material, and the mixture was dissolved in MeOH (0.125 mL). NaBH$_4$ (5 mg, excess) was slowly added with stirring. After 5 min, the reaction was quenched by the addition of NH$_4$Cl water solution (5.4 mg in 0.2 mL H$_2$O). A dry residue was obtained after complete evaporation of the reaction mixture. It was extracted with CHCl$_3$ to remove most inorganic admixtures. After vacuum drying of the CHCl$_3$ extract, the resulting dry material was dissolved in DMDS (0.1 mL), and a solution (0.025 mL) of iodine in diethyl ether (60 mg/mL) was added. The mixture was kept for 4 days at room temperature. Then the reaction was stopped by adding aqueous solution of Na$_2$S$_2$O$_3$ (5%, 0.2 mL), and the mixture was extracted with hexane (5 × 0.5 mL). The hexane extract was evaporated to dryness, and the resulting compounds were hydrolyzed in MeCN/HCl. Then, subfractions, containing sugar, peracetylated sphingoid bases (0.4 mg), and 2-acetyloxy FAs (0.3 mg), were obtained, as described above. The subfraction of peracetylated sphingoid bases consisted of peracetylated (13*S**,14*R**)-2-amino-13,14-methylene-eicosane-1,3,4-triol and DMDS derivatives of peracetylated (11*Z*)-2-aminoeicos-11-ene-1,3,4-triol and (13*Z*)-2-aminoeicos-13-ene-1,3,4-triol: ^{1}H-NMR (500 MHz, CDCl$_3$): 5.93 (d, *J* = 9.5 Hz, NH), 5.095 (dd, *J* = 3.1, 8.3 Hz, H-3), 4.94 (dt, *J* = 3.1, 9.9 Hz, H-4), 4.47 (m, H-2), 4.29 (dd, *J* = 4.8, 11.8 Hz, H-1b), 4.00 (dd, *J* = 3.1, 11.8 Hz, H-1a), 2.685 (m, –**CH**(SCH$_3$)–**CH**(SCH$_3$)–), 2.10 (s, –CH(S**CH$_3$**)–CH(S**CH$_3$**)–), 2.08 (s, −OCOCH$_3$ at C-3), 2.05 (two s, −OCOCH$_3$ at C-4 and at C-1), 2.025 (s, −NHCO**CH$_3$**), 1.84 (m; H-b, –**H$_2$**CCH(SCH$_3$)–CH(SCH$_3$)**CH$_2$**–), 1.64 (m, H$_2$C-5), 1.42−1.18 (m, –(CH$_2$)$_n$–), 1.32 (m, H-a, –**H$_2$**CCH(SCH$_3$)–CH(SCH$_3$)**CH$_2$**–), 0.88 (t, *J* = 6.9 Hz, H$_3$C-20 of *n*-t20:0), 0.89 (t, *J* = 6.9 Hz, H$_3$C-20 of *cis*-13,14-methylene-t21:0), 0.645 (m, H-13, H-14 of *cis*-13,14-methylene-t21:0), 0.56 (ddd, *J* = 4.2, 8.2, 8.2 Hz, H-21b of *cis*-13,14-methylene-t21:0), −0.33 (dt, *J* = 4.2, 5.4 Hz, H-21a of *cis*-13,14-methylene-t21:0). The DMDS derivative of per-acetylated (11*Z*)-2-aminoeicos-11-ene-1,3,4-triol MS, *m*/*z* (relative intensity, %): 557 [M − HSMe]$^+$ (1.4), 432 [M − H$_3$CSC$_9$H$_{18}$]$^+$ (24), 372 [M − H$_3$CSC$_9$H$_{18}$ − CH$_3$COOH]$^+$ (100), 330 (11), 312 (6), 173 [H$_3$CSC$_9$H$_{18}$]$^+$ (30), 102 [AcOCH$_2$CH(NHAc) − CH$_2$CO]$^+$ (25), 84 [AcOCH$_2$CH(NHAc) − AcOH]$^+$ (45). The DMDS derivative of per-acetylated (13*Z*)-2-aminoeicos-13-ene-1,3,4-triol MS, *m*/*z* (relative intensity, %): 557 [M − HSMe]$^+$ (2), 460 [M − H$_3$CSC$_7$H$_{14}$]$^+$ (24), 400 [M − H$_3$CSC$_7$H$_{14}$ − CH$_3$COOH]$^+$ (100), 358 (10), 340 (11), 145 [H$_3$CSC$_7$H$_{14}$]$^+$ (36), 102 [AcOCH$_2$CH(NHAc) − CH$_2$CO]$^+$ (36), 84 [AcOCH$_2$CH(NHAc) − AcOH]$^+$ (55). The 2-acetyloxy acids from FA subfraction were methylated using *N*-nitroso-*N*-methylurea. The resulting methyl esters were purified by column chromatography (SiO$_2$, 3.5 cm × 1.2 cm), eluting with 44 mL of hexane-ethyl acetate (10:1, *v*/*v*). The elution gave a mixture, containing the methyl esters of major mono(methylthio) and minor tris(methylthio) derivatives of 2-acetyloxy FAs (0.2 mg): ^{1}H-NMR ($CDCl_3$, 700 MHz): 5.40 (dt, *J* = 6.9, 15.2 Hz, –**HC**=CH–CH(SCH_3)–), 5.175 (dd, *J* = 9.0, 15.2 Hz, –HC=**CH**–CH(SCH_3)–), 4.99 (dd, *J* = 6.2, 6.9 Hz, H-2), 3.74 (s, H_3CO– at C-1), 2.99 (m, –HC=CH–**CH**(SCH_3)–), 2.135 (s, –$OCOCH_3$ at C-2), 2.04 (m, allylic CH_2), 1.965 (s, –HC=CH–CH(SC**H$_3$**)–), 1.82 (m, H_2C-3), 1.45–1.15 (m, –$(CH_2)_n$–), 0.88–0.89 (a series of triplets, $J \approx 7.0$ Hz), the methyl esters of the mono(methylthio) derivatives of 2-acetyloxy C_{23} acid MS: Figures S21–S24, the methyl esters of the mono(methylthio) derivatives of 2-acetyloxy C_{22} acid MS: Figures S25–S28, the methyl esters of the tris(methylthio) derivatives of 2-acetyloxy C_{23} acid MS: Figures S29 and S30, the methyl esters of the tris(methylthio) derivatives of 2-acetyloxy C_{22} acid MS: Figures S31 and S32. For the methyl esters of 2-acetyloxy C_{23} acids containing an allylic –SMe group, the order of elution was 15-methylthio-16-ene→17-methylthio-15-ene→16-methylthio-17-ene→18-methylthio-16-ene. The elution order for the methyl esters of 2-acetyloxy C_{22} acids containing an allylic –SMe group was 14-methylthio-15-ene→16-methylthio-14-ene→15-methylthio-16-ene→17-methylthio-15-ene. The mixture of these methylthio derivatives was hydrogenated over Adams' catalyst in AcOH at room temperature (12 h). The hydrogenation gave methyl 2-(acetyloxy)tricosanoate and methyl 2-(acetyloxy)docosanoate. The straight-chain structures of these compounds were clarified, using the GC-MS data on the methyl esters of straight-chain 2-acetyloxy tricosanoic and docosanoic acids, released from the non-oxidized cerebrosides of *Aulosaccus* sp. Then, methyl 2-(acetyloxy)tricosanoate and methyl 2-(acetyloxy)docosanoate, obtained by hydrodesulfurization, were de-esterified in MeCN (0.45 mL) and 5N HCl (0.05 mL) for 4 h at 73–75 °C to give free FAs. These FAs were converted into (2*S*)-oct-2-yl esters of 2-hydroxy FAs by treatment with 2% H_2SO_4 in (2*S*)-octan-2-ol (0.2 mL) for 4 h at 75 °C in a capped vial [8]. In GC analyses, the retention times of the resulting (2*S*)-oct-2-yl ester of 2-hydroxy tricosanoic and docosanoic acids were identical with those of the standard (2*S*)-oct-2-yl esters of (2*R*)-2-hydroxy tricosanoic and docosanoic acids, respectively.

The absolute configuration of glucose, released from oxidized cerebrosides, was determined by the GC analyses of per-acetylated (2*R*)-oct-2-yl glycosides according to the method of Leontein et al. [41]. The sugar (1.2 mg), (2*R*)-octan-2-ol (0.4 mL), and one drop of trifluoroacetic acid in a capped vial were kept for 7 h at 120 °C with stirring. Then, the mixture was concentrated in vacuo and acetylated with Ac_2O (0.4 mL) in pyridine (0.4 mL), overnight. The acetylated material was purified by column chromatography (SiO_2, 3.0 cm × 1.2 cm), eluting with a mixture of hexane/ethylacetate (5:1, *v*/*v*). The eluate was evaporated, yielding per-acetylated (2*R*)-oct-2-yl glucoside. Standards, D-Glc (1.0 mg) and L-Glc (1.0 mg), were treated and derivatized under the same conditions that were applied to the sugar subfractions, liberated from parts 1 and 2. The GC profiles (the retention times and intensities of GC peaks) of the derivatives of D-Glc and sugar, obtained from cerebrosides, were proven to be identical.

Supplementary Materials: The following are available online. Figure S1a: Base peak chromatogram from UPLC-MS analysis of oxidized cerebrosides (positive ion mode). Figure S1b: Extracted-ion chromatograms from UPLC-MS analysis of oxidized cerebrosides. Scheme S1: Hock fragmentation of some isomeric allylic hydroperoxides found in the present study. Figures S2, S2a–S2c: ^{1}H-NMR spectra (CD_3OD, 500 MHz) of the RP-HPLC fraction, containing oxidized cerebrosides. Figure S3 ^{13}C-NMR spectrum (CD_3OD, 125 MHz) of the RP-HPLC fraction containing oxidized cerebrosides. Figures S4–S7: Mass spectra of samples, enriched in the methyl esters of 2-acetyloxy C_{23} and C_{22} acids with a keto group in the (*n*–8) or (*n*–7) positions. Figures S8–S11: Mass spectra of samples, enriched in the methyl esters of 2-acetyloxy C_{23} and C_{22} acids with an additional acetyloxy group in the (*n*–8) or (*n*–7) positions. Figures S12, S13: Mass spectra of samples, enriched in the methyl esters of 2-acetyloxy C_{23} acids with a keto group in the (*n*–9) or (*n*–6) positions. Figure S14: Averaged mass spectrum for the four methyl ester of isomeric 2-acetyloxy keto C_{23} acids. Figures S15, S16: Mass spectra of samples, enriched in the methyl esters of 2-acetyloxy C_{22} acids with a keto group in the (*n*–9) or (*n*–6) positions. Figures S17–S20: Mass spectra of samples, enriched in the methyl esters of 2-acetyloxy C_{23} and C_{22} acids with an additional acetyloxy group in the (*n*–9) or (*n*–6) positions. Figures S21–S28: Mass spectra of samples, enriched in the methyl esters of 2-acetyloxy C_{23} and C_{22} acids with an allylic methylthio group. Figures S29–S32: Mass spectra of the methyl esters of 2-acetyloxy tris(methylthio) C_{23} and C_{22} acids. Figure S33: Mass spectrum of the bis(methylthio) adduct of methyl 11-hydroxy elaidate. Figure S34: Superimposed mass spectra of overlapping methyl 9-(methylthio)octadec-10-enoate and 11-(methylthio)octadec-9-enoate. Figure S35: Mass spectrum of methyl 9,10,11-tris(methylthio)octadecanoate. Figure S36: Superimposed mass spectra of overlapping methyl 10-(methylthio)octadec-8-enoate and 8-(methylthio)octadec-9-enoate. Figure S37: Mass spectrum of methyl 8,9,10-tris(methylthio)octadecanoate. A section before Figures S33–S37 provide experimental details. concerning allylic monohydroxylation of methyl oleate and methylthiolation of allylic hydroxy/acetyloxy elaidates.

Author Contributions: Investigation, data interpretation, methodology, writing, E.A.S. NMR analyses, V.A.D. ESI-MS analyses, validation, P.S.D. All authors have read and agreed to the published version of the manuscript.

Funding: This research was partly funded by the RFBR (Russian Foundation for Basic Research), grant NO. 20-03-00014, and by the Ministry of Science and Education (Russian Federation), grant NO. 13.1902.21.0012 (number of agreement: 075-15-2020-796).

Acknowledgments: The study was carried out on the equipment of the Collective Facilities Center "The Far Eastern Center for Structural Molecular Research (NMR/MS) of PIBOC FEB RAS". The authors express their gratitude to O.P. Moiseenko and L.P. Ponomarenko for their kind help in the GC-MS and GC analyses and V.A. Stonik for reading the manuscript and discussing its content. We thank D.N. Pelageev and O.S. Radchenko for the provision of some reagents.

Conflicts of Interest: The authors declare no conflict of interest. The funders had no role in the design of the study, in the collection, analyses, or interpretation of data, in the writing of the manuscript, or in the decision to publish the results.

Appendix A

The differences in δ_C (CD_3OD, Figure 5a,b and Figure A1a,b) of allylic –H**C**(OOH)– ($\Delta\delta_C = \delta_{C\ trans} - \delta_{C\ cis} = 88.2 - 82.5 = 5.7$) and –H**C**(OH)– ($\Delta\delta_C = \delta_{C\ trans} - \delta_{C\ cis} = 74.4 - 68.7 = 5.7$) were close to the corresponding values ($CDCl_3$) of *trans*-monoenoic and *cis*-monoenoic allylic hydroperoxides ($\Delta\delta_C = 86.9 - 81.1 = 5.8$) and allylic alcohols ($\Delta\delta_C = 73.1 - 67.5 = 5.6$), calculated in accordance with data in the literature [43]. Additionally, the signals of –HC(OH)– of *trans*-monoenoic allylic alcohols (δ_H 3.94, Figure 5b) were shifted upfield (by 0.42 ppm) compared to their *cis*-isomers (δ_H 4.36, Figure A1b), as described for *cis*/*trans*-isomeric methyl 12-hydroxyoctadec-10-enoate and 9-hydroxyoctadec-10-enoate in $CDCl_3$ ([44]: $\Delta\delta_H = \delta_{H\ cis} - \delta_{H\ trans} = 4.4 - 4.0 = 0.4$). Similarly, the signals of –HC(OOH)– of *trans*-monoenoic hydroperoxides (δ_H 4.16, Figure 5a) were shifted upfield (by 0.46 ppm) compared to their *cis*-isomers (δ_H 4.62, Figure A1a), as observed earlier ([45]: $\Delta\delta_H = \delta_{H\ cis} - \delta_{H\ trans} = 4.68 - 4.22 = 0.46$ for *cis*/*trans*-isomeric hydroperoxides of methyl oleate in $CDCl_3$).

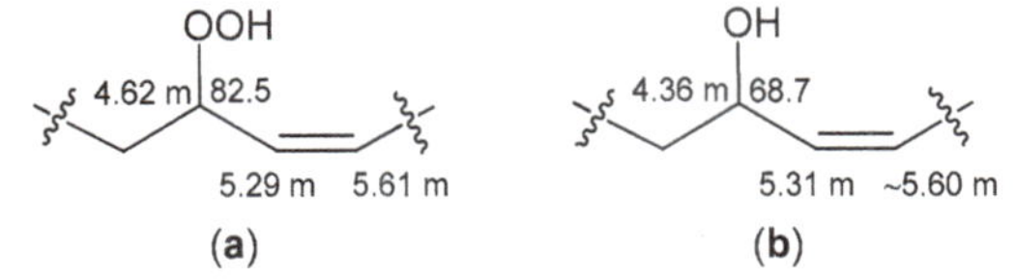

Figure A1. Some δ_H (500 MHz) and δ_C (125 MHz) values for trace *cis*-monoenes with allylic (**a**) hydroperoxy or (**b**) hydroxy groups (solvent: CD_3OD).

Appendix B

Although a single peak in the GC chromatogram represented overlapping esters of four isomeric keto acids, the mass spectra, obtained from different points in the GC peak, exhibited fragmentations of samples, enriched in the isomers with keto groups in the (*n*–9)→(*n*–8)→(*n*–7)→(*n*–6) positions (in this order of elution). Mass spectra of methyl esters of minor keto C_{23} and C_{22} acids displayed homologous pairs of *m*/*z* 141/156 and 99/114 fragments, containing methyl ends of the molecules with keto groups on the (*n*–9) or (*n*–6) carbon, respectively (Figures S12, S13, S15, and S16). The total mass spectra, obtained by averaging the spectra over the selected GC peaks, showed significant differences between relative abundances of several ions, characteristic of different isomers. For the overlapping esters of 15-keto, 16-keto, 17-keto, and 18-keto C_{23} acids, the abundance ratio of the characteristic ions of *m*/*z* 342, 356, 370, and 384 were about 1:3:3:1 (Figure S14). Accordingly, the mixture of C_{23} acid derivatives contained significant amounts of isomers with a keto group in the (*n*–8) and (*n*–7) positions and minor amounts of isomers with a keto group located at the (*n*–9) and (*n*–6) positions. The overlapping methyl esters of homologous keto C_{22} acids exhibited similar fragmentation and elution behavior.

GC-MS analysis also revealed the presence of clusters of closely overlapping chromatographic peaks for derivatives of regio-isomeric diacetyloxy C_{23} and C_{22} acids, including minor components. In particular, the α-fragments included the ions of *m/z* 371 and 413 (Figures S17 and S18) for the derivatives of 2,15-diacetyloxy and 2,18-diacetyloxy C_{23} acids, respectively, and the ions of *m/z* 357 and 399 (Figures S19 and S20) for the 2,14-diacetyloxy and 2,17-diacetyloxy C_{22} acid derivatives, respectively. The abundant ions of *m/z* 269 and 311 (Figures S17 and S18), which were detected in the mass spectra of the methyl esters of isomeric minor 2-acetyloxy C_{23} acids, confirmed the presence of second acetyloxy groups on C-15 and C-18, respectively. Similarly, *m/z* 255 and 297 ions in the mass spectra of the methyl esters of minor 2-acetyloxy C_{22} acids (Figures S19 and S20) indicated a second acetoxy group on C-14 and C-17, respectively. α-Fragmentation of isomers with an acetyloxy group in the (*n*–9), (*n*–8), (*n*–7), or (*n*–6) positions (in this order of elution) and subsequent loss of AcOH gave rise to *m/z* 125, 111, 97, and 83 ions, respectively, containing the methyl end of acyl chains. Among these α-fragments, the largest fragment of *m/z* 125 was much less abundant than the related ions. Unfortunately, the percentages of all the methyl esters of isomeric diacetylated FAs, which were only partially separated by GC, could not be accurately evaluated on the basis of mass spectral data because the presence of diagnostic α-fragments, formed from isomers with an acetyloxy group in the (*n*–9) positions, were masked by other peaks in averaged mass spectrum. However, for diacetyloxy C_{23} acid derivatives with an isolated acetyloxy group in the (*n*–8), (*n*–7), or (*n*–6) positions, the abundance ratio of the diagnostic α-fragments of *m/z* 385, 399, and 413, respectively, were about 3:3:1. The abundance ratio of homologous α-fragments (*m/z* 371, 385, and 399) were nearly the same for the derivatives of diacetyloxy C_{22} acids. Accordingly, the spectra, recorded at the top of the most intense overlapping GC peaks, exhibited fragmentation patterns of major isomers with an acetyloxy group in the (*n*–8) or (*n*–7) positions, while the mass spectra, recorded at lower points, belonged to minor isomers with an acetyloxy group on the (*n*–9) or (*n*–6) carbons.

We assumed that the abundance ratio of components with oxygen-containing groups on the (*n*–9), (*n*–8), (*n*–7), or (*n*–6) carbons (about 1:3:3:1, respectively) were nearly the same for isomeric hydrogenated products and initial hydroperoxy FA moieties. This suggestion was supported by (–)ESI-MS/MS study of isomeric allylic hydroperoxides (Figure 4a), showing two pairs of characteristic peaks with a 3:1 intensity ratio, including peaks at *m/z* 743.6/729.55 (fragments, formed from acyl chains with –OOH in the (*n*–8)/(*n*–9) positions) and *m/z* 784.6/798.6 (fragments, formed from acyl chains with –OOH in the (*n*–7)/(*n*–6) positions). Thus, the previously mentioned minor hydrogenated products were possibly derived from minor amide-linked FA with allylic hydroperoxy/hydroxy/keto groups in the (*n*–9) or (*n*–6) positions. However, complete structures of the corresponding oxidized cerebrosides could not be comprehensively elucidated due to their minor amounts. Therefore, only major cerebrosides with oxygen-containing groups in the (*n*–8) or (*n*–7) positions of acyl chains were the focus of our research.

Appendix C

Methyl esters of 11-hydroxy and 8-hydroxy elaidic acids were prepared from methyl oleate according to the method described by Li et al. [46]. Methyl 11-hydroxy elaidate was treated with DMDS to give a mixture, consisting of 59.2% of the expected DMDS adduct, 26.8% of isomeric allylic thioethers (9-(methylthio)octadec-10-enoate and 11-(methylthio)octadec-9-enoate), and 8.0% of 9,10,11-tris(methylthio)octadecenoate (Scheme A1). Two GC peaks that showed identical major fragments in GC-MS represented stereoisomeric *bis*(methylthio) adducts with a nearly 1:1 ratio. Assuming a specific anti-addition of DMDS [33], these stereoisomeric adducts could be racemic erythro-9,10-bis(methylthio)octadecanoates, differing in configuration only at the carbon atom linked to the –OH group. Based on the NOESY experiment, two pairs of singlets at δ_H 2.14, 2.22 and δ_H 2.06, 2.185 in the ^{1}H-NMR spectrum ($CDCl_3$) of the mixture were assigned to vicinal –SMe groups of different racemates. Two GC peaks (major and minor) for mono(methylthio)octadecenoates also had identical mass spectra. For major *trans*-allylic thioethers, the ^{1}H,^{1}H-COSY diagram ($CDCl_3$) showed a spin system, which consisted of the protons of two *trans*-olefinic CH (δ_H 5.17, dd, J = 9.2, 15.2 Hz, and 5.40, dt, J = 6.8, 15.2 Hz) and CH (δ_H 2.985, m), bearing an –SMe group (^{1}H-NMR spectrum: δ_H 1.965, s). Accordingly, the major GC peak for overlapping 9-(methylthio)octadec-10-enoates and 11-(methylthio)octadec-9-enoates represented isomers with a *trans*-double bond (23.0%), and the minor GC peak, observed before the major peak, likely represented their *cis*-isomers (3.8%). Similarly, two GC peaks (major and minor) for stereoisomeric 9,10,11-tris(methylthio)octadecanoates were observed. Upon MeCN/HCl hydrolysis of this methylthio derivative mixture, the stereoisomeric bis(methylthio) adducts were destroyed, so major allylic thioethers (9/11-(methylthio)octadec-10/9-enoic acids, 59.3%) and minor tris(methylthio) derivatives (stereoisomeric 9,10,11-tris(methylthio)octadecanoic acids, 12.4%), analogous to that shown in Scheme 3, along with octadecadienoic acid (11.7%), were found in the hydrolysate. Apparently, the bis(methilthio) adduct of allylic alcohol could lose an –OH group and one –SMe group during MeCN/HCl hydrolysis, giving rise to an additional amount of allylic thioethers.

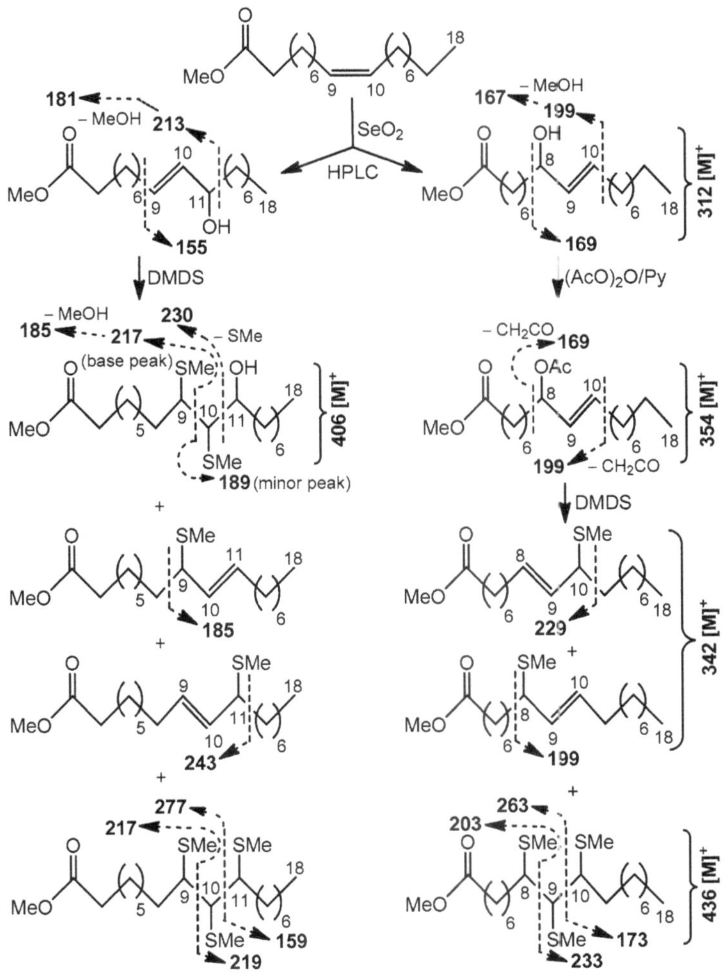

Scheme A1. Allylic mono-hydroxylation of methyl oleate, followed by transformations of allylic alcohols into S-methyl derivatives. GC-MS cleavage patterns for the derivatives, obtained from methyl oleate, are depicted.

The DMDS adduct of methyl oleate can be de-esterified in MeCN/HCl without detectable degradation of the –CH(SMe)–CH(SMe)– fragment, in contrast to the DMDS adduct of methyl 11-hydroxy elaidate. This was confirmed by the ^{1}H-NMR spectra, recorded before and after hydrolysis. In particular, the ^{1}H-NMR spectrum ($CDCl_3$) of the product, obtained after hydrolysis of the DMDS adduct of methyl oleate, showed the superimposed signals of two vicinal CH (δ_H 2.685, m), linked to –SMe groups (δ_H 2.10, s), and α-CH_2 (δ_H 1.845, m; 1.32, m) groups of *bis*(methylthio) oleic acid. Analogously, the *bis*(methylthio) derivatives of monoenoic sphingoid bases were obtained after MeCN/HCl hydrolysis of derivatized cerebrosides (part 2).

As for the allylic alcohol acetate, methyl 8-acetyloxy elaidate did not react appreciably with DMDS under the conditions used for methyl 11-hydroxy elaidate (room temperature, 24 h). However, with longer reaction times (4 days), methyl 8-acetyloxy elaidate reacted with DMDS to give major allylic ethers, 8-(methylthio)octadec-9-enoate and 10-(methylthio)octadec-8-enoate (61.5%, mainly *trans*-forms), and minor 8,9,10-tris(methylthio)octadecanoates (25.4%), formed after deacetylation of the starting compound. Surprisingly, the expected bis(methylthio)derivative of methyl 8-acetyloxy elaidate was not detected (Scheme A1). For experimental details and mass spectra (Figures S33–S37), see the supplementary materials.

References

1. Cyberlipid. Lipid Oxidation. Available online: http://cyberlipid.gerli.com/oxidation/generation-of-peroxidation-products/peroxidation-of-fatty-acids/ (accessed on 1 March 2020).
2. Miyamoto, S.; Ronsein, G.E.; Prado, F.M.; Uemi, M.; Corrêa, T.C.; Toma, I.N.; Bertolucci, A.; Oliveira, M.C.B.; Motta, F.D.; Medeiros, M.H.G.; et al. Biological Hydroperoxides and Singlet Molecular Oxygen Generation. *IUBMB Life* **2007**, *59*, 322–331. [CrossRef] [PubMed]
3. Yin, H.; Xu, L.; Porter, N.A. Free Radical Lipid Peroxidation: Mechanisms and Analysis. *Chem. Rev.* **2011**, *111*, 5944–5972. [CrossRef] [PubMed]
4. Xia, W.; Budge, S.M. Techniques for the Analysis of Minor Lipid Oxidation Products Derived from Triacylglycerols: Epoxides, Alcohols, and Ketones. *Compr. Rev. Food Sci. Food Saf.* **2017**, *16*, 735–758. [CrossRef]
5. Costantino, V.; Fattorusso, E.; Imperatore, C.; Mangoni, A.; Freigang, S.; Teyton, L. Corrugoside, a new immunostimulatory α-galactoglycosphingolipid from the marine sponge *Axinella corrugata*. *Bioorg. Med. Chem.* **2008**, *16*, 2077–2085. [CrossRef]
6. Couto, D.; Santinha, D.; Melo, T.; Ferreira-Fernandes, E.; Videira, R.A.; Campos, A.; Fardilha, M.; Domingues, P.; Domingues, M.R.M. Glycosphingolipids and oxidative stress: Evaluation of hydroxyl radical oxidation of galactosyl and lactosylceramides using mass spectrometry. *Chem. Phys. Lipids* **2015**, *191*, 106–114. [CrossRef]
7. Santinha, D.; Ferreira-Fernandes, E.; Melo, T.; Silva, E.M.P.; Maciel, E.; Fardilha, M.; Domingues, P.; Domingues, M.R.M. Evaluation of the photooxidation of galactosyl- and lactosyl- ceramide by electrospray ionization mass spectrometry. *Rapid Commun. Mass Spectrom.* **2014**, *28*, 2275–2284. [CrossRef]
8. Santalova, E.A.; Denisenko, V.A.; Dmitrenok, P.S.; Drozdov, A.L.; Stonik, V.A. Cerebrosides from a Far-Eastern glass sponge *Aulosaccus* sp. *Lipids* **2015**, *50*, 57–69. [CrossRef]
9. Adams, J.; Ann, Q. Structure determination of sphingolipids by mass spectrometry. *Mass Spectrom. Rev.* **1993**, *12*, 51–85. [CrossRef]
10. Cutignano, A.; De Palma, R.; Fontana, A. A chemical investigation of the Antarctic sponge *Lyssodendoryx flabellata*. *Nat. Prod. Res.* **2012**, *26*, 1240–1248. [CrossRef]
11. Nakagawa, K.; Kato, S.; Miyazawa, T. Determination of Phosphatidylcholine Hydroperoxide (PCOOH) as a Marker of Membrane Lipid Peroxidation. *J. Nutr. Sci. Vitaminol.* **2015**, *61*, S78–S80. [CrossRef]
12. Ito, J.; Mizuochi, S.; Nakagawa, K.; Kato, S.; Miyazawa, T. Tandem Mass Spectrometry Analysis of Linoleic and Arachidonic Acid Hydroperoxides via Promotion of Alkali Metal Adduct Formation. *Anal. Chem.* **2015**, *87*, 4980–4987. [CrossRef] [PubMed]
13. Ito, J.; Shimizu, N.; Kobayashi, E.; Hanzawa, Y.; Otoki, Y.; Kato, S.; Hirokawa, T.; Kuwahara, S.; Miyazawa, T.; Nakagawa, K. A novel chiral stationary phase LC-MS/MS method to evaluate oxidation mechanisms of edible oils. *Sci. Rep.* **2017**, *7*, 10026:1–10026:10. [CrossRef] [PubMed]

14. Hock, H.; Lang, S. Autoxydation von Kohlenwasserstoffen, IX. Mitteil.: Über Peroxyde von Benzol-Derivaten. *Ber. Dtsch. Chem. Ges. A* **1944**, *77*, 257–264. [CrossRef]
15. Wang, Z. *Comprehensive Organic Name Reactions and Reagents*; John Wiley & Sons: Hoboken, NJ, USA, 2010; pp. 1438–1442. ISBN 9780470638859. [CrossRef]
16. Mihailović, M.L.; Čeković, Ž.; Lorenc, L. Oxidations with Lead Tetraacetate. In *Organic Syntheses by Oxidation with Metal Compounds*; Mijs, W.J., De Jonge, C.R.H.I., Eds.; Springer: Boston, MA, USA, 1986; pp. 741–816, ISBN 978-1-4612-9248-7. [CrossRef]
17. Čeković, Ž. Reactions of carbon radicals generated by 1,5-transposition of reactive centers. *J. Serb. Chem. Soc.* **2005**, *70*, 287–318. [CrossRef]
18. Mićović, V.M.; Mamuzić, R.I.; Jeremić, D.; Mihailović, M.L. Reactions with lead tetraacetate. I. Oxidation of saturated aliphatic alcohols. Part 1. *Tetrahedron Lett.* **1963**, *4*, 2091–2095. [CrossRef]
19. Schaich, K.M. Challenges in elucidating lipid oxidation mechanisms: When, where, and how do products arise? In *Lipid Oxidation: Challenges in Food Systems*; Nienaber, U., Logan, A., Pan, X., Eds.; AOCS Press: Urbana, IL, USA, 2013; pp. 1–52, ISBN 978-0-9830791-6-3. [CrossRef]
20. MacMillan, D.K.; Murphycor, R.C. Analysis of lipid hydroperoxides and long-chain conjugated keto acids by negative ion electrospray mass spectrometry. *J. Am. Soc. Mass Spectrom.* **1995**, *6*, 1190–1201. [CrossRef]
21. Santalova, E.A.; Denisenko, V.A.; Dmitrenok, P.S. Structural Analysis of the Minor Cerebrosides from a Glass Sponge *Aulosaccus* sp. *Lipids* **2015**, *50*, 1209–1218. [CrossRef]
22. Santalova, E.A.; Denisenko, V.A. Analysis of the Configuration of an Isolated Double Bond in Some Lipids by Selective Homonuclear Decoupling. *Nat. Prod. Commun.* **2017**, *12*, 1913–1916. [CrossRef]
23. Porter, N.A.; Wujek, J.S. Allylic Hydroperoxide Rearrangement: β-Scission or Concerted Pathway? *J. Org. Chem.* **1987**, *52*, 5085–5089. [CrossRef]
24. Aveldaño, M.I.; Horrocks, L.A. Quantitative release of fatty acids from lipids by a simple hydrolysis procedure. *J. Lipid Res.* **1983**, *24*, 1101–1105.
25. Cobern, D.; Hobbs, J.S.; Lucas, R.A.; Mackenzie, D.J. Location of Hydroperoxide Groups in Monohydroperoxides formed by Chlorophyll-photosensitised Oxidation of Unsaturated Esters. *J. Chem. Soc. C* **1966**, *1966*, 1897–1902. [CrossRef]
26. Christie, W.W.; The LipidWeb. Mass Spectrometry of Methyl Esters. *Oxo (Keto) Fatty Acids*. Available online: http://www.lipidhome.co.uk/ms/methesters/me-oxo/index.htm (accessed on 5 May 2020).
27. Racovita, R.C.; Jetter, R. Identification of In-Chain-Functionalized Compounds and Methyl-Branched Alkanes in Cuticular Waxes of *Triticum aestivum* cv. Bethlehem. *PLoS ONE* **2016**, *11*, e0165827. [CrossRef]
28. Gemal, A.L.; Luche, J.-L. Lanthanoids in Organic Synthesis. 6. The Reductions of α-Enones by Sodium Borohydride in the Presence of Lanthanoid Chlorides: Synthetic and Mechanistic Aspects. *J. Am. Chem. Soc.* **1981**, *103*, 5454–5459. [CrossRef]
29. Zhang, X.; Rao, W.; Chan, P.W.H. Iodine-Catalyzed Allylic Alkylation of Thiols with Allylic Alcohols. *Synlett* **2008**, *2008*, 2204–2208. [CrossRef]
30. Tabarelli, G.; Godoi, M.; Canto, R.F.S.; Mora, J.R.; Nome, F.; Braga, A.L. Direct Synthesis of Allylic Thioethers under Greener Conditions: A Solvent- and Catalyst-Free Approach. *Synth. Commun.* **2014**, *44*, 3441–3449. [CrossRef]
31. Sugiyama, S.; Honda, M.; Komori, T. Biologically active glycosides from Asteroidea. XV. Asymmetric synthesis of phytosphingosine and phytosphingosine anhydro base: Assignment of the absolute stereochemistry. *J. Liebigs Ann. Chem.* **1988**, *1988*, 619–625. [CrossRef]
32. Santalova, E.A.; Denisenko, V.A. Fatty Acids from a Glass Sponge *Aulosaccus* sp. Occurrence of New Cyclopropane-Containing and Methyl-Branched Acids. *Lipids* **2017**, *52*, 73–82. [CrossRef]
33. Caserio, M.C.; Fisher, C.L.; Kim, J.K. Boron trifluoride catalyzed addition of disulfides to alkenes. *J. Org. Chem.* **1985**, *50*, 4390–4393. [CrossRef]
34. Scribe, P.; Guezennec, J.; Dagaut, J.; Pepe, C.; Saliot, A. Identification of the Position and the Stereochemistry of the Double Bond in Monounsaturated Fatty Acid Methyl Esters by Gas Chromatography/Mass Spectrometry of Dimethyl Disulfide Derivatives. *Anal. Chem.* **1988**, *60*, 928–931. [CrossRef]
35. Knothe, G.; Steidley, K.R. Fatty Acid Methyl Esters with Two Vicinal Alkylthio Side Chains and Their NMR Characterization. *J. Am. Oil Chem. Soc.* **2017**, *94*, 537–549. [CrossRef]
36. Leontein, K.; Lindberg, B.; Lönngren, J. Assignment of absolute configuration of sugars by g.l.c. of their acetylated glycosides formed from chiral alcohols. *Carbohydr. Res.* **1978**, *62*, 359–362. [CrossRef]

37. Santalova, E.A.; Denisenko, V.A. Steroids from a Far-Eastern Glass Sponge *Aulosaccus* sp. *Nat. Prod. Commun.* **2019**, *14*, 1–8. [CrossRef]
38. Taylor, M.W.; Radax, R.; Steger, D.; Wagner, M. Sponge-associated microorganisms: Evolution, ecology, and biotechnological potential. *Microbiol. Mol. Biol. Rev.* **2007**, *71*, 295–347. [CrossRef] [PubMed]
39. Bergstrand, L.H.; Cardenas, E.; Holert, J.; Van Hamme, J.D.; Mohn, W.W. Delineation of steroid degrading microorganisms through comparative genomic analysis. *mBio* **2016**, *7*, e00166-16:1–e00166-16:14. [CrossRef] [PubMed]
40. Kreit, J. Microbial catabolism of sterols: Focus on the enzymes that transform the sterol 3β-hydroxy-5-en into 3-keto-4-en. *FEMS Microbiol. Lett.* **2017**, *364*, fnx007:1–fnx007:9. [CrossRef] [PubMed]
41. Smith, A.G.; Brooks, C.J.W. Mass spectra of Δ^4- and 5α-3-ketosteroids formed during the oxidation of some 3β-hydroxysteroids by cholesterol oxidase. *Biomed. Mass Spectrom.* **1976**, *3*, 81–87. [CrossRef]
42. Christie, W.W.; The LipidWeb. Long-Chain (Sphingoid) Bases. Available online: http://www.lipidmaps.org/resources/lipidweb/index.php?page=lipids/sphingo/lcb/index.htm (accessed on 10 September 2020).
43. Garwood, R.F.; Khambay, B.P.S.; Weedon, B.C.L.; Frankel, E.N. Allylic Hydroperoxides from the Autooxidation of Methyl Oleate. *JCS Chem. Comm.* **1977**, 364–365. [CrossRef]
44. Frankel, E.N.; Garwood, R.F.; Khambay, B.P.S.; Moss, G.P.; Weedon, B.C.L. Stereochemistry of Olefin and Fatty Acid Oxidation. Part 3. The Allylic Hydroperoxides from the Autooxidation of Methyl Oleate. *J. Chem. Soc. Perkin Trans. 1* **1984**, *1984*, 2233–2240. [CrossRef]
45. Ahmed, R.; Siddiqui, H.; Choudhary, M.I.; Gerothanassis, I.P. ^{1}H-^{13}C HMBC NMR experiments as a structural and analytical tool for the characterization of elusive *trans/cis* hydroperoxide isomers from oxidized unsaturated fatty acids in solution. *Magn. Reson. Chem.* **2019**, *57*, S69–S74. [CrossRef]
46. Li, Z.; Tran, V.H.; Duke, R.K.; Ng, M.C.H.; Yang, D.; Duke, C. Synthesis and biological activity of hydroxylated derivatives of linoleic acid and conjugated linoleic acids. *Chem. Phys. Lipids* **2009**, *158*, 39–45. [CrossRef]

Sample Availability: Samples of the compounds are available from the authors.

Publisher's Note: MDPI stays neutral with regard to jurisdictional claims in published maps and institutional affiliations.

molecules

MDPI

Article

Isolation and Structure Determination of Echinochrome A Oxidative Degradation Products

Natalia P. Mishchenko [1,*], Elena A. Vasileva [1], Andrey V. Gerasimenko [2], Valeriya P. Grigorchuk [3], Pavel S. Dmitrenok [1] and Sergey A. Fedoreyev [1]

1 G.B. Elyakov Pacific Institute of Bioorganic Chemistry, Far Eastern Branch of the Russian Academy of Sciences, Vladivostok 690022, Russia; vasilieva_el_an@mail.ru (E.A.V.); paveldmitrenok@mail.ru (P.S.D.); fedoreev-s@mail.ru (S.A.F.)

2 Institute of Chemistry, Far Eastern Branch of the Russian Academy of Sciences, Vladivostok 690022, Russia; gerasimenko@ich.dvo.ru

3 Federal Scientific Center of the East Asia Terrestrial Biodiversity, Far-Eastern Branch of Russian Academy of Sciences, Vladivostok 690022, Russia; kera1313@mail.ru

* Correspondence: mischenkonp@mail.ru; Tel.: +8-423-231-40-50

Academic Editor: Marina DellaGreca

Received: 18 September 2020; Accepted: 16 October 2020; Published: 18 October 2020

Abstract: Echinochrome A (Ech A, **1**) is one of the main pigments of several sea urchin species and is registered in the Russian pharmacopeia as an active drug substance (Histochrome®), used in the fields of cardiology and ophthalmology. In this study, Ech A degradation products formed during oxidation by O_2 in air-equilibrated aqueous solutions were identified, isolated, and structurally characterized. An HPLC method coupled with diode-array detection (DAD) and mass spectrometry (MS) was developed and validated to monitor the Ech A degradation process and identify the appearing compounds. Five primary oxidation products were detected and their structures were proposed on the basis of high-resolution electrospray ionization mass spectrometry (HR-ESI-MS) as 7-ethyl-2,2,3,3,5,7,8-heptahydroxy-2,3-dihydro-1,4-naphthoquinone (**2**), 6-ethyl-5,7,8-trihydroxy-1,2,3,4-tetrahydronaphthalene-1,2,3,4-tetraone (**3**), 2,3-epoxy-7-ethyl-2,3-dihydro-2,3,5,6,8-pentahydroxy-1,4-naphthoquinone (**4**), 2,3,4,5,7-pentahydroxy-6-ethylinden-1-one (**5**), and 2,2,4,5,7-pentahydroxy-6-ethylindane-1,3-dione (**6**). Three novel oxidation products were isolated, and NMR and HR-ESI-MS methods were used to establish their structures as 4-ethyl-3,5,6-trihydroxy-2-oxalobenzoic acid (**7**), 4-ethyl-2-formyl-3,5,6-trihydroxybenzoic acid (**8**), and 4-ethyl-2,3,5-trihydroxybenzoic acid (**9**). The known compound 3-ethyl-2,5-dihydroxy-1,4-benzoquinone (**10**) was isolated along with products **7–9**. Compound **7** turned out to be unstable; its anhydro derivative **11** was obtained in two crystal forms, the structure of which was elucidated using X-ray crystallography as 7-ethyl-5,6-dihydroxy-2,3-dioxo-2,3-dihydrobenzofuran-4-carboxylic acid and named echinolactone. The chemical mechanism of Ech A oxidative degradation is proposed. The in silico toxicity of Ech A and its degradation products **2** and **7–10** were predicted using the ProTox-II webserver. The predicted median lethal dose (LD_{50}) value for product **2** was 221 mg/kg, and, for products **7–10**, it appeared to be much lower (≥2000 mg/kg). For Ech A, the predicted toxicity and mutagenicity differed from our experimental data.

Keywords: histochrome; echinochrome A; oxidative degradation; HPLC–DAD–MS; NMR

1. Introduction

Histochrome® is a solution of the sodium salt of naturally occurring quinone echinochrome A (7-ethyl-2,3,5,6,8-pentahydroxy-1,4-naphthoquinone) for intravenous injections and infusions, manufactured in ampoules (Supplementary Materials, Figure S1). Histochrome is registered in

Russia as an antioxidant drug. It is used in cardiology for the treatment of coronary heart disease and for restriction of the necrosis zone in myocardial infarction (state registration number P N002363/01), and in ophthalmology for the treatment of dystrophic diseases of the retina and cornea, macular degeneration, primary open-angle glaucoma, diabetic retinopathy, hemorrhages to vitreous humor, retina, and anterior chamber, and discirculatory disorders in the central artery and retinal vein (state registration number P N002363/02). Histochrome has no known analogues, and it simultaneously blocks a number of free-radical reactions, neutralizes reactive oxygen species (ROS), nitric oxide, and peroxide radicals, chelates metal ions, inhibits lipid peroxidation, and regulates antioxidant enzyme levels [1].

Echinochrome A (Ech A, **1**) is one of the main pigments of various sea urchin species [2–4] (Figure 1). Ech A isolated from the sand dollar *Scaphechinus mirabilis* (purity >98%) is registered in the Russian pharmacopeia as an active drug substance under the international non-patented name pentahydroxyethylnaphthoquinone (state registration number P N002362/01).

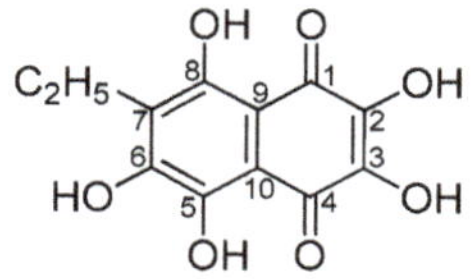

Figure 1. Chemical structure of echinochrome A (7-ethyl-2,3,5,6,8-pentahydroxy-1,4-naphthoquinone, Ech A).

Currently, more and more papers focused on elucidating the mechanisms underlying the diverse biological effects of Ech A are being published. Ech A was found to protect rat cardiomyoblasts and isolated cardiomyocytes from the effects of cardiotoxic compounds doxorubicin, *tert*-butyl hydroperoxide, and sodium nitroprusside, which cause an increase in ROS formation and depolarization of mitochondrial membranes [5]. In rat cardiomyoblast cells, Ech A dose-dependently increased the mass of mitochondria and the content of mitochondrial DNA and activated mitochondrial biogenesis, increasing the expression of the main regulators of the metabolic function of mitochondria [6]. Ech A was also found to activate mitochondrial biogenesis in skeletal muscle, increasing the endurance of rats during physical activity by increasing the number of mitochondria [7]. Being an inhibitor of the sarcoplasmic/endoplasmic reticulum Ca^{2+} ATPase 2a (SERCA2A) receptor, which is responsible for pumping calcium ions from the cytosol into the sarcoplasmic reticulum, Ech A prevented ischemic damage of the myocardium, reducing the area of myocardial infarction [8]. By reducing the level of intracellular ROS and regulating the expression of pro- and antiapoptotic proteins, Ech A protected human cardiac progenitor cells against oxidative stress [9]. This may be the basis for a simple and effective strategy to enhance myocardial regeneration by increasing the survival of transplanted cardiac cells under oxidative stress induced by ischemic damage. Ech A was found to be an effective agent for promoting cell proliferation and maintaining the stemness of hematopoietic stem and progenitor cells [10]. Ech A is also beneficial for human CD34+ progenitor cells from peripheral blood to maintain their self-renewal potential and function during ex vivo expansion. The efficacy of Ech A in a model of hemorrhagic and ischemic stroke in rats has been demonstrated [11,12]. It was found that the drug can cross the blood–brain barrier into the cerebrospinal fluid. Ech A also exhibited an antidiabetic effect due to antioxidant and hypoglycemic activities [13,14]. A study demonstrated that Ech A inhibits acetylcholinesterase and exhibits dose-dependent antiradical activity against nitric oxide, which opens up its possible use in the treatment of neurodegenerative diseases [15]. The therapeutic potential of Ech A in the treatment of various inflammatory diseases has been demonstrated in numerous studies: in a model of experimental colitis in mice [16], in an experimental model of bleomycin-induced pneumonia in immature rats [17], in children aged 7–12 years old with chronic inflammatory lung diseases [18,19], and in adolescents with erosive gastroduodenitis [20,21].

Since the search for active compounds and the creation of new drugs is a long, expensive, and risky process, one of the main modern pharmaceutical strategies is the use of registered drugs for a new medical application. Considering the above variety of activities possessed by Ech A with an established mechanism of action, the development of new dosage forms based on this substance seems promising.

To ensure the quality of active pharmaceutical substances and finished drug products, impurities must be monitored carefully during process development, optimization, and changeover. The isolation, characterization, and control of impurities in pharmaceutical substances are being reviewed with a greater focus on national regulatory and international guidelines [22]. According to International Conference on Harmonization (ICH) guidance Q3A(R2) and Q3B(R2), degradation products are impurities resulting from a chemical change in the drug substance during manufacture and/or storage of the drug product due to the effect of light, temperature, pH, water, or reaction with an excipient and/or the immediate container closure system. Due to the presence of a large number of phenolic hydroxyls, Ech A readily undergoes oxidative decomposition. The aim of this study was to isolate and determine the structure of Ech A degradation products formed during the oxidation of its preparation (Histochrome) by O_2 in air-equilibrated aqueous solutions.

2. Results and Discussion

2.1. Isolation and Structure Elucidation of Echinochrome A Oxidative Degradation Products

According to our observations, in dry form and in the absence of oxygen in solution, the pharmaceutical substance Ech A remains stable for several years. This was confirmed by us after a 3 year stability study of the Ech A substance and Histochrome preparation in ampoules closed under inert conditions. In aqueous solutions, the Ech A sodium derivative (Histochrome) readily hydrolyzes and oxidizes (Figure S2, Supplementary Materials). Therefore, to provide the opportunity to establish primary oxidation products, we did not use the onerous conditions recommended in the ICH guidelines for degradation product studies. In this work, we studied the oxidation of Ech A sodium derivative in air-equilibrated aqueous solutions.

An HPLC method coupled with diode-array detection (DAD) and mass spectrometry (MS) was developed and validated to monitor the degradation of Ech A and to support the peak identification procedure. The eluent system consisting of H_2O (A)/MeCN (B) with the addition of 0.2% AcOH in a gradient mode provided acceptable separation of Ech A and its oxidation products. The developed LC method demonstrated good linearity, and the correlation coefficient for Ech A was found to be 0.9987. The limit of detection (LOD) and limit of quantification (LOQ) of Ech A were found to be 22 and 72 ng/mL, respectively. The analytical area of this method was established by the range of experimental data satisfying the linear model. For Ech A (**1**), the corresponding range was determined to be 72–600 ng/mL. The accuracy and reproducibility of the quantification procedure was evaluated according to the results obtained for Ech A, shown in Table S1 (Supplementary Materials). The detection wavelength 254 nm was chosen on the basis of the fact that all target compounds display intense absorption in the region of 230–270 nm.

A solution of Ech A sodium derivative from an ampoule was diluted approximately 50-fold with distilled water saturated with atmospheric oxygen, pH 7.2. In this solution, the molar ratio of Ech A to dissolved O_2 was 3:1. HPLC–DAD–MS analysis showed that, after 1 h in the air-equilibrated aqueous solution, the first oxidation product of Ech A (**1**) was compound **2**, with a retention time of 7.79 min (Figure 2). The high-resolution electrospray ionization mass spectrum (HR-ESI-MS) of compound **2** presented a peak at *m/z* $[M - H]^-$ 299.0399 (calculated for $[C_{12}H_{11}O_9]^-$ 299.0409). The increase in molecular weight of 34 Da compared to Ech A indicates that compound **2** contained two additional hydroxyl groups in the molecule. In its absorption spectrum, there was no absorption band at 468 nm that is characteristic of Ech A, and absorption bands at 256, 321, and 391 nm were present, indicating a decrease in the length of the conjugation chain in the molecule and, therefore, that oxidative changes affected the quinonoid ring of Ech A. According to NMR data, the structure of compound **2** was previously established by us as 7-ethyl-2,2,3,3,5,6,8-heptahydroxy-2,3-dihydro-1,4-naphthoquinone (Figures S12–S17, Supplementary Materials) [23].

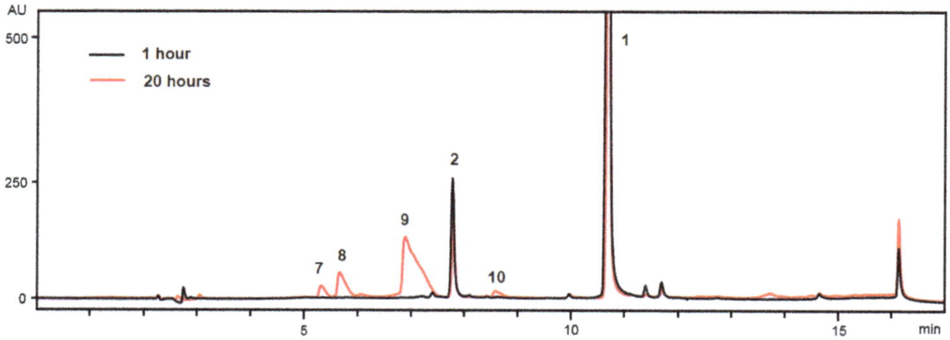

Figure 2. HPLC profiles of Histochrome (Ech A) oxidation products. Unmarked peaks are natural impurities of the Ech A substance [24].

The presence of four aliphatic hydroxyl groups was confirmed by the preparation of tetramethyl ether of compound **2** (*m/z* $[M - H]^-$ 355.1035, calculated for $C_{16}H_{19}O_9{}^-$ 355.1029) by methylation with methyl iodide according to the procedure in [25] (Figure S3, Table S2, Supplementary Materials), which also confirmed the structure of bis-*gem*-diol for compound **2**.

It is interesting to note that, in the mass spectrum of the chromatographic peak with a retention time of 7.79 min, along with the main peak at *m/z* $[M - H]^-$ 299, there were low-intensity peaks with mass values *m/z* $[M - H]^-$ 263, 281, 237, and 253. These peaks were observed in all cases when the ESI mass spectrum of compound **2** was obtained. On the basis of HR-ESI-MS (Table 1), structures of compounds **3–6** were predicted (Figure 3).

Table 1. High-resolution electrospray ionization mass spectrometry (HR-ESI-MS) characteristics of the chromatographic peaks of Ech A oxidation products at retention time 7.79 min.

Compound	Peak Intensity Relative to 2, (%)	Formula	Measured *m/z* $[M - H]^-$	Calculated *m/z* $[M - H]^-$
2	100	$C_{12}H_{12}O_9$	299.0399	299.0409
3	3	$C_{12}H_8O_7$	263.0192	263.0197
4	6	$C_{12}H_{10}O_8$	281.0294	281.0303
5	2	$C_{11}H_{10}O_6$	237.0397	237.0405
6	3	$C_{11}H_{10}O_7$	253.0356	253.0356

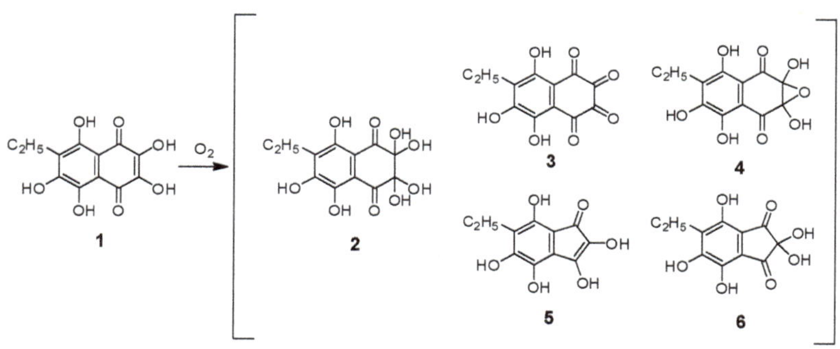

Figure 3. Primary oxidation products of Ech A (**1**); structures **3–6** were proposed on the basis of HR-ESI-MS data.

Prior to our studies, according to published data, the first oxidation product of Ech A was attributed to structures such as the dihydrate of 5,6,8-trihydroxy-7-ethyl-1,2,3,4-tetrahydronaphthalene-1,2,3,4-tetraone (**3**) [26] or monohydrate of 2,3-dihydro-2,3,5,6,8-pentahydroxy-2,3-epoxy-7-ethyl-1,4-naphthoquinone (**4**) [27]. These compounds were present in combination with compound **2**. A compound with the brutto-formula $C_{12}H_{10}O_8$ can exist both in the form of structure **4** and in the form of keto-*gem*-diols **4a** and **4b** (Figure 4). The presence of those compounds in the mixture of oxidation products was previously recorded by us using ^{1}H- and ^{13}C-NMR spectroscopy [23].

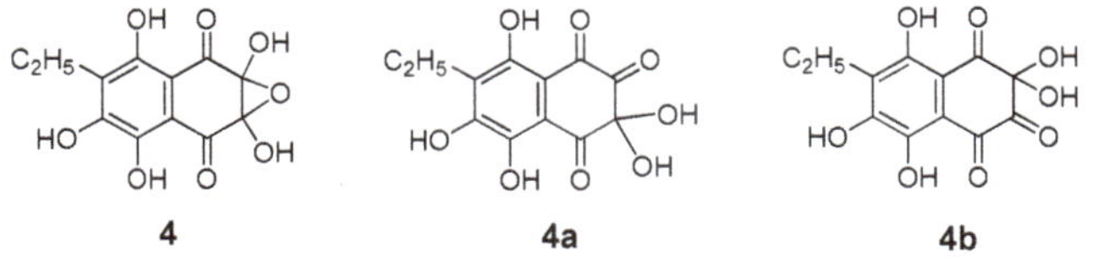

Figure 4. Structures of Ech A oxidation products with brutto-formula $C_{12}H_{10}O_8$.

For compounds **5** and **6**, structures 2,3,4,5,7-pentahydroxy-6-ethylinden-1-one and 2,2,4,5,7-pentahydroxy-6-ethylindane-1,3-dione were predicted, respectively. It is known that di- and polycarbonyl vicinal compounds are prone to hydration; therefore, they are often isolated in the form of *gem*-diols, and, in our case, compound **2** was predominant.

Additionally, structures of compounds **3–6** were confirmed by obtaining their methyl derivatives by methylation with both methyl iodide and diazomethane, the MS data for which are provided in Tables S2 and S3 (Supplementary Materials).

Compounds **3–6** were extremely reactive, as their formation was accompanied by the formation of an intermediate ene-diol radical, superoxide anion radical, hydroxyl radical, and hydrogen peroxide. It was not possible to isolate them from the mixture of Ech A oxidation products, and they continued the chain reaction of oxidation of bis-*gem*-diol **2**, even if O_2 was removed from the reaction medium. In the process of developing technology for the preparation of a Histochrome for injections (0.2 mg/mL Ech A), we observed that even a small amount of O_2 entering the drug solution in sealed ampoules led to the appearance of product **2**. Even the subsequent use of an inert medium (argon) did not stop the process of Ech A oxidation, which led to the formation of products **2–10** and continued until the complete discoloration of the red-brown Histochrome solution.

Twenty hours after the start of the reaction, the opaque dark red solution became transparent yellow red. HPLC–MS analysis showed that approximately 50% of Ech A was consumed during this time (Figure 2 and Figure S2, Supplementary Materials). Unreacted Ech A was removed from the aqueous solution by extraction with chloroform, and the oxidation products were extracted with ethyl acetate. Low-pressure reversed-phase chromatography on Toyopearl HW-40 gel of ethyl acetate extraction revealed five oxidation products of Ech A with retention times of 7.79 (**2**), 5.32 (**7**), 5.67 (**8**), 6.89 (**9**), and 8.56 (**10**) min (Table 2). The absorption spectra of compounds **7–9** contained absorption bands due to $\pi \rightarrow \pi^*$ transitions in the benzenoid core in the region of 310–370 nm, but there were no absorption bands associated with $\pi \rightarrow \pi^*$ transitions in the quinonoid core in the region of 460–540 nm, which indicated a cleavage of the quinonoid ring (Table 2, Figure 5).

Table 2. HPLC coupled with diode array detection (DAD) and MS parameters of Ech A (**1**) and its oxidation products **2** and **7–10**.

Compound	Rt (min)	Formula	Measured *m/z* [M − H]⁻	Calculated *m/z* [M − H]⁻	λ_{max} (nm)
1	10.71	$C_{12}H_{10}O_7$	265.0352	265.0348	254, 338, 471
2	7.79	$C_{12}H_{12}O_9$	299.0399	299.0409	256, 321, 391
7	5.32	$C_{11}H_{10}O_8$	269.0304	269.0297	219, 271, 320
8	5.67	$C_{10}H_{10}O_6$	225.0405	225.0399	270, 320
9	6.89	$C_9H_{10}O_5$	197.0452	197.0450	228, 251, 333
10	8.56	$C_8H_8O_4$	167.0343	167.0344	287

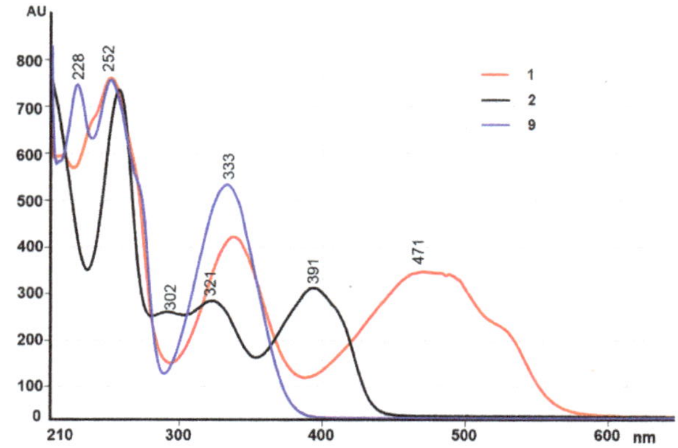

Figure 5. Absorption spectra of Ech A (**1**) and its oxidation products **2** and **9**.

It turned out that compound **7** was quite unstable in an acidic environment, and, after vacuum evaporation in fractions with compound **7**, product **11** was formed. According to the ESI-MS spectrum, product **11** had an *m/z* 251 $[M-H]^-$, which was 18 Da less than mass of compound **7**. The absorption band at 385 nm in the absorption spectrum of compound **11** indicated a longer $\pi \rightarrow \pi^*$ transition chain in its molecule compared to compound **7**. In the ^{1}H-NMR spectrum of compound **11**, we observed a triplet (δ_H 1.24) and a quartet (δ_H 2.78) of an ethyl substituent, a broadened singlet of two hydroxyl groups (δ_H 5.33), and a singlet of the hydroxyl group bound to carbonyl (δ_H 12.84) (Figure S18, Supplementary Materials). The ^{13}C-NMR spectrum of compound **11** contained 11 carbon signals: two signals for the ethyl group (δ_C 12.8 and 17.2), three quaternary carbons (δ_C 106.3, 108.2, and 121.4), and seven quaternary carbons bound to oxygen (δ_C 151.2, 158.0, 159.9, 161.6, 171.0, and 177.8) (Figure S19, Supplementary Materials). In the HMBC spectrum of compound **11**, protons of the ethyl group (δ_H 2.78) showed correlations with δ_C 121.4, 159.0, and 161.6 (Figure S20, Supplementary Materials). These spectral data were insufficient to establish the structure of **11**; however, it turned out that compound **11** easily formed crystals and, as such, X-ray analysis was used to establish its structure. Compound **11** showed polymorphism, with two types of crystals obtained from the same system of solvents (EtOH–$CHCl_3$ = 1:5 *v*/*v*). Recrystallization from these solvents simultaneously provided crystals as dark-red plates (α-form) and as orange prisms (β-form) (Figure 6). Red crystals were predominant (about 90%). Both crystal forms were a crystalline hydrate of **11** (Tables S4–S6, Supplementary Materials).

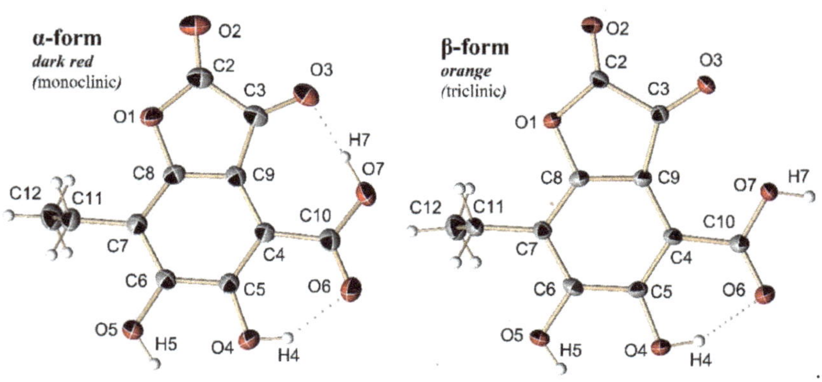

Figure 6. Molecular structure of echinolactone (**11**) in different crystal types (hydrogen bonds are shown as dotted lines).

The α-form of **11** crystallized as monoclinic system with the space group $P2_1/c$ and cell parameters a = 4.7823(6) Å, b = 7.9520(9) Å, c = 14.4705(17) Å, and Z = 4, and the final R-value was found to be 0.0512 (Table S4, Supplementary Materials). The β-form of **11** crystallized as a triclinic system with the space group $P\bar{1}$ and cell parameters a = 4.7823(6) Å, b = 7.9520(9) Å, c = 14.4705(17) Å, and Z = 2, and the final R-value was found to be 0.0407 (Table S4, Supplementary Materials).

In the α- and β-forms of molecule **11**, all atoms with the exception of the CH_3 carbon atom of the ethyl group C12 were located in the same plane (Figure 6). The deviation from the plane did not exceed 0.138(2) Å. The main difference between the two forms was the dimensional orientation of the hydrogen atom H7 of the carboxyl group, which allowed us to consider the molecules of α- and β-forms as stereoisomers (Figure 6); in the α-form, this atom participates in the formation of an intramolecular hydrogen bond, while, in the β-form, it participates in an intermolecular hydrogen bond. In both crystalline forms of **11**, all the corresponding C–C and C–O bonds had close values (Table S5, Supplementary Materials). The torsion angles of C6–C7–C11–C12 in the α- and β-forms were 96.3(3)° and 83.4(2)°, respectively. The intermolecular hydrogen bonds of molecule **11** with H_2O molecules played a decisive role in the formation of crystalline structures. In the α-form, the H_2O molecules were linked to each other by *Ow*–H···*Ow* hydrogen bonds in an infinite chain along the [0 0 1] direction, and they combined isolated **11** molecules into a three-dimensional framework (Figure S21, Supplementary Materials). The coordination number of O atoms of H_2O molecules in α-$C_{11}H_8O_7{\cdot}H_2O$ was 4. In the β-form, molecules of **11** were joined by O7–H7···O3 bonds in pairs into a centrosymmetric bimolecular associate, and H_2O molecules distributed their hydrogen bonds only between molecules of **11**, combining pairs into flat ribbons that were infinite along the [1 −1 0] direction (Figure S22, Supplementary Materials). Ribbons were packed in corrugated layers parallel to the plane [0 0 1] (Figure S23, Supplementary Materials). The coordination number of O atoms of H_2O molecules in β-$C_{11}H_8O_7{\cdot}H_2O$ was 3. The triclinic β-form of $C_{11}H_8O_7{\cdot}H_2O$ had a slightly higher density (1.664 g/cm^3) at a temperature of T = 173(2) K than the monoclinic α-form (1.642 g/cm^3) and could formally be considered as more stable.

On the basis of X-ray data, **11** was assigned the structure of 7-ethyl-5,6-dihydroxy-2,3-dioxo-2,3-dihydrobenzofuran-4-carboxylic acid, and this compound was named echinolactone.

Compounds **7** and **8** turned out to be unstable under conditions of repeated chromatographic separation; therefore, to establish their structures, their stable methyl derivatives with m/z [M − H]$^-$ 297 and 239, respectively, were obtained by methylation with diazomethane.

On the basis of the ^{1}H- and ^{13}C-NMR spectra of the dimethyl ether of compound **7**, methyl ether of compound **8**, and of compound **9** (Table 3, Figure 7), it could be concluded that these compounds retained the same substituents as the Ech A benzenoid ring: an ethyl substituent, a free hydroxyl group, and hydroxyl groups, the protons of which were bound by an intramolecular hydrogen bond with the corresponding carbonyl groups, but the carboxyl group appeared. The differences in the NMR spectra of the Ech A oxidation products consisted of the chemical shifts of the substituents next to the carboxyl group (Table 3); therefore, to establish the structure of compounds **7**–**9**, it was necessary to establish the nature of these substituents.

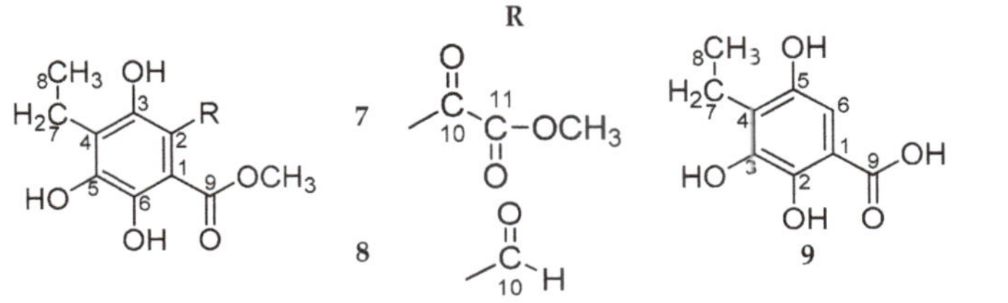

Figure 7. Structures of dimethyl ether of compound **7**, methyl ether of compound **8**, and of compound **9**.

Table 3. NMR data of dimethyl ether of compound **7** (500 MHz for ^{1}H and 126 MHz for ^{13}C, δ, ppm, *J*/Hz), methyl ether of compound **8**, and of compound **9** (700 MHz for ^{1}H and 176 MHz for ^{13}C, δ, ppm, *J*/Hz).

No.	Dimethyl Ether of 7 ($CDCl_3$)			Methyl Ether of 8 (Acetone-d_6)			9 (Acetone-d_6)		
	δ_C	δ_H	HMBC	δ_C	δ_H	HMBC	δ_C	δ_H	HMBC
1	107.7			107.2			109.4		
2	106.3			109.3			144.4	10.70 (1H, s, OH)	1, 2
3	158.4	11.30 (1H, s, OH)	2, 3, 4, 5	159.7	13.36 (1H, s, OH)	2, 3, 4	144.5	8.01 (1H, s, OH)	
4	124.0			124.2			126.3		
5	150.7	6.53 (1H, s, OH)	3, 4, 5, 6	150.7	6.56 (1H, s, OH)	4, 5, 6	148.2	7.54 (1H, s, OH)	
6	143.3	10.49 (1H, s, OH)	1, 5, 6	144.9	11.35 (1H, s, OH)	1, 5, 6	104.8	6.88 (1H, s, H)	1, 3, 4, 5, 9
7	16.2	2.76 (2H, q, J = 7.5, CH_2)	3, 4, 5, 8	16.5	2.77 (2H, q, J = 7.5, CH_2)	3, 4, 5, 8	17.7	2.73 (2H, q, J = 7.5, CH_2)	3, 4, 5, 8
8	12.6	1.16 (3H, t, J = 7.5, CH_3)	4, 7	12.5	1.17 (3H, t, J = 7.5, CH_3)	4, 7	13.2	1.12 (3H, t, J = 7.5, CH_3)	
9	169.1			170.6			172.7		
10	186.8			195.2	10.43 (1H, s, COH)	2, 3, 4			
11	162.9								
9-OCH_3	53.0	3.88 (3H, s, OCH_3)	1, 9	53.02	4.04 (3H, s, OCH_3)				
11-OCH_3	52.2	3.82 (3H, s, OCH_3)	10, 11						

In the dimethyl ether of compound **7**, the proton of the hydroxyl group at C-6 (δ_H 10.49) was hydrogen-bonded to the carbonyl of the ester group at C-9 (δ_H 169.1), and the proton of the hydroxyl group at C-3 (δ_H 11.30) was hydrogen-bonded to the carbonyl of the methylcarboxy group at C-10 (δ_C 186.8) (Figures S24–S26, Supplementary Materials). The signal at δ_C 162.9 ppm in the ^{13}C-NMR spectrum corresponded to the carbonyl of the ester group of the methylcarboxy fragment. Two singlets with an integrated intensity of 3H each at δ_H 3.82 and 3.88 in the ^{1}H-NMR spectrum corresponded to the protons of methoxy groups at C-9 and C-11. In the ^{13}C-NMR spectrum of compound **7** dimethyl ether, there were two corresponding signals at δ_C 52.2 and 53.0 ppm. According to an analysis of the NMR spectra of the dimethyl derivative, the structure of compound **7** was established as 4-ethyl-3,5,6-trihydroxy-2-oxalobenzoic acid.

The ^{13}C-NMR spectrum of methyl ether of compound **8** contained a signal in a low field at δ_C 195.2, the chemical shift of which was characteristic for the aldehyde carbon atom (Table 3, Figure S28, Supplementary Materials). The singlet at δ_H 10.43 in the ^{1}H-NMR spectrum of this compound corresponded to the proton of the aldehyde group. According to NMR data of its methyl derivative (Figures S27–S34, Supplementary Materials), the structure of compound **8** was established as 4-ethyl-2-formyl-3,5,6-trihydroxybenzoic acid.

The ^{1}H-NMR spectrum of compound **9** contained a singlet signal of the aromatic proton at C-6 (δ_H 6.88), which corresponded to a signal at δ_C 104.8 ppm in the ^{13}C-NMR spectrum (Table 3, Figures S35 and S36, Supplementary Materials). The chemical shift of the proton of the hydroxyl group at C-5 (δ_H 7.54) indicated that it was not bound by an intramolecular hydrogen bond as in compound **8**. Thus, the structure of compound **9** was established as 4-ethyl-2,3,5-trihydroxybenzoic acid (Figures S35–S39, Supplementary Materials).

On the basis of the NMR data of compound **10** and its methyl derivative (Figures S40–S48, Supplementary Materials), the structure of **10** was established as 3-ethyl-2,5-dihydroxy-1,4-benzoquinone. This compound was previously described by Moore et al. as a natural pigment of sea urchins of the genus *Echinothrix* [28]. However, it is likely that compound **10** was one of the most stable oxidation products of Ech A obtained by the authors during the storage and repeated chromatographic separation of sea urchin extracts. According to the conditions of our experiment, the oxidation process was stopped when half of the Ech A was oxidized; thus, a very small amount of **10** was isolated.

2.2. *Proposed Mechanism of Echinochrome A Oxidative Degradation*

Many bioactive natural and pharmaceutical compounds that are α-hydroxyketones such as oxolin, ascorbate, glyoxal, and cyclic ketones are susceptible to autooxidation [29]. It has been shown that α-hydroxyketones auto-oxidize under physiological conditions via the enediol tautomer [30,31]. One condition which favors the formation of the enediol is the presence of a vicinal carbonyl group.

The equilibrium is generally displaced in favor of the more thermodynamically stable ketol tautomer. The autooxidation of such enediols, as well as tetrahydroxy-1,4-benzoquinone, ascorbate, and Ech A, has been shown to involve the generation of intermediates such as carbon-centered free radicals and ROS including the superoxide radical anion, hydroxyl radical, and hydrogen peroxide [23,30,32,33]. The mechanisms of the primary attack of triplet and singlet oxygen molecules on Ech A (**1**), the result of which is bis-*gem*-diol **2** formation, were described in detail in previous studies [23,34]. Here, according to the structures of the isolated products, we propose the scheme of the further Ech A oxidative degradation process (Figure 8).

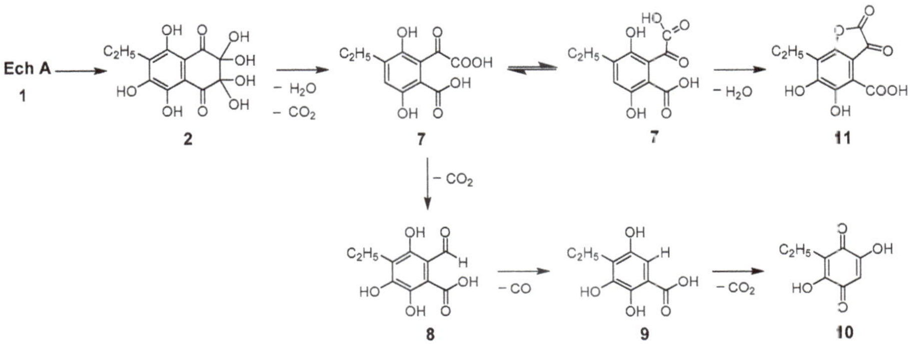

Figure 8. Proposed scheme of the Ech A oxidative degradation process.

Pracht et al. showed that splitting of the aromatic system of phenolic substances occurs only if the first oxidation stage includes the formation of *o*-quinone [35]. Subsequent cleavage of the ring structure can occur between two keto groups, as in our case, for example, in compounds **3**, **4**, **4a**, and **4b**, which always presented together with compound **2** (Figures 3 and 4). It can be assumed that ring rupture in the primary oxidation product **2** with loss of H_2O and CO_2 led to the formation of phthalonic acid derivative **7**. Compound **7**, resulting from keto–enol tautomerism, gave echinolactone **11** upon dehydration in an acidic medium and heating during the evaporation process. Establishment of the structure of **11** using X-ray analysis played an important role in the elucidation of the structure of the labile compound **7**. Further sequential decarboxylation of **7** resulted in the formation of **8**, a benzoic acid derivative with an aldehyde group, and a benzoic acid derivative **9**. Decarboxylation of **9** and subsequent hydration and oxidation led to the formation of stable benzoquinone **10**.

Thus, it was shown that the oxidative destruction of Ech A did not affect the benzenoid fragment of its molecule. All transformations occurred only in the quinonoid ring with the formation of bis-*gem*-diol **2**, further oxidation of which occurred upon cleavage of the dihydroquinonoid ring and led to the formation of derivatives of phthalonic (**7**) and benzoic (**8**, **9**) acids, as well as benzoquinone **10**.

2.3. Predicted Toxicity of Echinochrome A and Its Oxidation Products

Obtaining information on the toxicity of compounds and their impurities is an important part of the drug design development process. However, for impurities in particular, this information cannot be obtained experimentally. In this case, in silico studies assist in evaluating the results.

The potential toxicity of Ech A and its oxidative degradation products was assessed with the webserver ProTox-II. This virtual lab predicts the toxicity of small molecules on the basis of a total of 33 models for the prediction of various toxicity endpoints such as acute toxicity, hepatotoxicity, cytotoxicity, carcinogenicity, mutagenicity, immunotoxicity, adverse outcome (Tox21) pathways, and toxicity targets [36].

The results of the predicted toxicity of the original Ech A molecule and its oxidative degradation products are shown in Table 4.

Table 4. Oral toxicity prediction results for Ech A and its degradation products. LD_{50}, median lethal dose.

Classification and Target	Ech A and its Degradation Products						
	1 (Ech-B)	1 (Ech-Q)	2	7	8	9	10
Predicted toxicity class	II	IV	III	IV	IV	IV	V
Predicted LD_{50}, mg/kg	16	487	221	2000	2000	1800	2800
Organ Toxicity							
Hepatotoxicity	Inactive	Inactive	Inactive	Inactive	Inactive	Inactive	Inactive
Toxicity Endpoints							
Carcinogenicity	Inactive	Inactive	Inactive	Inactive	Inactive	Inactive	Inactive
Immunotoxicity	Inactive	Inactive	Inactive	Inactive	Inactive	Inactive	Inactive
Mutagenicity	Active	Active	Inactive	Inactive	Inactive	Inactive	Inactive
Cytotoxicity	Inactive	Inactive	Inactive	Inactive	Inactive	Inactive	Inactive

The predicted LD_{50} value for the first major oxidation product **2** was 221 mg/kg (toxicity class III), and the predicted acute toxicity for other oxidation products **7–10** was much lower (≥2000 mg/kg, toxicity class IV–V), suggesting they cannot lead to serious toxic effects.

As shown above, structures **2** and **7–10** were unambiguously defined as nontoxic and there was no doubt about the predicted toxicity results for these compounds. However, for derivatives of the naphthazarin (5,8-dihydroxy-1,4-naphthoquinone) structure, to which Ech A belongs, it is not so simple. Under various conditions, Ech A (**1**) can exist as a mixture of four tautomeric forms; however, only the 1,4-naphthoquinonoid forms Ech-B (ethyl in the benzene ring) and Ech-Q (ethyl in the quinonoid ring) are energetically favorable (Figure 9) [23,37,38].

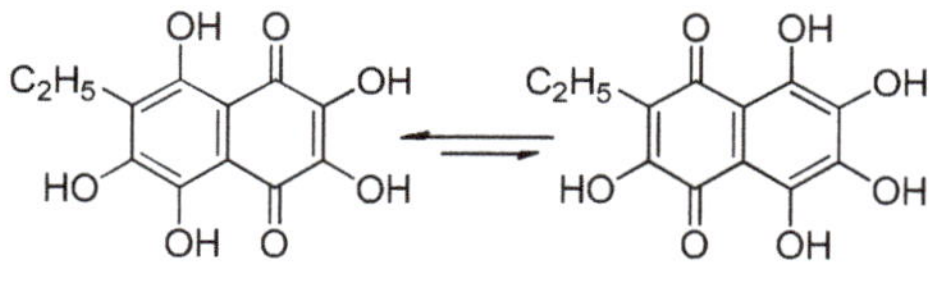

Figure 9. Tautomeric forms of Ech A: Ech-B (ethyl in the benzene ring) and Ech-Q (ethyl in the quinonoid ring).

As seen in Table 4, the Ech-B and Ech-Q formulas loaded in ProTox-II showed dramatically different results for acute toxicity. For the Ech-B form, a toxic LD_{50} of 16 mg/kg and toxicity class II were predicted; for Ech-Q, the LD_{50} was 487 mg/kg and the toxicity class IV was predicted. According to our experimental data for the determination of intraperitoneal acute toxicity of the Ech A substance in outbred mice, the LD_{50} was found to be 153.7 mg/kg (Table S7, Supplementary Materials) and toxicity class III was determined, which was somewhere between the predicted values for Ech-B and Ech-Q, confirming the benzenoid-quinonoid equilibrium of Ech A. However, the structure of bis-*gem*-diol **2** and of other products **3–6** indicated that only one of the possible tautomeric forms (Ech-B) of Ech A was involved in the oxidation process.

We established the experimental cytotoxicity for two compounds, Ech A (**1**) and bis-*gem*-diol **2**. The cytotoxicity of **1** and **2** was estimated by methylthiazolyltetrazolium bromide (MTT) assay using pig embryo kidney (PK) cells and African green monkey kidney (Vero) cells. For Ech A (**1**), 50% inhibition of cell viability was observed at 54.4 mkg/mL and 60.5 mkg/mL in PK and Vero cell lines, respectively [39]. For compound **2**, this value was found to be 140 mkg/mL in Vero cells; thus, it had a weaker toxic effect on normal cells.

With confidence scores of 0.77 and 0.82 for Ech-B and Ech-Q, respectively, it was predicted that these compounds had mutagenic activity (Table 4). However, as shown in a comprehensive study of the mutagenic properties of the Histochrome drug carried out in accordance with the requirements of the Pharmacological Committee of the Russian Ministry of Health, Histochrome in the range of 1.0–10 mg/kg does not have the ability to induce chromosomal damage in the bone marrow cells of C57BL/6 mice, nor does it lead to an increase in the level of spontaneous gene mutations in *Drosophila* or induce gene mutations in *Salmonella typhimurium* (Tables S8–S10, Supplementary Materials). These results allowed us to conclude that Ech A does not exhibit mutagenic activity, at least in the range of therapeutic doses. However, there is published evidence of mutagenic activity of Ech A [33]. In this publication, the source of the drug was not clearly indicated, and neither were its preparation method or purity. This once again confirms that the standardization of drug substances is very important. For the implementation of its medicinal properties, not only the structure of the active substance is important, but also the properties of the drug, determined by the technological process used for its production.

3. Materials and Methods

3.1. Materials

Drug substance echinochrome A and drug product Histochrome were produced by G.B. Elyakov Pacific Institute of Bioorganic Chemistry (Vladivostok, Russia). TSKgel Toyopearl HW-40 (TOYO SODA, Tokyo, Japan), and Sephadex LH-20 (GE Healthcare Bio-Sciences AB, Uppsala, Sweden) were used for column chromatography. HPLC-grade water and acetic acid were purchased from Panreac Quimica (Barcelona, Spain). MeCN grade 0 was sourced from Cryochrom (Saint Petersburg, Russia). Other solvents used in this study were of analytical grade. Deuterated solvents acetone-d_6, $CDCl_3$, CD_3CN, and Bruker® SampleJet NMR tubes WIMWG10007SJ (178 mm, cap, O.D. 5.0 mm) for NMR experiments were purchased from Sigma (St. Louis, MO, USA).

3.2. Instruments

Ultraviolet (UV) spectra were recorded using a UV 1800 spectrophotometer (Shimadzu USA Manufacturing Inc., Canby, OR, USA) and infrared (IR) spectra were obtained on an Equinox 55 Fourier-transform (FT) IR spectrophotometer (Bruker, Rheinstetten, Germany). HR-ESI-MS experiments were carried out using a Shimadzu hybrid ion trap time-of-flight mass spectrometer (Kyoto, Japan). The operating settings of the instrument were as follows: electrospray ionization (ESI) source potential, −3.8 and 4.5 kV for negative and positive polarity ionization, respectively; drying gas (N_2) pressure, 200 kPa; nebulizer gas (N_2) flow, 1.5 L/min; temperature for the curved desolvation line (CDL) and heat block, 200 °C; detector voltage, 1.5 kV; range of detection, 100–900 *m/z*. The mass accuracy was below 4 ppm. The data were acquired and processed using Shimadzu LCMS Solution software (v.3.60.361). The ^{1}H-, ^{13}C-, and two-dimensional (2D) NMR spectra were recorded using NMR Bruker AVANCE III DRX-700, AVANCE DRX-500, and AVANCE DPX-300 instruments (Bruker, Karlsruhe, Germany). The chemical shift values (δ) and the coupling constants (*J*) are given in parts per million and in Hz, respectively. HMBC spectra were optimized for 5 Hz coupling. The key NMR acquisition parameters are shown in the Supplementary Materials.

The X-ray experiments for single crystals were performed using a Bruker SMART-1000 Charge-Coupled Detector (CCD) diffractometer (MoK$_\alpha$-radiation, graphite monochromator). Intensity data were corrected for absorption using the multi-scan method. The structures were solved using the direct methods and refined by the least-squares calculation in anisotropic approximation for non-hydrogen atoms. Hydrogen atoms of ethyl groups after checking their presence in a difference map were placed in geometrically idealized positions and refined in the riding-model approximation. Hydrogen atoms of water molecules and hydroxyl groups were located in the difference Fourier maps and refined with $U_{iso}(H) = 1.5U_{eq}$ (O). Data collection and editing, as well as refinement of unit cell parameters, were performed with the SMART [40] and SAINT [41] program packages. All calculations on the determination and refinement of the structures were carried out using the SHELXTL/PC software [42,43].

3.3. HPLC–DAD–MS Analysis

HPLC–DAD–MS was performed using a system consisting of a CBM-20A system controller (Shimadzu USA Manufacturing Inc., Canby, OR, USA), two LC-20 CE pumps (Shimadzu USA Manufacturing Inc., Canby, OR, USA), a DGU-20A3 degasser (Shimadzu Corp., Kyoto, Japan), an SIL-20A autosampler (Shimadzu USA Manufacturing Inc., Canby, OR, USA), a diode-matrix SPD-M20A (Shimadzu USA Manufacturing Inc., Canby, OR, USA), and a mass-spectrometric detector LCMS-2020 (Shimadzu Corp., Kyoto, Japan). The separation was carried out on a Discovery HS C18 column (150 × 2.1 mm, 3 µm particle size, Supelco, Bellefonte, PA, USA) with a Supelguard Ascentis C18 pre-column (2 × 2.1 mm, 3 µm particle size, Supelco, Bellefonte, PA, USA) using a binary gradient of H_2O (A)/MeCN (B) with the addition of 0.2% AcOH at a flow rate of 0.2 mL/min and column temperature of 40 °C. The gradient was as follows: 0–6 min, 10–40% (B); 6–11 min, 40–100% (B); 11–12 min, 100% (B), 12–13 min, 100–10% (B); 13–17 min, 10% (B). The chromatograms were recorded at 254 nm. Mass spectra were taken in the electrospray ionization (ESI) mode at atmospheric pressure, recording negative ions (1.50 kV) in the *m/z* range of 100–900 with the following settings: drying gas, N_2 (10 L/min); nebulizer gas N_2 flow, 1.5 L/min; temperature for the curved desolvation line (CDL), 200 °C; temperature of heat block, 250 °C; interface voltage, 3.5 kV. Before analysis, samples were filtered through a 0.2 µm polytetrafluoroethylene (PTFE) syringe filter. The injection volume was 5 µL.

3.4. HPLC Method Validation

Linearity of the method was established by using methanolic Ech A solutions containing 50–1500 ng/mL. Each linearity sample was injected in triplicate. The calibration curve was constructed as linear regression analysis of the peak area versus concentration. The limits of detection (LOD) and quantification (LOQ) of Ech A were calculated as concentrations at which the signal-to-noise ratio was below 3 and 10, respectively. The accuracy of the method was established by recovery studies of Ech A samples (100, 250, 500 ng/mL); data are provided in the Supplementary Materials (Table S1). Accuracy was expressed as relative standard deviations (RSDs) and recoveries (%). Selectivity was confirmed through peak purity studies using a DAD detector.

3.5. Oxidation Products Preparation and Isolation

Histochrome from an ampoule (10 mg/mL Ech A, 5 mL) was diluted with distilled water (250 mL), balanced with atmospheric oxygen, to obtain a solution with an Ech A concentration of 0.2 mg/mL, pH 7.2. The oxidation was carried out in the light without stirring at ambient temperature. The process was monitored by HPLC–DAD–MS. When approximately 50% of the Ech A was consumed (after 20 h), the reaction was stopped by adding hydrochloric acid to pH 2. Unreacted Ech A was removed from solution by extraction with chloroform. The oxidation products were extracted with ethyl acetate, the solvent was removed under reduced pressure, and the residue was chromatographed on a Toyopearl HW-40 TSKgel column in a gradient of 10–50% ethanol containing 0.5% formic acid. From fractions eluted with 10–30% ethanol compounds were extracted with ethyl acetate and the solvent was evaporated. As a result, compounds **2**, **9**, and **10** were isolated. Fractions eluted with 30–50% ethanol were concentrated in vacuo at 50 °C; as a result, echinolactone **11** was obtained from the fraction of compound 7.

A small portion of oxidation products (10 mg) was dissolved in cold MeOH and treated with diazomethane in diethyl ether on ice as described [44]. The reaction process was controlled by HPLC–DAD–MS. After 1 h, the solvent was removed in vacuo, and the residue was subjected to column chromatography on Sephadex LH-20 eluting with $CHCl_3$/EtOH 8:1. As a result, dimethyl ether of **7** and methyl ether of **8** were obtained.

7-ethyl-2,2,3,3,5,6,8-heptahydroxy-2,3-dihydro-1,4-naphthoquinone (**2**): $C_{12}H_{12}O_9$; UV (ethanol) λ_{max} 256, 320, 391 nm; ESI-MS *m/z*: 299 $[M-H]^-$; HR-ESI-MS *m/z*: 299.0399 $[M-H]^-$ (calculated for $[C_{12}H_{11}O_9]^-$ 299.0409).

4-ethyl-3,5,6-trihydroxy-2-oxalobenzoic acid (**7**); $C_{11}H_{10}O_8$; UV (CH_3CN–H_2O) λ_{max} 219, 271, 320 nm; ESI-MS *m/z*: 269 $[M-H]^-$; HR-ESI-MS *m/z*: 269.0304 $[M-H]^-$ (calculated for $[C_{11}H_9O_8]^-$ 269.0297).

Dimethyl ether of compound **7**: light-yellow powder; $C_{13}H_{14}O_8$; UV (CH_3CN–H_2O) λ_{max} 224, 250, 324, 384 nm; IR ($CDCl_3$) ν_{max} 3520, 3156, 2975, 2956, 2937, 2878, 2855, 1734, 1683, 1632, 1595 cm^{-1}; ESI-MS *m/z*: 297 $[M-H]^-$; ^{1}H- and ^{13}C-NMR (see Table 3).

4-ethyl-2-formyl-3,5,6-trihydroxybenzoic acid (**8**); $C_{10}H_{10}O_6$; light-yellow powder; UV (CH_3CN–H_2O) λ_{max} 270, 320 nm. ESI-MS *m/z*: 225 $[M-H]^-$; HR-ESI-MS *m/z*: 225.0405 $[M-H]^-$ (calculated for $[C_{10}H_9O_6]^-$ 225.0399).

Methyl ether of compound **8**: $C_{11}H_{12}O_6$; light-yellow powder; UV (CH_3CN–H_2O) λ_{max} 238, 303, 369 nm; IR ν_{max} ($CDCl_3$) 3011, 2959, 2935, 2878, 2865, 1668, 1634, 1611 cm^{-1}. ESI-MS *m/z*: 239 $[M-H]^-$; HR-ESI-MS *m/z*: 239.0570 $[M-H]^-$ (calculated for $[C_{11}H_{11}O_6]^-$ 239.0555); ^{1}H- and ^{13}C-NMR (see Table 3).

4-ethyl-2,3,5-trihydroxybenzoic acid (**9**); $C_9H_{10}O_5$; light-yellow powder; UV (CH_3CN–H_2O) λ_{max} 228, 251, 333 nm; IR ν_{max} (CCl_4) 3352, 2960, 2928, 2875, 2855, 1642 cm^{-1}. ESI-MS *m/z*: 197 $[M-H]^-$; HR-ESI-MS *m/z*: 197.0452 $[M-H]^-$ (calculated for $[C_9H_9O_5]^-$ 197.0450); ^{1}H- and ^{13}C-NMR (see Table 3).

3-ethyl-2,5-dihydroxy-1,4-benzoquinone (**10**); $C_8H_8O_4$; light-yellow powder; UV (CH_3CN–H_2O) λ_{max} 228 nm, literature 228 nm [28]; IR ν_{max} ($CDCl_3$) 3366, 2976, 2935, 2878, 1731, 1641 cm^{-1}. ESI-MS *m/z*: 167 $[M-H]^-$; HR-ESI-MS *m/z*: 167.0343 $[M-H]^-$ (calculated for $[C_8H_7O_4]^-$ 167.0344). ^{1}H-NMR (700 MHz, $CDCl_3$) δ, ppm (*J*, Hz): 1.21 (3H, t, $J = 7.5$, CH_3), 2.49 (2H, q, $J = 7.5$, CH_2), 6.01 (1H, s, H), 7.54 (2H, s, OH); ^{1}H- and ^{13}C-NMR data of dimethyl ether of **10** (see Figures S43–S48, Supplementary Materials).

Echinolactone (7-ethyl-5,6-dihydroxy-2,3-dioxo-2,3-dihydrobenzofuran-4-carboxylic acid, **11**): $C_{11}H_8O_7$; UV (ethanol) λ_{max} (logε): 215 (3.4), 331 (2.8), 385 (2.8) nm; IR ν_{max} ($CDCl_3$) 3483, 1838, 1698, 1581 cm^{-1}; ESI-MS *m/z*: 251 $[M-H]^-$; EI-MS *m/z*: 252 $[M]^+$. ^{1}H-NMR (300 MHz, $CDCl_3$), δ, ppm: 1.24 (3H, t, $J = 7.5$, CH_3), 2.78 (2H, dd, $J = 7.5$, CH_2), 5.24, (1H, s, OH), 12.88 (1H, s, OH). ^{13}C-NMR (75 MHz, CD_3CN), δ, ppm: 13.3, 17.8, 106.9, 108.8, 121.9, 151.7, 158.6, 160.5, 162.2, 171.5, 178.4.

The main crystal data, data collection, and refinement parameters are presented in Table S4 (Supplementary Materials). The selected geometric parameters of **11** crystal forms are given in Table S5 (Supplementary Materials). The hydrogen-bond geometry parameters are listed in Table S6 (Supplementary Materials). Crystallographic data for the structures in this study were deposited to the Cambridge Crystallographic Data Center as supplementary publication No. CCDC 1,822,816 and 1822821; copies of the data can be obtained, free of charge, via an application to CCDC, 12 Union Road, Cambridge CB2 1EZ, UK, (fax: +44 1223 336,033 or e-mail: deposit@ccdc.cam.ac.uk).

3.6. In Silico Toxicity Studies

The potential acute toxicity, hepatotoxicity, carcinogenicity, immunotoxicity, mutagenicity, and cytotoxicity of Ech A and its degradation products via oral administration were assessed using ProTox-II [36].

4. Conclusions

In this study, Ech A degradation products formed during oxidation by O_2 in air-equilibrated aqueous solutions were identified, isolated, and structurally characterized. During the oxidation of Ech A, transformation was found to occur only in the quinonoid ring, with the formation of a hydrogenated quinonoid cycle with two carbonyl groups and two pairs of geminal hydroxyl groups. The further oxidation of bis-*gem*-diol occurred with the cleavage of the dihydroquinonoid ring and a chain of subsequent decarboxylations.

The HPLC method with DAD and MS detection was developed and validated to monitor the Ech A degradation process and to identify the appearing compounds. The structural studies and obtained

HPLC–MS parameters of the main Ech A oxidation products are of great interest from the point of view of investigation of the chemical properties of drug substances and for developing methods for monitoring the quality of drugs and food additives obtained from sea urchin pigments, in addition to their stability.

In recent years, there has also been a growing interest in the environmental significance of methods to characterize the destruction of pharmaceutical compounds and study their toxic properties in order to avoid the accumulation of these compounds in nature.

The in silico toxicity studies performed using ProTox-II webserver revealed that Ech A oxidative degradation products do not exhibit mutagenic properties, and their toxicity values were much lower than that of Ech A. This means that the spontaneous formation of these degradation products in preparations using Ech A would not be harmful to patients.

Supplementary Materials: The following are available online: Figure S1. The UV-Vis spectrum of Histochrome (blue) and Echinochrome A (black) in ethanol solution containing 1 mM HCl; Figure S2. Change in the concentration of echinochrome A during the oxidation of a 1% histochrome solution; Table S1. Accuracy and reproducibility of the quantification of echinochrome A (1) using HPLC method; Figure S3. HRESIMS (negative mode) data for methyl ethers of Ech A oxidation products obtained by methylation with methyl iodide; Table S2. HRESIMS (negative mode) data for methyl ethers of Ech A oxidation products obtained by methylation with methyl iodide; Table S3. ESIMS (negative mode) data for methyl ethers of Ech A oxidation products obtained by methylation with diazomethane; Figures S4 and S5. IR spectrum (CDCl3) of compound 7 dimethyl ether; Figures S6 and S7. IR spectrum (CDCl3) of compound 8 methyl ether; Figures S8 and S9. IR spectrum (CDCl3) of compound 10; Figures S10 and S11. IR spectrum (CDCl3) of compound 11; Figure S12. 1H NMR spectrum (300 MHz, acetone-d6) of 7-ethyl-2,2,3,3,5,6,8-heptahydroxy-2,3-dihydro-1,4-naphthoquinone (2); Figure S13. 13C NMR spectrum (75 MHz, acetone-d6) of 2; Figure S14. HMBC spectrum (300 MHz, acetone-d6) of 2; Figures S15–S17. HMBC correlations of 2 (enlarged); Figure S18. 1H NMR spectrum (300 MHz, CDCl3) of echinolactone (11); Figure S19. 13C NMR spectrum (75 MHz, CD3CN) of echinolactone (11); Figure S20. HMBC spectrum (300 MHz, CD3CN) of echinolactone (11); Table S4. Selected crystal data and refinement parameters for α- and β- forms of C11H8O7•H2O; Table S5. Selected geometric parameters (Å) for α- and β- forms of C11H8O7•H2O; Table S6. Hydrogen-bond geometry (Å, °) for α- and β- forms of C11H8O7•H2O; Figure S21. Overall packing for α-C11H8O7•H2O viewed along the a-axis direction; Figure S22. A plot of band for β-C11H8O7•H2O; Figure S23. Overall packing for β-C11H8O7•H2O viewed along the a-axis direction; Figure S24. 1H NMR spectrum (500 MHz, CDCl3) of dimethyl ether of compound 7; Figure S25. 13C NMR spectrum (126 MHz, CDCl3) of dimethyl ether of compound 7; Figure S26. HMBC spectrum (500 MHz, CDCl3) of dimethyl ether of compound 7; Figure S27. 1H NMR spectrum (700 MHz, CDCl3) of methyl ether of 4-ethyl-2-formyl-3,5,6-trihydroxybenzoic acid (8); Figure S28. 13C NMR spectrum (175 MHz, CDCl3) of methyl ether of 8; Figure S29. HMBC spectrum (700 MHz, CDCl3) of methyl ether of 8; Figures S30–S34. HMBC correlations of methyl ether of 8 (enlarged); Figure S35. 1H NMR spectrum (700 MHz, acetone-d6) of 4-ethyl-2,3,5-trihydroxybenzoic acid (9); Figure S36. 13C NMR spectrum (175 MHz, aceton-d6) of 9; Figure S37. HMBC spectrum (700 MHz, aceton-d6) of 9; Figures S38 and S39. HMBC correlations of 9 (enlarged); Figure S40. 1H NMR spectrum (700 MHz, CDCl3) of 3-ethyl-2,5-dihydroxy-1,4-benzoquinone (10); Figure S41. 13C NMR spectrum (175 MHz, CDCl3) of 10; Figure S42. HSQC spectrum (700 MHz, CDCl3) of 10; Figure S43. 1H NMR spectrum (700 MHz, CDCl3) of dimethyl ether of 10; Figure S44. 13C NMR spectrum (175 MHz, CDCl3) of dimethyl ether of 10; Figure S45. HMBC spectrum (700 MHz, CDCl3) of dimethyl ether of 10; Figures S46–S48. HMBC correlations of dimethyl ether of 10; Table S7. Histochrome toxicity values (intraperitoneal administration); Table S8. Accounting for chromosomal aberrations in mammalian bone marrow cells; Table S9. The results of a study of the mutagenic effect of the histochrome drug on indicator strains in the Ames test; Table S10. The results of the study of the ability of the histochrome drug to induce dominant lethal mutations in the germ cells of mice.

Author Contributions: Conceptualization, N.P.M.; formal analysis, V.P.G. and P.S.D.; investigation, E.A.V., N.P.M., A.V.G., V.P.G., and P.S.D.; project administration, N.P.M. and P.S.D.; resources, V.P.G.; software, A.V.G.; supervision, S.A.F.; writing—original draft, N.P.M. and E.A.V.; writing—review and editing, S.A.F. All authors have read and agreed to the published version of the manuscript.

Funding: This research received no external funding.

Acknowledgments: The study was carried out on the equipment of the Collective Facilities Center "Far Eastern Center for Structural Molecular Research (NMR/MS) PIBOC FEB RAS". The X-ray measurements were performed on the equipment of the Collective Facilities Center "Far Eastern Center of Structural Studies", Institute of Chemistry FEB RAS.

Conflicts of Interest: The authors report no conflicts of interest. The authors alone are responsible for the content and writing of this article.

References

1. Mishchenko, N.P.; Fedoreev, S.A.; Bagirova, V.L. Histochrome: A New Original Domestic Drug. *Pharm. Chem. J.* **2003**, *37*, 48–52. [CrossRef]
2. Mischenko, N.P.; Fedoreyev, S.A.; Pokhilo, N.D.; Anufriev, V.P.; Denisenko, V.A.; Glazunov, V.P. Echinamines A and B, First Aminated Hydroxynaphthazarins from the Sea Urchin *Scaphechinus mirabilis*. *J. Nat. Prod.* **2005**, *68*, 1390–1393. [CrossRef]
3. Vasileva, E.A.; Mishchenko, N.P.; Tran, V.T.T.; Vo, H.M.N.; Bui, L.M.; Denisenko, V.A.; Fedoreyev, S.A. Quinoid Pigments from the Sea Urchin *Astropyga radiata*. *Chem. Nat. Compd.* **2017**, *53*, 356–358. [CrossRef]
4. Hou, Y.; Vasileva, E.A.; Carne, A.; McConnell, M.; Bekhit, A.E.-D.A.; Hou, Y. Naphthoquinones of the spinochrome class: Occurrence, isolation, biosynthesis and biomedical applications. *RSC Adv.* **2018**, *8*, 32637–32650. [CrossRef]
5. Jeong, S.H.; Kim, H.K.; Song, I.-S.; Lee, S.J.; Ko, K.S.; Rhee, B.D.; Kim, N.; Mishchenko, N.P.; Fedoreyev, S.A.; Stonik, V.A.; et al. Echinochrome A Protects Mitochondrial Function in Cardiomyocytes against Cardiotoxic Drugs. *Mar. Drugs* **2014**, *12*, 2922–2936. [CrossRef]
6. Jeong, S.H.; Kim, H.K.; Song, I.-S.; Noh, S.J.; Marquez, J.; Ko, K.S.; Rhee, B.D.; Kim, N.; Mishchenko, N.P.; Fedoreyev, S.A.; et al. Echinochrome A Increases Mitochondrial Mass and Function by Modulating Mitochondrial Biogenesis Regulatory Genes. *Mar. Drugs* **2014**, *12*, 4602–4615. [CrossRef] [PubMed]
7. Seo, D.Y.; A McGregor, R.; Noh, S.J.; Choi, S.-J.; Mishchenko, N.P.; Fedoreyev, S.A.; Stonik, V.A.; Han, J. Echinochrome A Improves Exercise Capacity during Short-Term Endurance Training in Rats. *Mar. Drugs* **2015**, *13*, 5722–5731. [CrossRef]
8. Kim, H.K.; Youm, J.B.; Jeong, S.H.; Lee, S.R.; Song, I.-S.; Ko, T.H.; Pronto, J.R.; Ko, K.S.; Rhee, B.D.; Kim, N.; et al. Echinochrome A regulates phosphorylation of phospholamban Ser16 and Thr17 suppressing cardiac SERCA2A Ca2+ reuptake. *Pflügers Archiv. Eur. J. Physiol.* **2014**, *467*, 2151–2163. [CrossRef]
9. Park, J.H.; Lee, N.-K.; Lim, H.J.; Mazumder, S.; Rethineswaran, V.K.; Kim, Y.-J.; Jang, W.B.; Ji, S.T.; Kang, S.; Kim, D.Y.; et al. Therapeutic Cell Protective Role of Histochrome under Oxidative Stress in Human Cardiac Progenitor Cells. *Mar. Drugs* **2019**, *17*, 368. [CrossRef]
10. Park, G.-B.; Kim, M.-J.; Vasileva, E.A.; Mishchenko, N.P.; Fedoreyev, S.A.; Stonik, V.A.; Han, J.; Lee, H.S.; Kim, D.; Jeong, J.-Y. Echinochrome A Promotes Ex Vivo Expansion of Peripheral Blood-Derived CD34+ Cells, Potentially through Downregulation of ROS Production and Activation of the Src-Lyn-p110δ Pathway. *Mar. Drugs* **2019**, *17*, 526. [CrossRef]
11. Gusev, E.I.; Stonik, V.A.; Martynov, M.I.; Guseva, M.R.; Shchukin, I.A.; Agafonova, I.G.; Mishchenko, N.P.; Fedoreev, S.A. An influence of histochrome on the dynamics of neurological disorders and MRT-picture in experimental hemorrhagic stroke. *Zhurnal Nevrol. i psikhiatrii im. S.S. Korsakova* **2005**, *105*, 61–66.
12. Kim, R.; Hur, D.; Kim, H.K.; Han, J.; Mishchenko, N.P.; Fedoreyev, S.A.; Stonik, V.A.; Chang, W. Echinochrome A Attenuates Cerebral Ischemic Injury through Regulation of Cell Survival after Middle Cerebral Artery Occlusion in Rat. *Mar. Drugs* **2019**, *17*, 501. [CrossRef] [PubMed]
13. Mohamed, A.S.; Soliman, A.M.; Marie, M.A.S. Mechanisms of echinochrome potency in modulating diabetic complications in liver. *Life Sci.* **2016**, *151*, 41–49. [CrossRef]
14. Soliman, A.M.; Mohamed, A.S.; Marie, M.A.S. Comparative study between the hypoglycemic and antioxidant effects of echinochrome on type 1 and type 2 diabetes mellitus. *Res. J. Biol.* **2016**, *4*, 1–9.
15. Lee, S.R.; Pronto, J.R.D.; Sarankhuu, B.-E.; Ko, K.S.; Rhee, B.D.; Kim, N.; Mishchenko, N.P.; Fedoreyev, S.A.; Stonik, V.A.; Han, J. Acetylcholinesterase Inhibitory Activity of Pigment Echinochrome A from Sea Urchin Scaphechinus mirabilis. *Mar. Drugs* **2014**, *12*, 3560–3573. [CrossRef]
16. Oh, S.-J.; Seo, Y.; Ahn, J.-S.; Shin, Y.Y.; Yang, J.W.; Kim, H.K.; Han, J.; Hou, Y.; Fedoreyev, S.A.; Stonik, V.A.; et al. Echinochrome A Reduces Colitis in Mice and Induces In Vitro Generation of Regulatory Immune Cells. *Mar. Drugs* **2019**, *17*, 622. [CrossRef]
17. Lebed'ko, O.A.; Ryzhavskii, B.Y.; Tkach, O.V.; Kuznetsova, M.S.; Guseva, O.E. Influence of oral administration echinochrome A on structural-metabolic disorders, bleomycin-induced in the rat lung early stage postnatal ontogenesis. *Far East Med. J.* **2016**, *3*, 93–97.
18. Kozlov, V.K.; Kozlov, M.V.; Guseva, O.E.; Lebedko, O.A.; Morozova, N.V. Antioxidative activity of echinochrome A in case of chronic inflammatory lung diseases in children. *Pacific Med. J.* **2009**, *3*, 116–117.

19. Kozlov, V.K.; Kozlov, M.V.; Lebedko, O.A.; Yephimenko, M.V.; Guseva, O.E.; Morozova, N.V. Influence of echinochrome A on some parameters of systemic free-radical status and t-cell immunity under chronic inflammatory lung diseases in children at the period of remission. *Far East Med. J.* **2010**, *1*, 55–58.
20. Anufrieva, A.V.; Lebedko, O.A.; Berezina, G.P.; Kozlov, V.K. Effect of echinochrome A on biogenesis of free radicals in the gastric mucosa of adolescents with erosive gastroduodenitis. *Far East Med. J.* **2012**, *1*, 78–81.
21. Anufrieva, A.V.; Lebed'ko, O.A.; Berezina, G.P.; Kozlov, V.K. Correction of free radical disordes at erosive gastroduodenitis. *Health. Med. Ecol. Sci.* **2014**, *3*, 13–15.
22. Sengupta, P.; Chatterjee, B.; Tekade, R.K. Current regulatory requirements and practical approaches for stability analysis of pharmaceutical products: A comprehensive review. *Int. J. Pharm.* **2018**, *543*, 328–344. [CrossRef] [PubMed]
23. Novikov, V.L.; Shestak, O.P.; Mishchenko, N.P.; Fedoreev, S.A.; Vasileva, E.A.; Glazunov, V.P.; Artyukov, A.A. Oxidation of 7-ethyl-2,3,5,6,8-pentahydroxy-1,4-naphthoquinone (echinochrome A) by atmospheric oxygen 1. Structure of dehydroechinochrome. *Russ. Chem. Bull.* **2018**, *67*, 282–290. [CrossRef]
24. Mishchenko, N.P.; Fedoreev, S.A.; Glazunov, V.P.; Denisenko, V.A.; Krasovskaya, N.P.; Glebko, L.I.; Maslov, L.G.; Dmitrenok, P.S.; Bagirova, V.L. Isolation and Identification of Impurities in the Parent Substance of Echinochrome and in the Drug Histochrome. *Pharm. Chem. J.* **2004**, *38*, 50–53. [CrossRef]
25. Rao, P.S.; Reddy, P.P.; Seshadri, T.R. Methylation of hydroxy flavonols using methyl iodide and potassium carbonate. *Proc. Math. Sci.* **1940**, *12*, 495. [CrossRef]
26. Kuhn, R.; Wallenfels, K. Dehydroechinochrom. *Ber. Deutsch. Chem. Gesel. (A und B Ser.)* **1942**, *75*, 407–413. [CrossRef]
27. Malinovskaya, G.V.; Chizhova, A.Y.; Anufriev, V.P.; Glazunov, V.P.; Denisenko, V.A. The chemistry of naphthazarine derivatives. *Russ. Chem. Bull.* **1999**, *48*, 1587–1589. [CrossRef]
28. Moore, R.E.; Singh, H.; Scheuer, P.J. Isolation of Eleven New Spinochromes from Echinoids of the Genus Echinothrix. *J. Org. Chem.* **1966**, *31*, 3645–3650. [CrossRef]
29. Hovorka, S.W.; Schoneich, C. Oxidative Degradation of Pharmaceuticals: Theory, Mechanisms and Inhibition. *J. Pharm. Sci.* **2001**, *90*, 253–269. [CrossRef]
30. Hoffmann, M.E.; Ciampi, D.B.; Durán, N. The autooxidation of 2,3,5,6-tetrahydroxy-2,5-cyclohexadiene-1,4-dione under physiological conditions. *Cell. Mol. Life Sci.* **1987**, *43*, 217–220. [CrossRef]
31. Dántola, M.L.; Vignoni, M.; Capparelli, A.L.; Lorente, C.; Thomas, A.H. Stability of 7,8-Dihydropterins in Air-Equilibrated Aqueous Solutions. *Helvetica Chim. Acta* **2008**, *91*, 411–425. [CrossRef]
32. Herrmann, N.; Heinz, N.; Dolg, M.; Cao, X. Quantum chemical study of the autoxidation of ascorbate. *J. Comput. Chem.* **2016**, *37*, 1914–1923. [CrossRef] [PubMed]
33. Lebedev, A.V.; Levitskaya, E.L.; Tikhonova, E.V.; Ivanova, M.V. Antioxidant properties, autooxidation, and mutagenic activity of echinochrome a compared with its etherified derivative. *Biochem. (Moscow)* **2001**, *66*, 885–893. [CrossRef]
34. Novikov, V.L.; Shestak, O.P. Thermal and photochemical oxidation of 2-acetylcyclopentanone with atmospheric oxygen. *Russ. Chem. Bull.* **2012**, *61*, 1099–1110. [CrossRef]
35. Pracht, J.; Boenigk, J.; Isenbeck-Schröter, M.; Keppler, F.; Schöler, H.F. Abiotic Fe(III) induced mineralization of phenolic substances. *Chemosphere* **2001**, *44*, 613–619. [CrossRef]
36. Banerjee, P.; O Eckert, A.; Schrey, A.K.; Preissner, R. ProTox-II: A webserver for the prediction of toxicity of chemicals. *Nucleic Acids Res.* **2018**, *46*, W257–W263. [CrossRef]
37. Berdyshev, D.V.; Glazunov, V.P.; Novikov, V.L. 7-Ethyl-2,3,5,6,8-pentahydroxy-1,4-naphthoquinone (echinochrome A): A DFT study of the antioxidant mechanism. 1. Interaction of echinochrome A with hydroperoxyl radical. *Russ. Chem. Bull.* **2007**, *56*, 413–429. [CrossRef]
38. Glazunov, V.P.; Berdyshev, D.V.; Novikov, V.L. 7-Ethyl-2,3,5,6,8-pentahydroxy-1,4-naphthoquinone (echinochrome A): A DFT study of the antioxidant mechanism 2.* The structure of monosodium salts of echinochrome A and their reactions with the hydroperoxyl radical. *Russ. Chem. Bull.* **2010**, *59*, 43–54. [CrossRef]
39. Fedoreyev, S.A.; Krylova, N.V.; Mishchenko, N.P.; Vasileva, E.A.; Pislyagin, E.A.; Iunikhina, O.V.; Lavrov, V.F.; Svitich, O.A.; Ebralidze, L.K.; Leonova, G.N. Antiviral and Antioxidant Properties of Echinochrome A. *Mar. Drugs* **2018**, *16*, 509. [CrossRef]
40. *SMART*; Bruker AXS Inc.: Madison, Wisconsin, USA, 1998.
41. *SAINT*; Bruker AXS Inc.: Madison, Wisconsin, USA, 2003.

42. Sheldrick, G.M. SHELXT– Integrated space-group and crystal-structure determination. *Acta Crystallogr. Sect. A Found. Adv.* **2015**, *71*, 3–8. [CrossRef]
43. Sheldrick, G.M. Crystal structure refinement with SHELXL. *Acta Crystallogr. Sect. C Struct. Chem.* **2015**, *C71*, 3–8. [CrossRef] [PubMed]
44. Moore, R.; Singh, H.; Chang, C.; Scheuer, P. Polyhydroxynaphthoquinones: Preparation and hydrolysis of methoxyl derivatives. *Tetrahedron* **1967**, *23*, 3271–3305. [CrossRef]

Sample Availability: Samples of the compounds are not available.

molecules

MDPI

Article

Synthesis and Evaluation of Antimicrobial and Cytotoxic Activity of Oxathiine-Fused Quinone-Thioglucoside Conjugates of Substituted 1,4-Naphthoquinones

Yuri E. Sabutski [1,*], Ekaterina S. Menchinskaya [1], Ludmila S. Shevchenko [1], Ekaterina A. Chingizova [1], Artur R. Chingizov [1], Roman S. Popov [1], Vladimir A. Denisenko [1], Valery V. Mikhailov [1], Dmitry L. Aminin [1,2] and Sergey G. Polonik [1,*]

1 G.B. Elyakov Pacific Institute of Bioorganic Chemistry, Far-Eastern Branch of the Russian Academy of Sciences. Prospekt 100-let Vladivostoku 159, 690022 Vladivostok, Russia; ekaterinamenchinskaya@gmail.com (E.S.M.); lshev@piboc.dvo.ru (L.S.S.); martyyas@mail.ru (E.A.C.); chingizov84@gmail.com (A.R.C.); prs_90@mail.ru (R.S.P.); vladenis@piboc.dvo.ru (V.A.D.); vvmikhailov@inbox.ru (V.V.M.); daminin@piboc.dvo.ru (D.L.A.)

2 Department of Biomedical Science and Environmental Biology, Kaohsiung Medical University, Kaohsiung 80708, Taiwan

* Correspondence: alixar2006@gmail.com (Y.E.S.); sergpol@piboc.dvo.ru (S.G.P.); Tel.: +7-423-231-07-05 (Y.E.S. & S.G.P.); Fax: +7-423-231-40-50 (Y.E.S. & S.G.P.)

Academic Editor: Luisella Verotta
Received: 22 July 2020; Accepted: 5 August 2020; Published: 6 August 2020

Abstract: A series of new tetracyclic oxathiine-fused quinone-thioglycoside conjugates based on biologically active 1,4-naphthoquinones and 1-mercapto derivatives of per-*O*-acetyl D-glucose, D-galactose, D-xylose, and L-arabinose have been synthesized, characterized, and evaluated for their cytotoxic and antimicrobial activities. Six tetracyclic conjugates bearing a hydroxyl group in naphthoquinone core showed high cytotoxic activity with EC_{50} values in the range of 0.3 to 0.9 μM for various types of cancer and normal cells and no hemolytic activity up to 25 μM. The antimicrobial activity of conjugates was screened against Gram-positive bacteria (*Staphylococcus aureus*, *Bacillus cereus*), Gram-negative bacteria (*Pseudomonas aeruginosa* and *Escherichia coli*), and fungus *Candida albicans* by the agar diffusion method. The most effective juglone conjugates with D-xylose or L-arabinose moiety and hydroxyl group at C-7 position of naphthoquinone core at concentration 10 μg/well showed antimicrobial activity comparable with antibiotics vancomicin and gentamicin against Gram-positive bacteria strains. In liquid media, juglone-arabinosidic tetracycles showed highest activity with MIC 6.25 μM. Thus, a positive effect of heterocyclization with mercaptosugars on cytotoxic and antimicrobial activity for group of 1,4-naphthoquinones was shown.

Keywords: 1,4-naphthoquinones; quinoid compounds; thioglycosides; quinone-sugar conjugates; cytotoxic activity; antibiotic activity

1. Introduction

Naphthoquinones represent one of the largest families of natural products and are widespread in nature. They were isolated from plants, marine invertebrates, fungi, and bacteria [1]. Naphthoquinones revealed a diverse spectrum of activities: anticancer [2–4], antibacterial [5,6], anti-infective [7], antimalarial [8], and cardioprotective action [9]. The quinone ring contains a system of double bonds conjugated with carbonyl groups: it is easily susceptible to reduction, oxidation, and addition of *O*-, *N*-, and *S*-nucleophiles [10,11]. The high reactivity of naphthoquinones and the well-developed methods of its chemical modification make this group of compounds attractive for the profound development of new types of substances with high biological activity [12].

Studies are continuing on the antimicrobial activity of 1,4-naphthoquinones in relation to various microbial pathogens that are dangerous as sources of fatal diseases, epidemics, and nosocomial infections. In some cases, not only was the direct effect of new compounds on microbial cells investigated, but also their effect on the viability of biofilms formed by reproducing microorganisms. Thus, a series of new 2-hydroxy-3-phenylsulfanylmethyl-1,4-naphthoquinones were synthesized and evaluated against Gram-negative and Gram-positive bacterial strains and their biofilms to probe for potential lead structures. The structure modification applied in the series resulted in 12 new naphthoquinones with pronounced antimicrobial activity against *Escherichia coli* and *Pseudomonas aeruginosa*. Four molecules showed anti-biofilm activity and inhibited biofilm formation more than 60% with a better profile than standard antibacterial drug, ciprofloxacin [13,14].

Naphthoquinones often possess poor solubility which has hampered their practical use. The conjugation of naphthoquinones with non-toxic carbohydrates is one of the most successful ways for improving their solubility [15–19]. Moreover, the conjugation of naphthoquinones with carbohydrates led to novel structures with new types of biological activity [20,21]. Such naphthoquinone–carbohydrate conjugates include classical *O*- and *S*-glycosides (carbohydrates linked directly to naphthoquinone via glycosidic bond), non-glycosidic conjugates (connection with the carbohydrate component via not glycosidic ether linkage), and *N*-glycosyl triazoles (a triazolic ring connecting the carbohydrate moiety to naphthoquinone). In the course of our drug research project we developed an effective method for preparation of naphthoquinone acetylthioglucosides by the condensation of available substituted chloromethoxynaphthoquinone **1** with per-*O*-acetyl-1-thioderivatives of D-glucose **2a**, D-galactose **2b**, D-xylose **2c**, and L-arabinose **2d** and obtained related naphthoquinone acetylglucosides **3a–d** [22]. These acetylglycoside derivatives, **3a–d**, were readily deacetylated with MeONa/MeOH and immediately converted to the quinone–sugar tetracyclic conjugates **4a–d** in good yields (Scheme 1). The tetracyclic quinone–carbohydrate conjugates **4a–d** had a linear planar structure and retained the stereochemistry of the starting sugars.

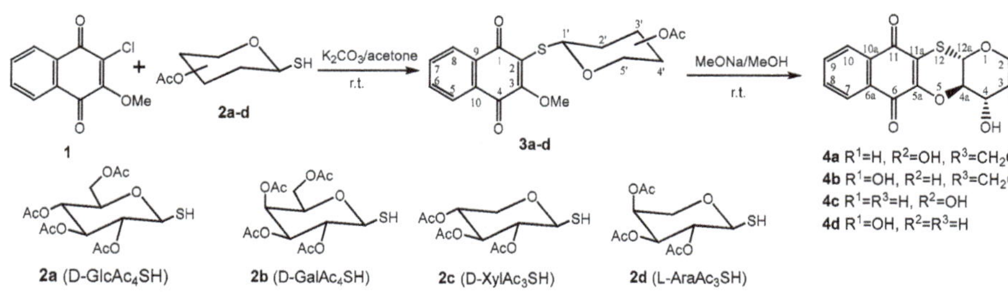

Scheme 1. Synthetic route for the synthesis of fused tetracyclic conjugates **4a–d**.

The obtained sugar–quinone tetracycles were converted to acetyl derivatives by treatment with Ac_2O/Py. Both synthesized tetracyclic quinone conjugates and their acetylated tetracyclic derivatives were active in vitro against human promyelocytic leukemia HL-60 in 1.0–5.0 μM concentrations, while starting acyclic acetylglycosides were approximately 10–100 times less active [23].

2. Results and Discussion

The synthesis of tetracyclic oxathiine-fused glycoside naphthoquinone conjugates can provide bioactive compounds, and the variation of naphthoquinone and carbohydrate moieties allows a structure–activity relationship (SAR) study. Therefore, this work aimed to conjugate per-*O*-acetylated 1-thiosugars **2a–d** with various substituted 1,4-naphthoquinones. The key intermediates, per-*O*-acetylated 1-thiosugars **2a–d**, were prepared from the respective peracetylated glycosyl halides (D-glucose, D-galactose, D-xylose, and L-arabinose) using reducing cleavage of its thiouronium salt with sodium metabisulfite according to literature procedures [24–26]. Sugar thioderivatives comprise two pairs of structurally related carbohydrates: hexopyranoses of D-glucose and D-galactose, as well as pentopyranoses of D-xylose and L-arabinose, which differ in the configuration of the C_4-OH group.

2.1. Preparation of Starting Chloro(bromo)methoxynaphthoquinones

The starting substituted 2-chloro-3,5,8-trimethoxy-1,4-naphthoquinones **5** and **6** were prepared by treatment of appropriated 2,3-dichloro-5,8-dimethoxynaphtalene-1,4-dione and 2,3,6,7-tetrachloro-5,8-dimethoxynaphtalene-1,4-dione with AcONa in dry methanol at reflux as described earlier [27] (Figure 1). Bromination of available hydroxyjuglone derivatives **7** and **8** in chloroform solution according to the work in [28] led to bromoquinones **9** and **10** in good yields, 92%. Subsequent treatment of quinones **9** and **10** in 1,4-dioxane with 0.2 M ethereal diazomethane solution and crystallization from MeOH gave a second pair of initial bromomethoxyjuglones **11** and **12** in yields 85% (Scheme 2), which were identical to such compounds described in [29].

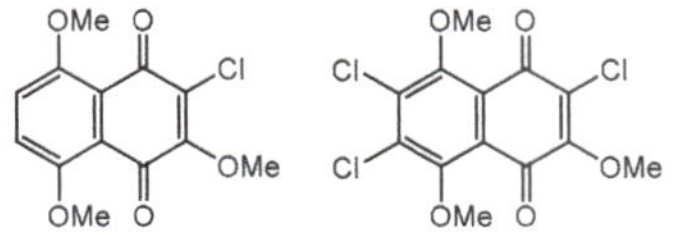

5 **6**

Figure 1. Starting derivatives of 5,8-dimethoxy-1,4-naphthoquinone.

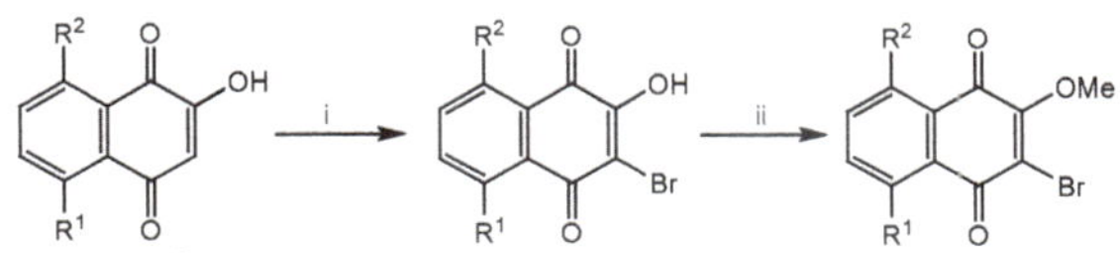

7 R^1 =OH, R^2 = H
8 R^1 = H, R^2 = OH

9 R^1 = OH, R^2 = H
10 R^1 = H, R^2 = OH

11 R^1 = OH, R^2 = H
12 R^1 = H, R^2 = OH

Scheme 2. Reagents and conditions for synthesis juglone bromomethoxyderivatives **11** and **12**: (i) Br_2, H_2O_2, $CHCl_3$, rt; (ii) CH_2N_2/1,4-dioxane.

2.2. Synthesis of 3-Acetylthioglycosides of 2-Methoxynaphthoquinones **13a–d–16a–d**

Then, four substituted 1,4-naphthoquinones—**5**, **6**, **11**, and **12**—were condensed in acetone with per-*O*-acetylated 1-thiosugars **2a–d** at equimolar ratio of quinone: thiosugar under base conditions in presence of K_2CO_3 according the procedure described in our previous work [22] (Scheme 1). This condensation led to acetylated thioglycosides of 1,4-naphthoquinones **13a–d–16a–d** in good yields, 71–91% (Figure 2). The structures of new compounds were proved by NMR, IR spectroscopy, and HR mass spectrometry. The 1′,2′-*trans*-configuration of glycosidic bond was confirmed by the value of anomeric proton doublets ($J_{1',2'}$ = 7.5–10.2 Hz) in ^{1}H-NMR spectra. The other spectral characteristics of the naphthoquinone methoxyderivatives **13a–d–16a–d** were in a good agreement with their proposed structures (see also Supplementary Materials file).

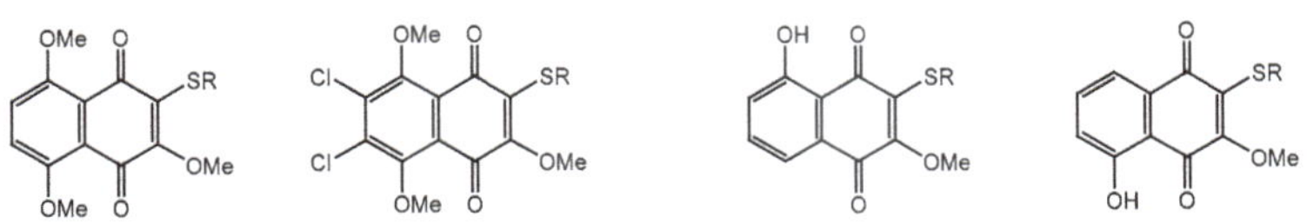

13a R = GlcAc$_4$ (90%)
13b R = GalAc$_4$ (82%)
13c R = XylAc$_3$ (85%)
13d R = AraAc$_3$ (87%)

14a R = GlcAc$_4$ (82%)
14b R = GalAc$_4$ (89%)
14c R = XylAc$_3$ (86%)
14d R = AraAc$_3$ (91%)

15a R = GlcAc$_4$ (74%)
15b R = GalAc$_4$ (71%)
15c R = XylAc$_3$ (83%)
15d R = AraAc$_3$ (79%)

16a R = GlcAc$_4$ (87%)
16b R = GalAc$_4$ (82%)
16c R = XylAc$_3$ (73%)
16d R = AraAc$_3$ (76%)

Figure 2. Acetylated thioglycosides of substituted methoxy-1,4-naphthoquinones.

2.3. Preparation of Naphthoquinone–Sugar Tetracyclic Conjugates

Under the base treatment by MeONa/MeOH thioglycosides **13a–d–16a–d** were readily converted in tetracycles **17a–d–20a–d** in good yields 82–97% (Figure 3). It is evident that tetracyclic quinone–glucoside conjugates **17a–d–20a–d** were formed from methoxyglucosides **13a–d–16a–d** through deacetylation stage and intramolecular nucleophilic substitution of the methoxy group by sugar C_2-OH group.

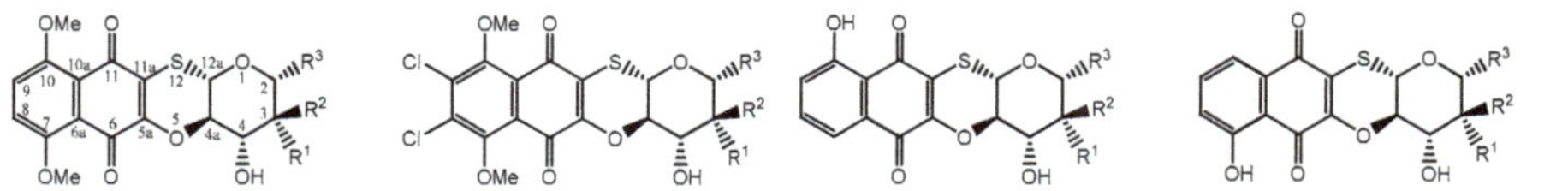

17a R^1=H, R^2=OH, R^3=CH_2OH (86%) **17b** R^1=OH, R^2=H, R^3=CH_2OH (93%) **17c** R^1=R^3=H, R^2=OH (84%) **17d** R^1=OH, R^2=R^3=H (92%)

18a R^1=H, R^2=OH, R^3=CH_2OH (83%) **18b** R^1=OH, R^2=H, R^3=CH_2OH (82%) **18c** R^1=R^3=H, R^2=OH (91%) **18d** R^1=OH, R^2=R^3=H (86%)

19a R^1=H, R^2=OH, R^3=CH_2OH (94%) **19b** R^1=OH, R^2=H, R^3=CH_2OH (82%) **19c** R^1=R^3=H, R^2=OH (90%) **19d** R^1=OH, R^2=R^3=H (92%)

20a R^1=H, R^2=OH, R^3=CH_2OH (87%) **20b** R^1=OH, R^2=H, R^3=CH_2OH (82%) **20c** R^1=R^3=H, R^2=OH (83%) **20d** R^1=OH, R^2=R^3=H (97%)

Figure 3. Synthesized 1,4-naphthoquinone-thioglycoside tetracyclic conjugates.

This process proceeds with retention of the configuration of all asymmetric centers of the carbohydrate portion and formation of linear tetracyclic structure. The alternative angular structure for the obtained tetracycles **17a–d–20a–d** was rejected based upon on the direct comparison with the spectral data of similar angular tetracycles, which we obtained earlier [30].

2.4. Biological Evaluation

2.4.1. Cytotoxic Activity

Tetracyclic conjugates **17a–d–20a–d** and 5-hydroxy-1,4-naphthoquinone (juglone) were examined for cytotoxicity against three cancer cell lines and two normal cell lines such as human cervical cancer (HeLa), mouse neuroblastoma (Neuro 2a), mouse ascites Ehrlich carcinoma, mouse normal epithelial cell line (JB6 Cl 41-5a), and mouse blood erythrocytes. Known cytotoxic agent cucumarioside A_2-2 [31] was used as positive control. The results are presented in Table 1.

Conjugates **17a,c,d** with a 7,10-dimethoxynaphthoquinone core were inactive for all cell lines at EC_{50} value ≤ 25 μM. Galactoside derivative **17b** had poor solubility in DMSO, which did not allow for the measurement its activity. Introduction of two chlorine atoms in a 7,10-dimethoxynaphthoquinone core led to tetracyclic 8,9-dichloroderivatives **18a–d** with better solubility and promising activity in EC_{50} values ranging from 1.1 to 10.9 μM. Among the substances of this group, galactoside derivative **18b** showed the best activity against ascites Ehrlich carcinoma cell line with EC_{50} value 1.1 μM.

The following two groups of tetracycles, **19a–d** and **20a–d**, are conjugates of thiosaccharides **2a–d** with derivatives of 5-hydroxy-1,4-naphthoquinone **11–12**. Among them, six substances had promising values, EC_{50} < 1 μM. In the group of tetracycles **19a–d**, bearing a hydroxyl group at position 10, only hexapyranoside derivatives D-Glc **19a** and D-Gal **19b** showed the high cytotoxic activity with EC_{50} values in the range of 0.6 to 0.9 μM, while all tetracycles **20a–d**, bearing hydroxyl group at position 7, revealed highly toxic compounds with EC_{50} 0.3–0.7 μM for various types of cells. Mouse epithelial Jb6 cells were more susceptible to the action of juglone tetracyclic derivatives **20a–d**. As evidenced from Table 1, the presence of a hydroxy group in the naphthoquinone scaffold led to the formation of naphthoquinone tetracycles **19a–b–20a–b** with promising cytotoxicity. However, unsubstituted 5-hydroxy-1,4-naphthoquinone (juglone) did not show any cytotoxic activity up to 100 μM. This fact proves the positive effect of heterocyclization on tetracycles cytotoxicity. Moreover, it was shown that all tested compounds did not cause lysis of murine erythrocytes up to 25 μM.

For all cytotoxic compounds **18a–d–20a–d** their selectivity index (SI) was calculated (Table 2). Among the tested tetracycles, the most selective substance was **19a** in relation to all studied lines of tumor cells. In comparison to non-tumor mouse epithelial Jb6 Cl 41-5a cells, the selectivity index for

Ehrlich carcinoma cells was 9.3, for HeLa—1.5, and for Neuro-2a cell cuture—1.4. Compounds **18b** and **19d** also showed rather high values of the selectivity index for Ehrlich carcinoma cells with SI = 2.6 and SI = 2.1, respectively. On HeLa cells, the most active compounds were **19a**, **19d**, and **20d**, with SI > 1.

Table 1. Cytotoxicity (EC_{50}) of oxathiine fused 1,4-naphthoquinone tetracycles on cancer and non-cancer cell lines (μM).

Compound	Tested Cell Lines			
	Neuro 2a *	HeLa *	Ascites Ehrlich Carcinoma **	Jb6 Cl 41-5a *
17a	>25	>25	>25	>25
17b	- ***	-	-	-
17c	>25	>25	>25	>25
17d	>25	>25	>25	>25
18a	2.4	10.9	8.8	1.5
18b	2.9	8	1.1	2.9
18c	4.1	10.1	4.5	3.8
18d	2.8	8.7	4.1	4.3
19a	4.1	3.7	0.6	5.6
19b	2.3	2.7	0.7	0.9
19c	13.3	4.9	7.4	5.9
19d	2.2	1.3	1.1	2.3
20a	0.7	6.3	2.7	0.7
20b	1.2	4.7	1.1	0.5
20c	2.2	2.3	0.7	0.3
20d	1.4	0.4	1.2	0.7
Juglone	>100	>100	>100	>100
Cucumarioside A_2-2	37.5	42.4	18.5	15.1

* Cytotoxicity evaluation with MTT reagent; ** cytotoxicity evaluation with with FDA; *** non-tested due to poor solubility.

Table 2. Tumor cell selectivity (Selectivity Index (SI)) of tested tetracycles **18a–d–20a–d**.

Compound	Neuro-2a	HeLa	Ascitic Erclich Carcinoma
18a	0.6	0.1	0.2
18b	1	0.4	2.6
18c	0.9	0.4	0.8
18d	1.5	0.5	1
19a	1.4	1.5	9.3
19b	0.4	0.3	1.3
19c	0.4	1.2	0.8
19d	1	1.8	2.1
20a	1	0.1	0.3
20b	0.4	0.1	0.5
20c	0.1	0.1	0.4
20d	0.5	1.75	0.6
Cucumarioside A_2-2	0.4	0.4	0.8

2.4.2. Antimicrobial Activity

All synthesized tetracyclic conjugates **17a–b–20a–b** and 5-hydroxy-1,4-naphthoquinone (juglone) were screened by the agar diffusion method for antibacterial activity against two strains of Gram-positive bacteria (*S. aureus* ATCC 21027; *B. cereus* ATCC 10702), two strains of Gram-negative bacteria (*P. aeruginosa* ATCC 27853; *E. coli* K-12), and antifungal activity against fungus *C. albicans* KMM 453 from the Collection of Marine Microorganisms (KMM) of the G.B. Elyakov Pacific Institute of Bioorganic Chemistry. Commercial antibacterial drugs (vancomicin and gentamicin) and antifungal (clotrimazol) drugs were used as positive controls. Compound concentrations and diameter of inhibition zone are presented in Table 3.

Table 3. Antimicrobial activity (zone inhibition) of oxathiine fused 1,4-naphthoquinone-thioglucoside tetracycles.

Compound	Compound Concentration (μg/well)	Diameter of Zone Inhibition (mm)				
		Gram-Positive Strains		Gram-Negative Strains		Fungus
		S. aureus	*B. cereus*	*P. aeruginosa*	*E. coli*	*C. albicans*
17a	100	0	0	0	0	0
17b	100	0	+	0	0	0
17c	100	0	0	0	0	0
17d	100	0	0	0	0	8 ± 2
18a	100	+	0	0	0	0
18b	100	0	0	0	0	0
18c	100	10 ± 2	16 ± 3	0	0	0
18d	100	10 ± 2	16 ± 2	0	0	0
19a	100	0	0	0	0	0
19b	100	0	8 ± 2	0	0	0
19c	100	8 ± 1	10 ± 2	0	0	0
19d	100	20 ± 3	25 ± 2	0	0	0
20a	100	30 ± 2	22 ± 1	0	0	0
	10	10 ± 2	10 ± 2	–	–	–
	1	0	0	–	–	–
20b	100	30 ± 2	20 ± 1	0	0	0
	10	0	10 ± 2	–	–	–
	1	0	0	–	–	–
20c	100	30 ± 2	25 ± 1	0	0	8 ± 1
	10	25 ± 3	22 ± 2	–	–	–
	1	+	+	–	–	–
20d	100	40 ± 2	30 ± 3	0	0	0
	10	25 ± 3	23 ± 2	–	–	–
	1	0	+	–	–	–
Juglone	100	0	0	12 ± 2	0	–
Gentamicin (disk)	10	25 ± 2	25 ± 1	18 ± 2	22 ± 3	–
Vancomicin (disk)	30	20 ± 2	20 ± 3	0	0	–
Clotrimazol (disk)	10	–	–	–	–	15 ± 2

Zones within 20–40 mm mean strong antibiotic activity, zones up to 20 mm—moderate, ≤10 mm—weak, +—insignificant (– means that the study was not conducted).

Conjugates **17a–c** with a 7,10-dimethoxynaphthoquinone core were inactive to Gram-positive and Gram-negative bacteria in concentrations of 100 μg/well. Only one of these conjugates—**17d**, based on L-arabinose, revealed weak activity against *C. albicans*. Conjugates **18c,d** with two chlorine atoms in their naphthoquinone skeleton showed moderate activity against Gram-positive strains *S. aureus* and *B. cereus* at a concentration of 100 μg/well, but also were not active to Gram-negative bacteria and fungus *C. albicans*. As evidenced from Table 3, tetracycles **19b–d**, bearing a hydroxyl group at C-10 atom of naphthoquinone core, showed various antimicrobial activity levels to Gram-positive bacteria from weak activity for tetracycle **19b**, to strong activity for compound **19d**.

Sugar tetracylic conjugates **20a–d**, with a hydroxyl group at the C-7 atom of their naphthoquinone core, constituted the most effective set of antimicrobials among tested compounds **17a–d–20a–d**. All these compounds, **20a–d**, have an inhibition zone diameter of 22–40 mm at concentration 100 μg/well and kept the value of inhibition zone in the range of 10 to 25 mm for the concentration 10 μg/well. At a concentration of 10 μg/well, the most effective conjugates with D-xylose **20c** and L-arabinose **20d** showed antimicrobial activity comparable with antibiotics vancomicin and gentamicin, and these ones retained residual effect upon dilution to 1 μg/well for Gram-positive strains. Surprisingly, 5-hydroxy-1,4-naphthoquinone (juglone) showed moderate activity against only *P. aeruginosa*. Thus, heterocyclization with 1-thiosugars leads to an increase of antibiotic activity, especially against the *S. aureus* strain.

The antimicrobial activity for the most active compounds, **19d** and **20a–d**, was also determined against *S. aureus* by the minimum inhibitory concentration (MIC) using the broth microdilution method. As it follows from Table 4, juglone derivatives **19d** and **20d** with D-arabinose moiety showed highest antibacterial activity with a MIC of 6.25 μM. The worse activity for **19d** in the agar medium test is probably associated with interactions with agar molecules in the medium. In comparison to non-tumor mouse epithelial Jb6 Cl 41-5a cells, no selective inhibitory activity on bacterial cells was observed, as SI < 1.

Table 4. Minimum inhibitory concentration (MIC, μM) and selectivity index (SI) of testing compounds against *S. aureus*.

Compound	MIC	SI
19d	6.25	0.4
20a	25.0	-
20b	25.0	-
20c	12.5	0.02
20d	6.25	0.1
Juglone	>100.0	-

3. Materials and Methods

3.1. General Information

All reagents were obtained from commercial suppliers and were used without additional purification. All solvents were distilled before use. Melting points were determined by using a Boetius apparatus (VEB Analytic, Dresden, Germany) and are uncorrected. IR spectra were recorded in KBr pellets or in $CHCl_3$ by using a Bruker Equinox 55 spectrophotometer (Bruker Optik GmbH, Ettlingen, Germany). ^{1}H-NMR spectra were recorded on a Bruker Avance III-500 HD (500 MHz) or a Bruker Avance III-700 (700 MHz) spectrometer (Bruker Corporation, Bremen, Germany) with $CDCl_3$ or DMSO-d_6 as the solvent and TMS as the internal standard. ^{13}C-NMR spectra were recorded on a Bruker Avance III HD-500 spectrometer at 125 MHz or a Bruker Avance III-700 spectrometer at 176 MHz. ESI mass spectra were recorded on an Agilent 6510 Q-TOF LC/MS instrument (Agilent, Santa Clara, CA, USA). Silufol UV–Vis TLC plates (Sklarny Kavalier, Votitsa, Czech Republic) treated with HCl vapor were used for analytical TLC. Preparative TLC was performed on silica gel 60 Merck

(40–60 μm). TLC plates were developed in system A (hexane–benzene–acetone (2:1:1 *v*/*v*)), system B (hexane–benzene–acetone (2:1:2 *v*/*v*)), or system C (benzene–EtOAc–MeOH (7:4:2 *v*/*v*)).

3.2. *Synthesis*

3.2.1. General Procedure for Synthesis of Acetylated Thioglycosides **13a–d–16a–d**

2-Chloro(bromo)-3-methoxyquinone, **5**, **6**, **11**, and **12** (0.50 mM), and per-*O*-acetylated 1-mercaptosugar derivatives **2a–d** (0.55 mM) were dissolved in acetone (30 mL) and 76 mg (0.55 mM) of dry finely powdered K_2CO_3 was added. The resulting mixture was stirred for 2 h at room temperature until the consumption of thioglicose and conversion of starting quinone. Precipitate of inorganic salt was filtered, the filtrate was dried in a vacuum, and the residue was subjected to preparative TLC (system B for **13a–d** and system A for others). The main fraction was washed off from silica gel with acetone, dried, and recrystallized from MeOH to give pure thioglycoside **13a–d–16a–d**.

*2-(2,3,4,6-Tetra-O-acetyl-β-*D*-glucopyranosyl-1-thio)-3,5,8-trimethoxy-1,4-naphthoquinone* (**13a**); yield 276 mg (90%), orange solid, R_f 0.37 (B), m.p. 162–165 °C. ^{1}H-NMR (500 MHz, $CDCl_3$), δ: 1.97 (s, 3H, COC$\underline{H}_3$), 2.00 (s, 3H, COC$\underline{H}_3$), 2.01 (s, 3H, COC$\underline{H}_3$), 2.06 (s, 3H, COC$\underline{H}_3$), 3.70 (m, 1H, H-5′), 3.93 (s, 3H, -OC$\underline{H}_3$), 3.94 (s, 3H, -OC$\underline{H}_3$), 3.99 (dd, 1H, *J* = 12.4, 1.9 Hz, H-6′a), 4.14 (s, 3H, -OC$\underline{H}_3$), 4.20 (dd, 1H, *J* = 12.4, 4.5, Hz, H-6′b), 5.10 (t, 1H, *J* = 9.6 Hz, H-2′), 5.12 (t, 1H, *J* = 9.7 Hz, H-4′), 5.26 (t, 1H, *J* = 9.2 Hz, H-3′), 5.61 (d, 1H, *J* = 10.1 Hz, H-1′), 7.23 (d, 1H, *J* = 9.5 Hz, Ar-H), 7.27 (d, 1H, *J* = 9.5 Hz, Ar-H). ^{13}C-NMR(125 MHz, $CDCl_3$), δ: 20.6 (3 × CO$\underline{C}H_3$), 20.7 (CO$\underline{C}H_3$), 56.9 (-O$\underline{C}H_3$), 57.1 (-O$\underline{C}H_3$), 61.1 (-O$\underline{C}H_3$), 61.8 (C-6′), 68.2 (C-4′), 71.2 (C-2′), 74.1 (C-3′), 75.7 (C-5′), 80.7 (C-1′), 119.4, 120.5, 120.7, 122.2, 126.9 (C-2), 153.0, 153.7, 158.3 (C-3), 169.3 ($\underline{C}OCH_3$), 169.4 ($\underline{C}OCH_3$), 170.2 ($\underline{C}OCH_3$), 170.6 ($\underline{C}OCH_3$), 178.0, 181.4. IR ($CHCl_3$): 3050, 2943, 2842, 1755, 1663, 1597, 1571, 1479, 1463, 1435, 1413, 1368, 1334, 1270, 1243, 1210, 1193 cm^{-1}. HRMS (ESI): *m*/*z* $[M - H]^-$ calcd. for $C_{27}H_{29}O_{14}S$ 609.1284, found 609.1281.

*2-(2,3,4,6-Tetra-O-acetyl-β-*D*-galactopyranosyl-1-thio)-3,5,8-trimethoxy-1,4-naphthoquinone* (**13b**); yield 253 mg (82.5%), orange solid, R_f 0.37 (B), m.p. 99–101 °C. ^{1}H-NMR (500 MHz, $CDCl_3$), δ: 1.94 (s, 3H, COC$\underline{H}_3$), 1.98 (s, 3H, COC$\underline{H}_3$), 2.08 (s, 3H, COC$\underline{H}_3$), 2.13 (s, 3H, COC$\underline{H}_3$), 3.91 (m, 1H, H-5′), 3.93 (s, 3H, -OC$\underline{H}_3$), 3.95 (s, 3H, -OC$\underline{H}_3$), 4.02 (m, 2H, H-6′a, H-6′b), 4.14 (s, 3H, -OC$\underline{H}_3$), 5.10 (dd, 1H, *J* = 9.9, 3.5 Hz, H-3′), 5.31 (t, 1H, *J* = 10.0 Hz, H-2′), 5.42 (m, 1H, H-4′), 5.59 (d, 1H, *J* = 10.2 Hz, H-1′), 7.23 (d, 1H, *J* = 9.5 Hz, Ar-H), 7.27 (d, 1H, *J* = 9.5 Hz, Ar-H). ^{13}C-NMR(125 MHz, $CDCl_3$), δ: 20.5 (CO$\underline{C}H_3$), 20.6 (2 × CO$\underline{C}H_3$), 20.8 (CO$\underline{C}H_3$), 56.9 (-O$\underline{C}H_3$), 57.0 (-O$\underline{C}H_3$), 61.0 (C-6′), 61.1 (-O$\underline{C}H_3$), 67.2 (C-4′), 68.4 (C-2′), 72.0 (C-3′), 74.3 (C-5′), 81.5 (C-1′), 119.4, 120.5, 120.6, 122.1, 126.9 (C-2), 153.1, 153.7, 158.5 (C-3), 169.6 ($\underline{C}OCH_3$), 170.0 (2 × $\underline{C}OCH_3$), 170.2 ($\underline{C}OCH_3$), 178.1, 181.3. IR ($CHCl_3$): 3054, 3006, 2941, 2842, 1750, 1663, 1597, 1571, 1479, 1463, 1435, 1413, 1372, 1334, 1270, 1251, 1185, 1155, 1085, 1060, 1022 cm^{-1}. HRMS (ESI): *m*/*z* $[M - H]^-$ calcd. for $C_{27}H_{29}O_{14}S$ 609.1284, found 609.1283.

*2-(2,3,4-Tri-O-acetyl-β-*D*-xylopyranosyl-1-thio)-3,5,8-trimethoxy-1,4-naphthoquinone* (**13c**); yield 230 mg (85%), orange solid, R_f 0.39 (B), m.p. 197–199 °C. ^{1}H-NMR (500 MHz, $CDCl_3$), δ: 2.04 (s, 3H, COC$\underline{H}_3$), 2.06 (s, 3H, COC$\underline{H}_3$), 2.08 (s, 3H, COC$\underline{H}_3$), 3.39 (dd, 1H, *J* = 11.9, 8.3 Hz, H-5′a), 3.93 (s, 3H, -OC$\underline{H}_3$), 3.95 (s, 3H, -OC$\underline{H}_3$), 4.15 (s, 3H, -OC$\underline{H}_3$), 4.21 (dd, 1H, *J* = 11.9, 4.7 Hz, H-5′b), 4.94 (m, 1H, H-4′), 5.04 (t, 1H, *J* = 8.0 Hz, H-2′), 5.20 (t, 1H, *J* = 8.0 Hz, H-3′), 5.60 (d, 1H, *J* = 8.0 Hz, H-1′), 7.24 (d, 1H, *J* = 9.5 Hz, Ar-H), 7.27 (d, 1H, *J* = 9.5 Hz, Ar-H). ^{13}C-NMR(125 MHz, $CDCl_3$), δ: 20.7 (3 × CO$\underline{C}H_3$), 56.9 (-O$\underline{C}H_3$), 57.1 (-O$\underline{C}H_3$), 61.2 (-O$\underline{C}H_3$), 64.9 (C-5′), 68.6 (C-4′), 70.6 (C-2′), 71.7 (C-3′), 81.6 (C-1′), 119.4, 120.6, 120.8, 122.0, 126.6 (C-2), 153.3, 153.8, 159.4 (C-3), 169.4 ($\underline{C}OCH_3$), 169.7 ($\underline{C}OCH_3$), 169.8 ($\underline{C}OCH_3$), 178.3, 181.3. IR ($CDCl_3$): 3054, 3018, 3006, 2942, 2842, 1752, 1663, 1597, 1571, 1478, 1463, 1435, 1413, 1371, 1334, 1272, 1248, 1210, 1185, 1062, 1024 cm^{-1}. HRMS (ESI): *m*/*z* $[M - H]^-$ calcd. for $C_{24}H_{25}O_{12}S$ 537.1072, found 537.1073.

*2-(2,3,4-Tri-O-acetyl-α-*L*-arabinopyranosyl-1-thio)-3,5,8-trimethoxy-1,4-naphthoquinone* (**13d**); yield 236 mg (87.2%), dark orange solid, R_f 0.39 (B), m.p. 92–94 °C. ^{1}H-NMR (500 MHz, $CDCl_3$), δ: 2.07 (s, 3H, COC$\underline{H}_3$), 2.10 (s, 3H, COC$\underline{H}_3$), 2.11 (s, 3H, COC$\underline{H}_3$), 3.64 (dd, 1H, *J* = 12.6, 2.3 Hz, H-5′a), 3.92 (s, 3H,

-OC$\underline{H}_3$), 3.94 (s, 3H, -OC$\underline{H}_3$), 4.08 (dd, 1H, J = 12.6, 4.3 Hz, H-5′b), 4.15 (s, 3H, -OC$\underline{H}_3$), 5.14 (dd, 1H, J = 8.2, 3.4 Hz, H-3′), 5.27 (m, 1H, H-4′), 5.31 (t, 1H, J = 7.9 Hz, H-2′), 5.60 (d, 1H, J = 7.9 Hz, H-1′), 7.23 (d, 1H, J = 9.5 Hz, Ar-H), 7.26 (d, 1H, J = 9.5 Hz, Ar-H). ^{13}C-NMR(125 MHz, CDCl$_3$), δ: 20.7 (CO$\underline{C}$H$_3$), 20.8 (CO$\underline{C}$H$_3$), 20.9 (CO$\underline{C}$H$_3$), 56.9 (-O$\underline{C}$H$_3$), 57.2 (-O$\underline{C}$H$_3$), 61.2 (-O$\underline{C}$H$_3$), 65.3 (C-5′), 67.6 (C-4′), 69.2 (C-2′), 70.3 (C-3′), 81.8 (C-1′), 119.4, 120.6, 120.7, 122.1, 127.1 (C-2), 153.2, 153.7, 159.1 (C-3), 169.5 ($\underline{C}$OCH$_3$), 169.8 ($\underline{C}$OCH$_3$), 170.2 ($\underline{C}$OCH$_3$), 178.3, 181.4. IR (CHCl$_3$): 3054, 3005, 2941, 2841, 1747, 1661, 1597, 1570, 1478, 1463, 1435, 1412, 1372, 1335, 1272, 1250, 1185, 1159, 1105, 1087, 1060, 1022 cm^{-1}. HRMS (ESI): m/z [M − H]$^-$ calcd. for $C_{24}H_{25}O_{12}S$ 537.1072, found 537.1070.

2-(2,3,4,6-Tetra-O-acetyl-β-D-glucopyranosyl-1-thio)-6,7-dichloro-3,5,8-trimethoxy-1,4-naphthoquinone (**14a**); yield 279 mg (82%), yellow solid, R_f 0.48 (A), m.p. 103–105 °C. ^{1}H-NMR (500 MHz, CDCl$_3$), δ: 1.96 (s, 3H, COC$\underline{H}_3$), 2.01 (s, 3H, COC$\underline{H}_3$), 2.02 (s, 3H, COC$\underline{H}_3$), 2.08 (s, 3H, COC$\underline{H}_3$), 3.67 (m, 1H, H-5′), 3.98 (s, 6H, 2 × -OC$\underline{H}_3$), 4.01 (dd, 1H, J = 12.6, 2.2 Hz, H-6′a), 4.15 (s, 3H, -OC$\underline{H}_3$), 4.18 (dd, 1H, J = 12.6, 4.7 Hz, H-6′b), 5.10 (m, 2H, H-2′, H-4′), 5.27 (t, 1H, J = 9.3 Hz, H-3′), 5.60 (d, 1H, J = 10.2 Hz, H-1′). ^{13}C-NMR(125 MHz, CDCl$_3$), δ: 20.5 (2 × CO$\underline{C}$H$_3$), 20.6 (2 × CO$\underline{C}$H$_3$), 61.3 (-O$\underline{C}$H$_3$), 61.8 (C-6′), 62.3 (-O$\underline{C}$H$_3$), 62.4 (-O$\underline{C}$H$_3$), 68.1 (C-4′), 71.1 (C-2′), 74.0 (C-3′), 76.0 (C-5′), 80.4 (C-1′), 124.1, 125.7, 127.6 (C-2), 136.2, 136.9, 152.6, 153.0, 158.3 (C-3), 169.3 ($\underline{C}$OCH$_3$), 169.4 ($\underline{C}$OCH$_3$), 170.1 ($\underline{C}$OCH$_3$), 170.5 ($\underline{C}$OCH$_3$), 176.4, 179.7. IR (CHCl$_3$): 3050, 2944, 2857, 1756, 1670, 1581, 1547, 1459, 1440, 1380, 1326, 1304, 1241, 1207, 1194, 1030 cm^{-1}. HRMS (ESI): m/z [M − H]$^-$ calcd. for $C_{27}H_{27}Cl_2O_{14}S$ 677.0504, found 677.0504.

2-(2,3,4,6-Tetra-O-acetyl-β-D-galactopyranosyl-1-thio)-6,7-dichloro-3,5,8-trimethoxy-1,4-naphthoquinone (**14b**); yield 302 mg (88.8%), orange solid, R_f 0.48 (A), m.p. 80–81 °C. ^{1}H-NMR (500 MHz, CDCl$_3$), δ: 1.95 (s, 3H, COC$\underline{H}_3$), 1.99 (s, 3H, COC$\underline{H}_3$), 2.09 (s, 3H, COC$\underline{H}_3$), 2.14 (s, 3H, COC$\underline{H}_3$), 3.90 (m, 1H, H-5′), 3.98 (s, 6H, 2 × -OC$\underline{H}_3$), 4.03 (m, 2H, H-6′a, H-6′b), 4.16 (s, 3H, -OC$\underline{H}_3$), 5.11 (dd, 1H, J = 9.9, 3.5 Hz, H-3′), 5.30 (t, 1H, J = 10.0 Hz, H-2′), 5.42 (m, 1H, H-4′), 5.57 (d, 1H, J = 10.2 Hz, H-1′). ^{13}C-NMR(125 MHz, CDCl$_3$), δ: 20.5 (2 × CO$\underline{C}$H$_3$), 20.6(CO$\underline{C}$H$_3$), 20.7 (CO$\underline{C}$H$_3$), 61.2 (C-6′), 61.7 (-O$\underline{C}$H$_3$), 62.3 (-O$\underline{C}$H$_3$), 62.4 (-O$\underline{C}$H$_3$), 67.2 (C-4′), 68.3 (C-2′), 71.9 (C-3′), 74.6 (C-5′), 81.3 (C-1′), 124.1, 125.6, 127.7 (C-2), 136.2, 136.9, 152.7, 153.0, 158.4 (C-3), 169.6 ($\underline{C}$OCH$_3$), 170.0 ($\underline{C}$OCH$_3$), 170.2 ($\underline{C}$OCH$_3$), 170.3 ($\underline{C}$OCH$_3$), 176.5, 179.7. IR (CHCl$_3$): 3053, 3007, 2944, 2856, 1751, 1670, 1580, 1459, 1440, 1380, 1326, 1304, 1245, 1191, 1113, 1086, 1059, 1028 cm^{-1}. HRMS (ESI): m/z [M − H]$^-$ calcd. for $C_{27}H_{27}Cl_2O_{14}S$ 677.0504, found 677.0501.

2-(2,3,4-Tri-O-acetyl-β-D-xylopyranosyl-1-thio)-6,7-dichloro-3,5,8-trimethoxy-1,4-naphthoquinone (**14c**); yield 263 mg (86.5%), dark orange solid, R_f 0.52 (A), m.p. 132–134 °C. ^{1}H-NMR (500 MHz, CDCl$_3$), δ: 2.04 (s, 3H, COC$\underline{H}_3$), 2.07 (s, 3H, COC$\underline{H}_3$), 2.10 (s, 3H, COC$\underline{H}_3$), 3.36 (dd, 1H, J = 11.8, 8.6 Hz, H-5′a), 3.97 (s, 3H, -OC$\underline{H}_3$), 3.98 (s, 3H, -OC$\underline{H}_3$), 4.15 (s, 3H, -OC$\underline{H}_3$), 4.18 (dd, 1H, J = 11.8, 5.0 Hz, H-5′b), 4.95 (m, 1H, H-4′), 5.04 (t, 1H, J = 8.2 Hz, H-2′), 5.22 (t, 1H, J = 8.2 Hz, H-3′), 5.57 (d, 1H, J = 8.2 Hz, H-1′). ^{13}C-NMR(125 MHz, CDCl$_3$), δ: 20.7 (3 × CO$\underline{C}$H$_3$), 61.3 (-O$\underline{C}$H$_3$), 62.3 (2×-O$\underline{C}$H$_3$), 65.2 (C-5′), 68.5 (C-4′), 70.7 (C-2′), 71.9 (C-3′), 81.3 (C-1′), 124.1, 125.4, 127.6 (C-2), 136.2, 136.9, 152.8, 153.0, 159.1 (C-3), 169.4 ($\underline{C}$OCH$_3$), 169.7 ($\underline{C}$OCH$_3$), 169.9 ($\underline{C}$OCH$_3$), 176.7, 179.7. IR (CHCl$_3$): 3054, 3019, 3006, 2944, 2857, 1754, 1671, 1580, 1547, 1459, 1440, 1380, 1326, 1304, 1246, 1190, 1114, 1067, 1029 cm^{-1}. HRMS (ESI): m/z [M − H]$^-$ calcd. for $C_{24}H_{23}Cl_2O_{12}S$ 605.0293, found 605.0291.

2-(2,3,4-Tri-O-acetyl-α-L-arabinopyranosyl-1-thio)-6,7-dichloro-3,5,8-trimethoxy-1,4-naphthoquinone (**14d**); yield 276 mg (90.8%), dark orange solid, R_f 0.50 (A), m.p. 79–81 °C. ^{1}H-NMR (500 MHz, CDCl$_3$), δ: 2.07 (s, 3H, COC$\underline{H}_3$), 2.11 (s, 6H, 2 × COC$\underline{H}_3$), 3.63 (dd, 1H, J = 12.7, 1.9 Hz, H-5′a), 3.96 (s, 3H, -OC$\underline{H}_3$), 3.98 (s, 3H, -OC$\underline{H}_3$), 4.06 (dd, 1H, J = 12.7, 4.2 Hz, H-5′b), 4.16 (s, 3H, -OC$\underline{H}_3$), 5.16 (dd, 1H, J = 8.3, 3.4 Hz, H-3′), 5.28 (m, 1H, H-4′), 5.31 (t, 1H, J = 8.0 Hz, H-2′), 5.57 (d, 1H, J = 8.0 Hz, H-1′). ^{13}C-NMR(125 MHz, CDCl$_3$), δ: 20.7 (CO$\underline{C}$H$_3$), 20.8 (CO$\underline{C}$H$_3$), 20.8 (CO$\underline{C}$H$_3$), 61.3 (-O$\underline{C}$H$_3$), 62.3 (2 × -O$\underline{C}$H$_3$), 65.6 (C-5′), 67.5 (C-4′), 69.2 (C-2′), 70.3 (C-3′), 81.6 (C-1′), 124.1, 125.5, 128.0 (C-2), 136.1, 136.9, 152.8, 153.0, 158.9 (C-3), 169.4 ($\underline{C}$OCH$_3$), 169.9 ($\underline{C}$OCH$_3$), 170.2 ($\underline{C}$OCH$_3$), 176.7, 179.8. IR (CHCl$_3$): 3054, 3027, 3005,

2943, 2856, 1748, 1670, 1580, 1546, 1459, 1440, 1380, 1326, 1304, 1247, 1225, 1190, 1113, 1087, 1060, 1028 cm^{-1}. HRMS (ESI): *m/z* $[M - H]^-$ calcd. for $C_{24}H_{23}Cl_2O_{12}S$ 605.0293, found 605.0289.

2-(2,3,4,6-Tetra-O-acetyl-β-D-glucopyranosyl-1-thio)-8-hydroxy-3-methoxy-1,4-naphthoquinone (**15a**); yield 210 mg (73.9%), orange solid, R_f 0.46 (A), m.p. 158–161 °C. ^{1}H-NMR (700 MHz, $CDCl_3$): δ 1.93 (s, 3H, $COC\underline{H}_3$), 2.02 (s, 3H, $COC\underline{H}_3$), 2.03 (s, 3H, $COC\underline{H}_3$), 2.09 (s, 3H, $COC\underline{H}_3$), 3.72 (m, 1H, H-5′), 4.08 (dd, 1H, *J* = 12.4, 2.0 Hz, H-6′a), 4.15 (dd, 1H, *J* = 12.4, 5.5 Hz, H-6′b), 4.29 (s, 3H, $-OC\underline{H}_3$), 5.08 (t, 1H, *J* = 9.7 Hz, H-4′), 5.11 (t, 1H, *J* = 10.1 Hz, H-2′), 5.27 (t, 1H, *J* = 9.3 Hz, H-3′), 5.50 (d, 1H, *J* = 10.1 Hz, H-1′), 7.26 (dd, 1H, *J* = 7.4, 1.2 Hz, H-7), 7.59 (dd, 1H, *J* = 8.0, 7.4 Hz, H-6), 7.61 (dd, 1H, *J* = 8.0, 1.2 Hz, H-5), 12.03 (s, 1H, $C^8O\underline{H}$). ^{13}C-NMR(176 MHz, $CDCl_3$): δ 20.4 ($CO\underline{C}H_3$), 20.5 ($CO\underline{C}H_3$), 20.6 ($CO\underline{C}H_3$), 20.7 ($CO\underline{C}H_3$), 62.1 (C-6′, $-O\underline{C}H_3$), 68.3 (C-4′), 71.3 (C-2′), 73.9 (C-3′), 76.0 (C-5′), 81.7 (C-1′), 114.3 (C-9), 119.7 (C-5), 125.0 (C-7), 126.0 (C-2), 131.3 (C-10), 135.8 (C-6), 160.5 (C-3), 161.2 (C-8), 169.3 ($\underline{C}OCH_3$), 169.4 ($\underline{C}OCH_3$), 170.2 ($\underline{C}OCH_3$), 170.5 ($\underline{C}OCH_3$), 178.4 (C-4), 187.7 (C-1). IR ($CHCl_3$): 3053, 3007, 2953, 1756, 1669, 1630, 1580, 1559, 1458, 1369, 1313, 1250, 1191, 1162, 1077, 1050 cm^{-1}. HRMS (ESI): *m/z* $[M - H]^-$ calcd. for $C_{25}H_{25}O_{13}S$ 565.1021, found 565.1016.

2-(2,3,4,6-Tetra-O-acetyl-β-D-galactopyranosyl-1-thio)-8-hydroxy-3-methoxy-1,4-naphthoquinone (**15b**); yield 203 mg (71.4%), orange solid, R_f 0.46 (A), m.p. 190–193 °C. ^{1}H-NMR (500 MHz, $CDCl_3$), δ: 1.91 (s, 3H, $COC\underline{H}_3$), 1.99 (s, 3H, $COC\underline{H}_3$), 2.10 (s, 3H, $COC\underline{H}_3$), 2.15 (s, 3H, $COC\underline{H}_3$), 3.92 (m, 1H, H-5′), 4.06 (m, 2H, H-6′a, H-6′b), 4.30 (s, 3H, $-OC\underline{H}_3$), 5.10 (dd, 1H, *J* = 9.9, 3.5 Hz, H-3′), 5.32 (t, 1H, *J* = 10.0 Hz, H-2′), 5.43 (m, 1H, H-4′), 5.46 (d, 1H, *J* = 10.0 Hz, H-1′), 7.26 (dd, 1H, *J* = 7.7, 1.5 Hz, H-7), 7.58 (dd, 1H, *J* = 7.7, 7.5 Hz, H-6), 7.61 (dd, 1H, *J* = 7.5, 1.5 Hz, H-5), 12.04 (s, 1H, $C^8O\underline{H}$). ^{13}C-NMR(125 MHz, $CDCl_3$), δ: 20.4 ($CO\underline{C}H_3$), 20.5 ($CO\underline{C}H_3$), 20.6 ($CO\underline{C}H_3$), 20.8 ($CO\underline{C}H_3$), 61.6 (C-6′), 62.1 ($-O\underline{C}H_3$), 67.3 (C-4′), 68.4 (C-2′), 71.9 (C-3′), 74.8 (C-5′), 82.5 (C-1′), 114.3 (C-9), 119.6 (C-5), 125.0 (C-7), 126.2 (C-2), 131.4 (C-10), 135.8 (C-6), 160.5 (C-3), 161.3 (C-8), 169.6 ($\underline{C}OCH_3$), 170.0 ($\underline{C}OCH_3$), 170.2 ($\underline{C}OCH_3$), 170.3 ($\underline{C}OCH_3$), 178.5 (C-4), 187.7 (C-1). IR ($CHCl_3$): 3055, 3019, 2954, 1751, 1669, 1630, 1580, 1559, 1458, 1441, 1370, 1312, 1249, 1193, 1162, 1080, 1053 cm^{-1}. HRMS (ESI): *m/z* $[M - H]^-$ calcd. for $C_{25}H_{25}O_{13}S$ 565.1021, found 565.1021.

2-(2,3,4-Tri-O-acetyl-β-D-xylopyranosyl-1-thio)-8-hydroxy-3-methoxy-1,4-naphthoquinone (**15c**); yield 205 mg (82.6%), orange solid, R_f 0.50 (A), m.p. 133–135 °C. ^{1}H-NMR (500 MHz, $CDCl_3$), δ: 2.04 (s, 3H, $COC\underline{H}_3$), 2.08 (s, 3H, $COC\underline{H}_3$), 2.11 (s, 3H, $COC\underline{H}_3$), 3.41 (dd, 1H, *J* = 11.9, 8.1 Hz, H-5′a), 4.25 (dd, 1H, *J* = 11.9, 4.8 Hz, H-5′b), 4.29 (s, 3H, $-OC\underline{H}_3$), 4.95 (m, 1H, H-4′), 5.05 (t, 1H, *J* = 7.8 Hz, H-2′), 5.21 (t, 1H, *J* = 7.8 Hz, H-3′), 5.51 (d, 1H, *J* = 7.9 Hz, H-1′), 7.26 (dd, 1H, *J* = 8.0, 1.7 Hz, H-7), 7.57 (dd, 1H, *J* = 8.0, 7.5 Hz, H-6), 7.60 (dd, 1H, *J* = 7.5, 1.5 Hz, H-5), 12.05 (s, 1H, $C^8O\underline{H}$). ^{13}C-NMR(125 MHz, $CDCl_3$), δ: 20.7 ($3 \times CO\underline{C}H_3$), 62.1 ($-O\underline{C}H_3$), 64.9 (C-5′), 68.3 (C-4′), 70.6 (C-2′), 71.5 (C-3′), 82.7 (C-1′), 114.4 (C-9), 119.6 (C-5), 125.0 (C-7), 126.0 (C-2), 131.4 (C-10), 135.8 (C-6), 161.3 (C-8), 161.4 (C-3), 169.4 ($\underline{C}OCH_3$), 169.7 ($\underline{C}OCH_3$), 169.8 ($\underline{C}OCH_3$), 178.8 (C-4), 187.7 (C-1). IR ($CHCl_3$): 3056, 2953, 1752, 1671, 1630, 1580, 1558, 1458, 1371, 1313, 1249, 1240, 1208, 1163, 1076 cm^{-1}. HRMS (ESI): *m/z* $[M - H]^-$ calcd. for $C_{22}H_{21}O_{11}S$ 493.0810, found 493.0805.

2-(2,3,4-Tri-O-acetyl-α-L-arabinopyranosyl-1-thio)-8-hydroxy-3-methoxy-1,4-naphthoquinone (**15d**); yield 197 mg (79.4%), orange solid, R_f 0.50 (A), m.p. 84–86 °C. ^{1}H-NMR (500 MHz, $CDCl_3$), δ: 2.10 (s, 3H, $COC\underline{H}_3$), 2.12 (s, 6H, $2 \times COC\underline{H}_3$), 3.65 (dd, 1H, *J* = 12.5, 2.5 Hz, H-5′a), 4.14 (dd, 1H, *J* = 12.5, 4.9 Hz, H-5′b), 4.30 (s, 3H, $-OC\underline{H}_3$), 5.17 (dd, 1H, *J* = 7.9, 3.4 Hz, H-3′), 5.28 (m, 1H, H-4′), 5.32 (t, 1H, *J* = 7.5 Hz, H-2′), 5.48 (d, 1H, *J* = 7.5 Hz, H-1′), 7.26 (dd, 1H, *J* = 7.8, 1.6 Hz, H-7), 7.57 (dd, 1H, *J* = 7.8, 7.5 Hz, H-6), 7.60 (dd, 1H, *J* = 7.5, 1.6 Hz, H-5), 12.06 (s, 1H, $C^8O\underline{H}$). ^{13}C-NMR(125 MHz, $CDCl_3$), δ: 20.7 ($CO\underline{C}H_3$), 20.8 ($CO\underline{C}H_3$), 20.9 ($CO\underline{C}H_3$), 62.1 ($-O\underline{C}H_3$), 64.7 (C-5′), 67.2 (C-4′), 69.4 (C-2′), 70.0 (C-3′), 82.9 (C-1′), 114.4 (C-9), 119.6 (C-5), 125.0 (C-7), 126.5 (C-2), 131.3 (C-10), 135.8 (C-6), 161.2 (C-3), 161.3 (C-8), 169.4 ($\underline{C}OCH_3$), 169.8 ($\underline{C}OCH_3$), 170.1 ($\underline{C}OCH_3$), 178.8 (C-4), 187.7 (C-1). IR ($CHCl_3$): 3054, 3006, 1748, 1671, 1629, 1580, 1558, 1458, 1372, 1313, 1250, 1162, 1106, 1078, 1061 cm^{-1}. HRMS (ESI): *m/z* $[M - H]^-$ calcd. for $C_{22}H_{21}O_{11}S$ 493.0810, found 493.0808.

2-(2,3,4,6-Tetra-O-acetyl-β-D-glucopyranosyl-1-thio)-5-hydroxy-3-methoxy-1,4-naphthoquinone (**16a**); yield 248 mg (87.3%), red solid, R_f 0.46 (A), m.p. 159–161 °C. ^{1}H-NMR (500 MHz, $CDCl_3$), δ: 1.93 (s, 3H, $COC\underline{H}_3$), 2.01 (s, 3H, $COC\underline{H}_3$), 2.02 (s, 3H, $COC\underline{H}_3$), 2.08 (s, 3H, $COC\underline{H}_3$), 3.73 (m, 1H, H-5′), 4.06 (dd, 1H, *J* = 12.4, 2.3 Hz, H-6′a), 4.16 (dd, 1H, *J* = 12.4, 5.2 Hz, H-6′b), 4.22 (s, 3H, -$OC\underline{H}_3$), 5.09 (dd, 1H, *J* = 10.1, 4.2 Hz, H-4′), 5.11 (dd, 1H, *J* = 10.1, 4.1 Hz, H-2′), 5.27 (t, 1H, *J* = 9.3 Hz, H-3′), 5.67 (d, 1H, *J* = 10.1 Hz, H-1′), 7.24 (dd, 1H, *J* = 8.1, 1.4 Hz, H-6), 7.59 (dd, 1H, *J* = 8.1, 7.5 Hz, H-7), 7.62 (dd, 1H, *J* = 7.5, 1.4 Hz, H-8), 11.72 (s, 1H, $C^5O\underline{H}$). ^{13}C-NMR(125 MHz, $CDCl_3$), δ: 20.5 (2 × $CO\underline{C}H_3$), 20.6 (2 × $CO\underline{C}H_3$), 62.0 (-$O\underline{C}H_3$, C-6′), 68.3 (C-4′), 71.2 (C-2′), 74.0 (C-3′), 75.9 (C-5′), 81.1 (C-1′), 114.1 (C-10), 119.7 (C-8), 124.4 (C-6), 130.1 (C-2), 132.3 (C-9), 136.5 (C-7), 158.6 (C-3), 161.8 (C-5), 169.3 ($\underline{C}OCH_3$), 169.4 ($\underline{C}OCH_3$), 170.1 ($\underline{C}OCH_3$), 170.5 ($\underline{C}OCH_3$), 181.3 (C-1), 183.5 (C-4). IR ($CHCl_3$): 3050, 3004, 2950, 1756, 1661, 1636, 1579, 1560, 1458, 1369, 1240, 1228, 1212, 1192, 1046 cm^{-1}. HRMS (ESI): *m/z* [M − H]$^-$ calcd. for $C_{25}H_{25}O_{13}S$ 565.1021, found 565.1017.

2-(2,3,4,6-Tetra-O-acetyl-β-D-galactopyranosyl-1-thio)-5-hydroxy-3-methoxy-1,4-naphthoquinone (**16b**); yield 233 mg (82.0%), dark orange solid, R_f 0.46 (A), m.p. 83–86 °C. ^{1}H-NMR (500 MHz, $CDCl_3$), δ: 1.91 (s, 3H, $COC\underline{H}_3$), 1.99 (s, 3H, $COC\underline{H}_3$), 2.09 (s, 3H, $COC\underline{H}_3$), 2.15 (s, 3H, $COC\underline{H}_3$), 3.93 (m, 1H, H-5′), 4.05 (m, 2H, H-6′a, H-6′b), 4.23 (s, 3H, -$OC\underline{H}_3$), 5.11 (dd, 1H, *J* = 9.9, 3.4 Hz, H-3′), 5.31 (t, 1H, *J* = 10.0 Hz, H-2′), 5.43 (m, 1H, H-4′), 5.64 (d, 1H, *J* = 10.2 Hz, H-1′), 7.24 (dd, 1H, *J* = 8.2, 1.3 Hz, H-6), 7.59 (dd, 1H, *J* = 8.2, 7.5 Hz, H-7), 7.63 (dd, 1H, *J* = 7.5, 1.3 Hz, H-8), 11.72 (s, 1H. $C^5O\underline{H}$). ^{13}C-NMR(125 MHz, $CDCl_3$), δ: 20.5 ($CO\underline{C}H_3$), 20.55 ($CO\underline{C}H_{3'}$), 20.6 ($CO\underline{C}H_3$), 20.7 ($CO\underline{C}H_3$), 61.4 (C-6′), 62.0 (-$C\underline{C}H_3$), 67.3 (C-4′), 68.3 (C-2′), 71.9 (C-3′), 74.6 (C-5′), 81.9 (C-1′), 114.1 (C-10), 119.7 (C-8), 124.3 (C-6), 130.2 (C-2), 132.3 (C-9), 136.5 (C-7), 158.6 (C-3), 161.8 (C-5), 169.6 ($\underline{C}OCH_3$), 170.0 ($\underline{C}OCH_3$), 170.2 ($\underline{C}OCH_3$), 170.3 ($\underline{C}OCH_3$), 181.2 (C-1), 183.5 (C-4). IR ($CHCl_3$): 3056, 3006, 1751, 1661, 1636, 1579, 1563, 1458, 1371, 1255, 1191, 1171, 1154, 1085, 1049 cm^{-1}. HRMS (ESI): *m/z* [M − H]$^-$ calcd. for $C_{25}H_{25}O_{13}S$ 565.1021, found 565.1019.

2-(2,3,4-Tri-O-acetyl-β-D-xylopyranosyl-1-thio)-5-hydroxy-3-methoxy-1,4-naphthoquinone (**16c**); yield 181 mg (73.0%), orange solid, R_f 0.50 (A), m.p. 135–137 °C. ^{1}H-NMR (500 MHz, $CDCl_3$), δ: 2.04 (s, 3H, $COC\underline{H}_3$), 2.07 (s, 3H, $COC\underline{H}_3$), 2.10 (s, 3H, $COC\underline{H}_3$), 3.40 (dd, 1H, *J* = 11.8, 8.4 Hz, H-5′a), 4.20 (dd, 1H, *J* = 11.8, 5.0 Hz, H-5′b), 4.22 (s, 3H, -$OC\underline{H}_3$), 4.96 (m, 1H, H-4′), 5.05 (t, 1H, *J* = 8.2 Hz, H-2′), 5.22 (t, 1H, *J* = 8.1 Hz, H-3′), 5.67 (d, 1H, *J* = 8.2 Hz, H-1′), 7.23 (dd, 1H, *J* = 8.1, 1.3 Hz, H-6), 7.59 (dd, 1H, *J* = 8.1, 7.5 Hz, H-7), 7.63 (dd, 1H, *J* = 7.5, 1.3 Hz, H-8), 11.71 (s, 1H, $C^5O\underline{H}$). ^{13}C-NMR(125 MHz, $CDCl_3$), δ: 20.7 (3 × $CO\underline{C}H_3$), 62.1 (-$O\underline{C}H_3$), 65.1 (C-5′), 68.5 (C-4′), 70.6 (C-2′), 71.8 (C-3′), 82.0 (C-1′), 114.2 (C-10), 119.8 (C-8), 124.3 (C-6), 130.0 (C-2), 132.3 (C-9), 136.6 (C-7), 159.4 (C-3), 161.8 (C-5), 169.4 ($\underline{C}OCH_3$), 169.7 ($\underline{C}OCH_3$), 169.8 ($\underline{C}OCH_3$), 181.3 (C-1), 183.7 (C-4). IR ($CHCl_3$): 3053, 3007, 2949, 1754, 1661, 1637, 1579, 1563, 1458, 1371, 1248, 1192, 1171, 1070, 1048 cm^{-1}. HRMS (ESI): *m/z* [M − H]$^-$ calcd. for $C_{22}H_{21}O_{11}S$ 493.0810, found 493.0811.

2-(2,3,4-Tri-O-acetyl-α-L-arabinopyranosyl-1-thio)-5-hydroxy-3-methoxy-1,4-naphthoquinone (**16d**); yield 189 mg (76.2%), dark orange solid, R_f 0.50 (A), m.p. 97–99 °C. ^{1}H-NMR(700 MHz, $CDCl_3$), δ: 2.08 (s, 3H, $COC\underline{H}_3$), 2.11 (s, 3H, $COC\underline{H}_3$), 2.12 (s, 3H, $COC\underline{H}_3$), 3.66 (dd, 1H, *J* = 12.6, 2.3 Hz, H-5′a), 4.09 (dd, 1H, *J* = 12.6, 4.4 Hz, H-5′b), 4.23 (s, 3H, -$OC\underline{H}_3$), 5.16 (dd, 1H, *J* = 8.2, 3.4 Hz, H-3′), 5.29 (m, 1H, H-4′), 5.31 (t, 1H, *J* = 7.7 Hz, H-2′), 5.66 (d, 1H, *J* = 7.7 Hz, H-1′), 7.23 (dd, 1H, *J* = 8.3, 1.0 Hz, H-6), 7.59 (dd, 1H, *J* = 8.3, 7.5 Hz, H-7), 7.63 (dd, 1H, *J* = 7.5, 1.0 Hz, H-8), 11.72 (s, 1H, $C^5O\underline{H}$). ^{13}C-NMR (176 MHz, $CDCl_3$), δ: 20.7 ($CO\underline{C}H_3$), 20.8 ($CO\underline{C}H_3$), 20.9 ($CO\underline{C}H_3$), 62.0 (-$O\underline{C}H_3$), 65.3 (C-5′), 67.4 (C-4′), 69.2 (C-2′), 70.2 (C-3′), 82.2 (C-1′), 114.2 (C-10), 119.8 (C-8), 124.3 (C-6), 130.5 (C-2), 132.3 (C-9), 136.6 (C-7), 159.3 (C-3), 161.8 (C-5), 169.5 ($\underline{C}OCH_3$), 169.8 ($\underline{C}OCH_3$), 170.2 ($\underline{C}OCH_3$), 181.4 (C-1), 183.8 (C-4). IR ($CHCl_3$): 3056, 2947, 1748, 1661, 1636, 1579, 1562, 1457, 1371, 1240, 1193, 1170, 1105, 1087, 1066, 1024 cm^{-1}. HRMS (ESI): *m/z* [M − H]$^-$ calcd. for $C_{22}H_{21}O_{11}S$ 493.0810, found 493.0807.

3.2.2. General Procedure for Cyclization of Acetylated Thioglycosides to Tetracyclic Conjugates **17a–d–20a–d**

Acetylated thioglycoside, **13a–d–16a–d** (0.25 mM), was suspended in 15 mL of dried MeOH and 0.9 mL MeONa solution (0.5 N) was added. The mixture was stirred at room temperature for 1 h until TLC analysis indicated complete consumption of initial thioglycoside. During the reaction, the formation of a new polar compound precipitate was also observed. The reaction mixture was acidified with 0.25 mL HCl solution (2 N) and the precipitate was filtered off, washed with water, cold MeOH, and gave high purity quinone-tioglycosidic conjugates **17a–d–20a–d**.

(2R,3S,4S,4aR,12aS)-3,4-Dihydroxy-2-hydroxymethyl-7,10-dimethoxy-3,4,4a,12a-tetrahydro-2H-naphtho[2,3-b]pyrano[2,3-e][2,5]oxathiine-6,11-dione (**17a**); yield 88 mg (86.1%), red solid, R_f 0.35 (C), m.p. 332–335 °C[27]. ^{1}H-NMR(700 MHz, DMSO-d_6), δ: 3.30 (m, 1H, H-3), 3.47 (m, 3H, H-2, H-4a, H-13a), 3.57 (m, 1H, H-4), 3.74 (m,1H, H-13b), 3.84 (s, 3H, -OC$\underline{H_3}$), 3.85 (s, 3H, -OC$\underline{H_3}$), 4.72 (br s, 1H, C^{13}O$\underline{H}$), 4.92 (d, 1H, *J* = 8.3 Hz, H-12a), 5.37 (br s, 1H, C^{3}O$\underline{H}$), 5.58 (br s, 1H, C^{4}O$\underline{H}$), 7.52 (d, 1H, *J* = 9.6 Hz, Ar-H), 7.54 (d, 1H, *J* = 9.6 Hz, Ar-H). ^{13}C-NMR (176 MHz, DMSO-d_6), δ: 56.7 (-O$\underline{C}$H$_3$), 56.8 (-O$\underline{C}$H$_3$), 60.8 (C-13), 70.5 (C-3), 73.7 (C-12a), 73.9 (C-4), 79.2 (C-4a), 82.1 (C-2), 118.6, 118.9, 121.7, 122.0, 122.4, 149.7, 153.3, 153.8, 174.7, 179.9. IR (KBr): 3444, 1638, 1614, 1561, 1476, 1405, 1266, 1181, 1075, 935 cm^{-1}. HRMS (ESI): *m/z* [M + Na]$^+$ calcd. for $C_{18}H_{18}NaO_9S$ 433.0564, found 433.0562.

(2R,3R,4S,4aR,12aS)-3,4-Dihydroxy-2-hydroxymethyl-7,10-dimethoxy-3,4,4a,12a-tetrahydro-2H-naphtho[2,3-b]pyrano[2,3-e][2,5]oxathiine-6,11-dione (**17b**); yield 95 mg (93%), orange solid, R_f 0.30 (C), m.p. > 350 °C. ^{1}H-NMR(500 MHz, DMSO-d_6), δ: 3.55 (m, 2H, H-13), 3.72 (t, 1H, *J* = 6.0 Hz, H-2), 3.77 (m, 2H, H-4, H-4a), 3.83 (m, 1H, H-3), 3.84 (s, 3H, -OC$\underline{H_3}$), 3.85 (s, 3H, -OC$\underline{H_3}$), 4.74 (t, 1H, *J* = 5.6 Hz, C^{13}O$\underline{H}$), 4.88 (d, 1H, *J* = 7.6 Hz, H-12a), 4.89 (d, 1H, *J* = 4.6 Hz, C^{3}O$\underline{H}$), 5.31 (d, 1H, *J* = 6.3 Hz, C^{4}O$\underline{H}$), 7.52 (d, 1H, *J* = 9.7 Hz, Ar-H), 7.55 (d, 1H, *J* = 9.7 Hz, Ar-H). ^{13}C-NMR (125 MHz, DMSO-d_6), δ: 56.7 (-O$\underline{C}$H$_3$), 56.8 (-O$\underline{C}$H$_3$), 60.5 (C-13), 69.4 (C-3), 70.4 (C-4), 74.4 (C-12a), 77.7 (C-4a), 80.6 (C-2), 118.6, 118.9, 121.7, 122.0, 122.4, 150.3, 153.3, 153.8, 174.4, 179.8. IR (KBr): 3487, 3289, 3016, 2968, 2935, 2838, 1649, 1611, 1581, 1562, 1476, 1434, 1407, 1384, 1349, 1325, 1267, 1213, 1196, 1180, 1110, 1082, 1056, 1041, 1019, 1008, 936, 904, 873, 824, 803, 755 cm^{-1}. HRMS (ESI): *m/z* [M + Na]$^+$ calcd. for $C_{18}H_{18}NaO_9S$ 433.0564, found 433.0565.

(3R,4S,4aR,12aS)-3,4-Dihydroxy-7,10-dimethoxy-3,4,4a,12a-tetrahydro-2H-naphtho[2,3-b]pyrano[2,3-e][2,5] oxathiine-6,11-dione (**17c**); yield 80 mg (83.9%), orange solid, R_f 0.47 (C), m.p. 329–332 °C. ^{1}H-NMR(700 MHz, DMSO-d_6), δ: 3.37 (m, 1H, H-2a), 3.48 (m, 1H, H-4a), 3.52 (m, 2H, H-3, H-4), 3.84 (s, 3H, -OC$\underline{H_3}$), 3.85 (s, 3H, -OC$\underline{H_3}$), 3.91 (dd, 1H, *J* = 11.1, 4.8 Hz, H-2b), 4.85 (d, 1H, *J* = 8.1 Hz, H-12a), 5.38 (d, 1H, *J* = 4.8 Hz, C^{3}O$\underline{H}$), 5.61 (d, 1H, *J* = 5.3 Hz, C^{4}O$\underline{H}$), 7.52 (d, 1H, *J* = 9.6 Hz, Ar-H), 7.54 (d, 1H, *J* = 9.6 Hz, Ar-H). ^{13}C-NMR (176 MHz, DMSO-d_6), δ: 56.7 (-O$\underline{C}$H$_3$), 56.8 (-O$\underline{C}$H$_3$), 70.0 (C-3), 70.3 (C-2), 74.1 (C-4), 74.6 (C-12a), 79.2 (C-4a), 118.6, 118.9, 121.7, 122.1, 122.3, 149.7, 153.4, 153.9, 174.6, 179.8. IR (KBr): 3502, 3470, 3296, 3013, 2981, 2939, 2877, 2836, 1661, 1636, 1611, 1581, 1562, 1478, 1459, 1431, 1408, 1359, 1281, 1267, 1253, 1223, 1196, 1164, 1125, 1095, 1060, 1042, 1023, 975, 935, 887, 818, 797, 755, 718 cm^{-1}. HRMS (ESI): *m/z* [M + Na]$^+$ calcd. for $C_{17}H_{16}NaO_8S$ 403.0458, found 403.0459.

(3S,4S,4aR,12aS)-3,4-Dihydroxy-7,10-dimethoxy-3,4,4a,12a-tetrahydro-2H-naphtho[2,3-b]pyrano[2,3-e][2,5] oxathiine-6,11-dione (**17d**); yield 88 mg (92.3%), yellow solid, R_f 0.43 (C), m.p. 317–320 °C. ^{1}H-NMR(700 MHz, DMSO-d_6), δ: 3.77 (m, 3H, H-2a, H-4, H-4a), 3.84 (m, 1H, H-3), 3.85 (s, 3H, -OC$\underline{H_3}$), 3.86 (s, 3H, -OC$\underline{H_3}$), 3.88 (dd, 1H, *J* = 12.2, 1.7 Hz, H-2b), 4.82 (d, 1H, *J* = 7.7 Hz, H-12a), 5.02 (d, 1H, *J* = 4.1 Hz, C^{3}O$\underline{H}$), 5.31 (d, 1H, *J* = 6.6 Hz, C^{4}O$\underline{H}$), 7.52 (d, 1H, *J* = 9.6 Hz, Ar-H), 7.55 (d, 1H, *J* = 9.6 Hz, Ar-H). ^{13}C-NMR (176 MHz, DMSO-d_6), δ: 56.7 (-O$\underline{C}$H$_3$), 56.8 (-O$\underline{C}$H$_3$), 69.4 (C-3), 69.8 (C-4), 71.4 (C-2), 74.8 (C-12a), 77.6 (C-4a), 118.6, 118.9, 121.7, 122.0, 122.4, 150.2, 153.4, 153.8, 174.6, 179.8. IR (KBr): 3496, 3449, 3094, 3012, 2979, 2932, 2864, 2837, 1659, 1642, 1613, 1479, 1459, 1433, 1409, 1352, 1338, 1325, 1280, 1259, 1206, 1188, 1166, 1123, 1097, 1080, 1069, 1048, 1022, 1007, 955, 934, 910, 866, 833, 815, 800, 747 cm^{-1}. HRMS (ESI, *m/z*): [M + Na]$^+$ calcd. for $C_{17}H_{16}NaO_8S$ 403.0458, found 403.0453.

(2R,3S,4S,4aR,12aS)-8,9-Dichloro-3,4-dihydroxy-2-hydroxymethyl-7,10-dimethoxy-3,4,4a,12a-tetrahydro-2H-naphtho[2,3-b]pyrano[2,3-e][2,5]oxathiine-6,11-dione (**18a**); yield 100 mg (83.4%), orange solid, R_f 0.51 (C), m.p. 222–224 °C[27]. ^{1}H-NMR(500 MHz, DMSO-d_6), δ: 3.32 (m, 1H, H-3), 3.50 (m, 2H, H-2, H-13a), 3.54 (m, 1H, H-4a), 3.60 (m, 1H, H-4), 3.75 (m, 1H, H-13b), 3.82 (s, 3H, -OC$\underline{H}_3$), 3.83 (s, 3H, -OC$\underline{H}_3$), 4.74 (t, 1H, J = 5.3 Hz, C^{13}O$\underline{H}$), 4.96 (d, 1H, J = 8.1 Hz, H-12a), 5.40 (d, 1H, J = 5.7 Hz, C^{3}O$\underline{H}$), 5.67 (d, 1H, J = 5.8 Hz, C^{4}O$\underline{H}$). ^{13}C-NMR (125 MHz, DMSO-d_6), δ: 60.7 (C-13), 61.5 (2 × -O$\underline{C}$H$_3$), 70.4 (C-3), 73.6 (C-12a), 73.9 (C-4), 79.3 (C-4a), 82.2 (C-2), 123.1, 123.4, 123.9, 135.1, 135.3, 149.9, 152.3, 152.9, 173.4, 178.9. IR (KBr): 3432, 2941, 1653, 1603, 1625, 1458, 1381, 1333, 1275, 1209, 1131, 1025, 951 cm^{-1}. HRMS (ESI): m/z [M + Na]$^+$ calcd. for $C_{18}H_{16}Cl_2NaO_9S$ 500.9784, found 500.9784.

(2R,3R,4S,4aR,12aS)-8,9-Dichloro-3,4-dihydroxy-2-(hydroxymethyl)-7,10-dimethoxy-3,4,4a,12a-tetrahydro-2H-naphtho[2,3-b]pyrano[2,3-e][2,5]oxathiine-6,11-dione (**18b**); yield 98 mg (81.7%), orange solid, R_f 0.47 (C), m.p. 276–279 °C. 1H-NMR(700 MHz, DMSO-*d6*), δ: 3.56 (m, 2H, H-13), 3.75 (t, 1H, J = 6.0 Hz, H-2), 3.79 (dd, 1H, J = 9.6, 3.3 Hz, H-4), 3.82 (s, 3H, -OC$\underline{H}_3$), 3.83 (s, 3H, -OC$\underline{H}_3$), 3.85 (m, 2H, H-3, H-4a), 4.80 (br s, 2H, 2 × O$\underline{H}$), 4.92 (d, 1H, J = 7.9 Hz, H-12a), 5.40 (br s, 1H, O$\underline{H}$). ^{13}C-NMR (176 MHz, DMSO-d_6), δ: 60.5 (C-13), 61.5 (2 × -O$\underline{C}$H$_3$), 69.5 (C-3), 70.4 (C-4), 74.3 (C-12a), 77.9 (C-4a), 80.8 (C-2), 123.2, 123.4, 123.9, 135.1, 135.2, 150.5, 152.3, 152.9, 173.4, 178.9. IR (KBr): 3413, 2940, 2855, 1652, 1603, 1542, 1524, 1457, 1380, 1336, 1274, 1236, 1207, 1121, 1090, 1024, 945, 876, 839, 805, 762 cm^{-1}. HRMS (ESI): m/z [M + Na]$^+$ calcd. for $C_{18}H_{16}Cl_2NaO_9S$ 500.9784, found 500.9783.

(3R,4S,4aR,12aS)-8,9-Dichloro-3,4-dihydroxy-7,10-dimethoxy-3,4,4a,12a-tetrahydro-2H-naphtho[2,3-b]pyrano [2,3-e][2,5]oxathiine-6,11-dione (**18c**); yield 102 mg (90.7%), orange solid, R_f 0.59 (C), m.p. 277–279 °C. 1H-NMR(700 MHz, DMSO-*d6*), δ: 3.40 (m, 1H, H-2a), 3.54 (m, 3H, H-4a, H-4, H-3), 3.81 (s, 3H, -OC$\underline{H}_3$), 3.83 (s, 3H, -OC$\underline{H}_3$), 3.93 (dd, 1H, J = 11.2, 4.6 Hz, H-2b), 4.90 (d, 1H, J = 7.5 Hz, H-12a), 5.42 (d, 1H, J = 4.4 Hz, C^{3}O$\underline{H}$), 5.72 (d, 1H, J = 4.9 Hz, C^{4}O$\underline{H}$). ^{13}C-NMR (176 MHz, DMSO-d_6), δ: 61.5 (2 × -O$\underline{C}$H$_3$), 69.9 (C-3), 70.4 (C-2), 74.0 (C-4), 74.5 (C-12a), 79.4 (C-4a), 123.0, 123.4, 123.8, 135.1, 135.3, 149.9, 152.3, 152.9, 173.4, 178.9. IR (KBr): 3482, 3385, 3003, 2938, 2880, 2855, 1651, 1604, 1544, 1524, 1458, 1382, 1330, 1307, 1272, 1237, 1225, 1208, 1124, 1090, 1074, 1061, 1054, 1024, 975, 948, 898, 869, 838, 816, 802, 763, 741 cm^{-1}. HRMS (ESI): m/z [M + Na]$^+$ calcd. for $C_{17}H_{14}Cl_2NaO_8S$ 470.9679, found 470.9673.

(3S,4S,4aR,12aS)-8,9-Dichloro-3,4-dihydroxy-7,10-dimethoxy-3,4,4a,12a-tetrahydro-2H-naphtho[2,3-b]pyrano [2,3-e][2,5]oxathiine-6,11-dione (**18d**); yield 97 mg (86.3%), orange solid, R_f 0.56 (C), m.p. 286–289 °C. ^{1}H-NMR(500 MHz, DMSO-d_6), δ: 3.80 (m, 2H, H-2a, H-4), 3.82 (s, 3H, -OC$\underline{H}_3$), 3.83 (s, 3H, -OC$\underline{H}_3$), 3.85 (m, 2H, H-3, H-4a), 3.90 (dd, 1H, J = 12.2, 1.7 Hz, H-2b), 4.86 (d, 1H, J = 8.1 Hz, H-12a), 5.05 (br s, 1H, C^{3}O$\underline{H}$), 5.40 (br s, 1H, C^{4}O$\underline{H}$). ^{13}C-NMR (125 MHz, DMSO-d_6), δ: 61.5 (2 × -O$\underline{C}$H$_3$), 69.4 (C-3), 69.8 (C-4), 71.5 (C-2), 74.7 (C-12a), 77.8 (C-4a), 123.2, 123.4, 123.9, 135.1, 135.2, 150.4, 152.3, 152.9, 173.4, 178.9. IR (KBr): 3391, 2980, 2940, 2856, 1654, 1603, 1543, 1525, 1458, 1381, 1334, 1305, 1272, 1237, 1211, 1128, 1092, 1048, 1026, 967, 944, 913, 876, 839, 805, 757 cm^{-1}. HRMS (ESI): m/z [M + Na]$^+$ calcd. for $C_{17}H_{14}Cl_2NaO_8S$ 470.9679, found 470.9675.

(2R,3S,4S,4aR,12aS)-3,4,10-Trihydroxy-2-hydroxymethyl-3,4,4a,12a-tetrahydro-2H-naphtho[2,3-b]pyrano[2,3-e][2,5]oxathiine-6,11-dione (**19a**); yield 86 mg (93.7%), orange solid, R_f 0.56 (C), decomposition > 342 °C. ^{1}H-NMR(700 MHz, DMSO-d_6), δ: 3.32 (m, 1H, H-3), 3.50 (m, 2H, H-2, H-13a), 3.60 (m, 2H, H-4, H-4a), 3.75 (m, 1H, H-13b), 4.74 (t, 1H, J = 5.5 Hz, C^{13}O$\underline{H}$), 5.01 (d, 1H, J = 7.7 Hz, H-12a), 5.41 (d, 1H, J = 5.8 Hz, C^{3}O$\underline{H}$), 5.69 (d, 1H, J = 5.7 Hz, C^{4}O$\underline{H}$), 7.30 (dd, 1H, J = 8.4, 0.7 Hz, H-9), 7.54 (dd, 1H, J = 7.4, 0.7 Hz, H-7), 7.70 (dd, 1H, J = 8.4, 7.4 Hz, H-8), 11.52 (s, 1H, C^{10}O$\underline{H}$). ^{13}C-NMR (176 MHz, DMSO-d_6), δ: 60.7 (C-13), 70.4 (C-3), 73.5 (C-12a), 73.9 (C-4), 79.4 (C-4a), 82.3 (C-2), 113.8 (C-10a), 119.2 (C-7), 122.9 (C-11a), 124.2 (C-9), 130.8 (C-6a), 136.5 (C-8), 151.0 (C-5a), 159.8 (C-10), 175.1 (C-6), 186.1 (C-11). IR (KBr): 3464, 3351, 3233, 2955, 2881, 1646, 1616, 1580, 1517, 1477, 1455, 1416, 1383, 1356, 1324, 1297, 1280, 1252, 1229, 1219, 1194, 1165, 1148, 1133, 1092, 1055, 1037, 1000, 973, 889, 863, 836, 828, 816, 755, 728 cm^{-1}. HRMS (ESI): m/z [M + Na]$^+$ calcd. for $C_{16}H_{14}NaO_8S$ 389.0302, found 389.0300.

(2R,3R,4S,4aR,12aS)-3,4,10-Trihydroxy-2-hydroxymethyl-3,4,4a,12a-tetrahydro-2H-naphtho[2,3-b]pyrano-[2,3-e][2,5]oxathiine-6,11-dione (**19b**); yield 75 mg (81.7%), dark orange solid, R_f 0.52 (C), decompose above > 346 °C. ^{1}H-NMR(500 MHz, DMSO-d_6), δ: 3.57 (m, 2H, H-13), 3.76 (t, 1H, *J* = 6.0 Hz, H-2), 3.80 (m, 1H, H-4), 3.85 (m, 1H, H-3), 3.90 (t, 1H, *J* = 9.5, 8.0 Hz, H-4a), 4.77 (t, 1H, *J* = 5.6 Hz, C^{13}O$\underline{H}$), 4.96 (d, 1H, *J* = 4.8 Hz, C^{3}O$\underline{H}$), 4.98 (d, 1H, *J* = 8.0 Hz, H-12a), 5.43 (d, 1H, *J* = 6.8 Hz, C^{4}O$\underline{H}$), 7.30 (dd, 1H, *J* = 8.4, 0.9 Hz, H-9), 7.54 (dd, 1H, J = 7.4, 0.9 Hz, H-7), 7.70 (dd, 1H, *J* = 8.4, 7.4 Hz, H-8), 11.53 (s, 1H, C^{10}O$\underline{H}$). ^{13}C-NMR (125 MHz, DMSO-d_6), δ: 60.5 (C-13), 69.5 (C-3), 70.4 (C-4), 74.1 (C-12a), 78.1 (C-4a), 80.8 (C-2), 113.8 (C-10a), 119.2 (C-7), 122.9 (C-11a), 124.2 (C-9), 130.8 (C-6a), 136.5 (C-8), 151.5 (C-5a), 159.8 (C-10), 175.1 (C-6), 186.1 (C-11). IR (KBr): 3456, 3338, 2984, 2938, 2888, 1644, 1616, 1580, 1476, 1456, 1431, 1401, 1384, 1327, 1296, 1279, 1250, 1208, 1165, 1136, 1105, 1085, 1056, 1024, 979, 903, 879, 860, 833, 813, 754, 734, 690 cm^{-1}. HRMS (ESI, *m*/*z*): [M + Na]$^+$ calcd. for $C_{16}H_{14}NaO_8S$ 389.0302, found 389.0295.

(3R,4S,4aR,12aS)-3,4,10-Trihydroxy-3,4,4a,12a-tetrahydro-2H-naphtho[2,3-b]pyrano[2,3-e][2,5]oxathiine-6,11-dione (**19c**); yield 76 mg (90.1%), orange solid, R_f 0.67 (C), m.p. 348–350 °C with decomposition. ^{1}H-NMR(700 MHz, DMSO-d_6), δ: 3.41 (m, 1H, H-2), 3.55 (m, 2H, H-3, H-4), 3.60 (m, 1H, H-4a), 3.94 (dd, 1H, *J* = 11.1, 4.7 Hz, H-2′), 4.95 (d, 1H, *J* = 8.0 Hz, C-12a), 5.42 (d, 1H, *J* = 4.7 Hz, C^{3}O$\underline{H}$), 5.73 (d, 1H, *J* = 5.3 Hz, C^{4}O$\underline{H}$), 7.30 (dd, 1H, *J* = 8.4, 0.9 Hz, H-9), 7.54 (dd, 1H, *J* = 7.4, 0.9 Hz, H-7), 7.70 (dd, 1H, *J* = 8.4, 7.4 Hz, H-8), 11.51 (s, 1H, C^{10}O$\underline{H}$). ^{13}C-NMR (176 MHz, DMSO-d_6), δ: 70.0 (C-3), 70.4 (C-2), 74.0 (C-4), 74.3 (C-12a), 79.4 (C-4a), 113.8 (C-10a), 119.2 (C-7), 122.8 (C-11a), 124.2 (C-9), 130.8 (C-6a), 136.6 (C-8), 151.0 (C-5a), 159.8 (C-10), 175.1 (C-6), 186.1 (C-11). IR (KBr): 3381, 3300, 2940, 2899, 2867, 1652, 1625, 1621, 1581, 1516, 1475, 1462, 1454, 1417, 1378, 1323, 1306, 1295, 1239, 1222, 1200, 1168, 1146, 1136, 1074, 1056, 1044, 1035, 1002, 976, 962, 900, 868, 833, 818, 789, 756 cm^{-1}. HRMS (ESI): *m*/*z* [M + Na]$^+$ calcd. for $C_{15}H_{12}NaO_7S$ 359.0196, found 359.0197.

(3S,4S,4aR,12aS)-3,4,10-Trihydroxy-3,4,4a,12a-tetrahydro-2H-naphtho[2,3-b]pyrano[2,3-e][2,5]oxathiine-6,11-dione (**19d**); yield 78 mg (92.5%), orange solid, R_f 0.62 (C), m.p. 319–321 °C. ^{1}H-NMR(500 MHz, DMSO-d_6), δ: 3.79 (m, 2H, H-4, H-2a), 3.86 (m, 1H, H-3), 3.90 (m, 2H, H-4a, H-2b), 4.91 (d, 1H, *J* = 8.0 Hz, H-12a), 5.10 (d, 1H, *J* = 4.0 Hz, C^{3}O$\underline{H}$), 5.44 (d, 1H, *J* = 7.0 Hz, C^{4}O$\underline{H}$), 7.30 (dd, 1H, *J* = 8.4, 1.0 Hz, H-9), 7.53 (dd, 1H, *J* = 7.5, 1.0 Hz, H-7), 7.70 (dd, 1H, *J* = 8.4, 7.5 Hz, H-8), 11.54 (s, 1H, C^{10}O$\underline{H}$). ^{13}C-NMR (125 MHz, DMSO-d_6), δ: 69.4 (C-3), 69.7 (C-4), 71.5 (C-2), 74.5 (C-12a), 77.9 (C-4a), 113.8 (C-10a), 119.2 (C-7), 122.9 (C-11a), 124.2 (C-9), 130.7 (C-6a), 136.5 (C-8), 151.4 (C-5a), 159.8 (C-10), 175.1 (C-6), 186.1 (C-11). IR (KBr): 3511, 3196, 2987, 2925, 2892, 1654, 1624, 1578, 1475, 1463, 1444, 1378, 1356, 1333, 1308, 1279, 1237, 1177, 1164, 1139, 1109, 1084, 1071, 1022, 1004, 982, 938, 911, 837, 861, 835, 815, 754 cm^{-1}. HRMS (ESI, *m*/*z*): [M + Na]$^+$ calcd. for $C_{15}H_{12}NaO_7S$ 359.0196, found 359.0192.

(2R,3S,4S,4aR,12aS)-3,4,7-Trihydroxy-2-(hydroxymethyl)-3,4,4a,12a-tetrahydro-2H-naphtho[2,3-b]pyrano-[2,3-e][2,5]oxathiine-6,11-dione (**20a**); yield 80 mg (87.1%), orange solid, R_f 0.56 (C), m.p. 274–276 °C. ^{1}H-NMR(700 MHz, DMSO-d_6), δ: 3.33 (m, 1H, H-3), 3.50 (m, 2H, H-2, H-13a), 3.60 (m, 2H, H-4, H-4a), 3.75 (m, 1H, H-13b), 4.73 (t, 1H, *J* = 5.6 Hz, C^{13}O$\underline{H}$), 5.01 (d, 1H, *J* = 7.7 Hz, H-12a), 5.41 (d, 1H, *J* = 5.7 Hz, C^{3}O$\underline{H}$), 5.68 (d, 1H, *J* = 5.7 Hz, C^{4}O$\underline{H}$), 7.32 (dd, 1H, *J* = 8.5, 0.9 Hz, H-8), 7.51 (dd, 1H, *J* = 7.4, 0.9 Hz, H-10), 7.70 (dd, 1H, *J* = 8.5, 7.4 Hz, H-9), 11.72 (s, 1H, C^{7}O$\underline{H}$). ^{13}C-NMR (176 MHz, DMSO-d_6), δ: 60.7 (C-13), 70.4 (C-3), 73.6 (C-12a), 73.9 (C-4), 79.2 (C-4a), 82.2 (C-2), 113.4 (C-6a), 118.7 (C-10), 124.1 (C-11a), 124.5 (C-8), 131.2 (C-10a), 136.5 (C-9), 150.0 (C-5a), 160.7 (C-7), 180.4 (C-11), 180.5 (C-6). IR (KBr): 3493, 3474, 3415, 3365, 3264, 2959, 2931, 2892, 1658, 1628, 1585, 1481, 1457, 1403, 1457, 1380, 1296, 1280, 1246, 1223, 1198, 1156, 1143, 1124, 1096, 1077, 1048, 1035, 1009, 952, 895, 869, 834, 816, 796, 747 cm^{-1}. HRMS (ESI): *m*/*z* [M + Na]$^+$ calcd. for $C_{16}H_{14}NaO_8S$ 389.0302, found 389.0297.

(2R,3R,4S,4aR,12aS)-3,4,7-Trihydroxy-2-hydroxymethyl-3,4,4a,12a-tetrahydro-2H-naphtho[2,3-b]pyrano[2,3-e][2,5]oxathiine-6,11-dione (**20b**); yield 75 mg (81.7%), orange solid, R_f 0.52 (C), m.p. 293–295 °C. ^{1}H-NMR(500 MHz, DMSO-d_6), δ: 3.56 (m, 2H, H-13), 3.76 (t, 1H, *J* = 6.0 Hz, H-2), 3.80 (dd, 1H, *J* = 9.5, 3.2 Hz, H-4), 3.85 (dd, 1H, *J* = 3.3, 1.2 Hz, H-3), 3.89 (d, 1H, *J* = 9.5, 7.9 Hz, H-4a), 4.83 (br s, 3H, C^{3}O$\underline{H}$,

$C^4O\underline{H}$, $C^{13}O\underline{H}$), 4.97 (d, 1H, *J* = 7.9 Hz, H-12a), 7.32 (dd, 1H, *J* = 8.5, 1.0 Hz, H-8), 7.51 (dd, 1H, *J* = 7.5, 1.0 Hz, H-10), 7.70 (dd, 1H, *J* = 8.5, 7.5 Hz, H-9), 11.72 (s, 1H, $C^7O\underline{H}$). ^{13}C-NMR (125 MHz, DMSO-d_6), δ: 60.5 (C-13), 69.5 (C-3), 70.4 (C-4), 74.2 (C-12a), 77.9 (C-4a), 80.8 (C-2), 113.4 (C-6a), 118.7 (C-10), 124.1 (C-11a), 124.5 (C-8), 131.3 (C-10a), 136.5 (C-9), 150.6 (C-5a), 160.7 (C-7), 180.4 (C-11), 180.5 (C-6). IR (KBr): 3412, 2941, 1627, 1581, 1475, 1454, 1385, 1297, 1282, 1241, 1208, 1192, 1155, 1129, 1074, 1049, 986, 954, 937, 870, 832, 813, 745, 698 cm^{-1}. HRMS (ESI): *m/z* [M + Na]$^+$ calcd. for $C_{16}H_{14}NaO_8S$ 389.0302, found 389.0296.

(3R,4S,4aR,12aS)-3,4,7-Trihydroxy-3,4,4a,12a-tetrahydro-2H-naphtho[2,3-b]pyrano[2,3-e][2,5]oxathiine-6,11-dione (**20c**); yield 70 mg (83.0%), orange solid, R_f 0.67 (C), m.p. 334–336 °C. ^{1}H-NMR(500 MHz, DMSO-d_6), δ: 3.41 (m, 1H, H-2a), 3.56 (m, 3H, H-3, H-4, H-4a), 3.94 (dd, 1H, *J* = 11.0, 4.7 Hz, H-2b), 4.95 (d, 1H, *J* = 7.5 Hz, H-12a), 5.42 (d, 1H, *J* = 4.8 Hz, $C^3O\underline{H}$), 5.73 (d, 1H, *J* = 5.1 Hz, $C^4O\underline{H}$), 7.32 (dd, 1H, *J* = 8.5, 1.0 Hz, H-8), 7.51 (dd, 1H, *J* = 7.5, 1.0 Hz, H-10), 7.70 (dd, 1H, *J* = 8.5, 7.5 Hz, H-9), 11.71 (s, 1H, $C^7O\underline{H}$). ^{13}C-NMR (125 MHz, DMSO-d_6), δ: 69.4 (C-3), 69.7 (C-4), 71.5 (C-2), 74.5 (C-12a), 77.9 (C-4a), 113.8 (C-10a), 119.2 (C-7), 122.9 (C-11a), 124.2 (C-9), 130.7 (C-6a), 136.5 (C-8), 151.4 (C-5a), 159.8 (C-10), 180.4 (C-11), 180.5 (C-6). IR (KBr): 3401, 2887, 1653, 1622, 1579, 1476, 1454, 1407, 1390, 1297, 1277, 1251, 1220, 1207, 1194, 1156, 1134, 1090, 1063, 1048, 1008, 976, 952, 902, 840, 831, 818, 745 cm^{-1}. HRMS (ESI): *m/z* [M + Na]$^+$ calcd. for $C_{15}H_{12}NaO_7S$ 359.0196, found 359.0194.

(3S,4S,4aR,12aS)-3,4,7-Trihydroxy-3,4,4a,12a-tetrahydro-2H-naphtho[2,3-b]pyrano[2,3-e][2,5]oxathiine-6,11-dione (**20d**); yield 82 mg (97.3%), dark orange solid, R_f 0.62 (C), m.p. 284–286 °C. ^{1}H-NMR(500 MHz, DMSO-d_6), δ: 3.79 (dd, 1H, *J* = 12.7, 3.4 Hz, H-2a), 3.80 (m, 1H, H-3), 3.86 (m, 1H, H-4), 3.89 (t, 1H, *J* = 8.0 Hz, H-4a), 3.91 (dd, 1H, *J* = 12.7, 1.8 Hz, H-2b), 4.90 (d, 1H, *J* = 8.0 Hz, H-12a), 5.08 (d, 1H, *J* = 4.0 Hz, $C^3O\underline{H}$), 5.41 (d, 1H, *J* = 6.9 Hz, $C^4O\underline{H}$), 7.32 (dd, 1H, *J* = 8.4, 1.0 Hz, H-8), 7.51 (dd, 1H, *J* = 7.5, 1.0 Hz, H-10), 7.69 (dd, 1H, *J* = 8.4, 7.5 Hz, H-9), 11.72 (s, 1H, $C^7O\underline{H}$). ^{13}C-NMR (125 MHz, DMSO-d_6), δ: 69.4 (C-3), 69.7 (C-4), 71.5 (C-2), 74.6 (C-12a), 77.7 (C-4a), 113.4 (C-6a), 118.7 (C-10), 124.1 (C-11a), 124.5 (C-8), 131.2 (C-10a), 136.4 (C-9), 150.5 (C-5a), 160.6 (C-7), 180.3 (C-11), 180.4 (C-6). IR (KBr): 3541, 3268, 3167, 2998, 2936, 1648, 1627, 1621, 1618, 1581, 1474, 1453, 1405, 1384, 1330, 1299, 1284, 1245, 1215, 1191, 1173, 1153, 1115, 1088, 1068, 1049, 1037, 1007, 950, 915, 881, 869, 841, 831, 812, 790, 744, 729 cm^{-1}. HRMS (ESI): *m/z* [M + Na]$^+$ calcd. for $C_{15}H_{12}NaO_7S$ 359.0196, found 359.0191.

3.3. Biology

3.3.1. Cell Culture

Human adenocarcinoma cell line HeLa, mouse neuroblastoma cell line Neuro-2a, and mouse epithelial cells Jb6 Cl 41-5a were obtained from ATCC (Manassas, VA, USA). Mouse ascites Ehrlich carcinoma was provided by the N.N. Blokhin NMRCO (Ministry of Health of the Russian Federation, Moscow, Russia).

HeLa and Neuro-2a cells were cultured in DMEM medium containing 10% fetal bovine serum (FBS) (Biolot, Russia) and 1% penicillin/streptomycin (Biolot, Russia). Jb6 Cl 41-5a cells were cultivated in MEM medium containing 5% fetal bovine serum (FBS) (Biolot, Russia) and 1% penicillin/streptomycin (Biolot, Russia). All cell lines were incubated at 37 °C in a humidified atmosphere with 5% (*v*/*v*) CO_2.

The cells of Ehrlich carcinoma were injected into the peritoneal cavity of 18–20 g albino CD-1 mice (male and female). Cells for experiments were collected 7 days after inoculation. The mice were killed by perivisceral dislocation and the ascitic fluid containing tumor cells was collected with a syringe. The cells were washed three times by centrifugation at 2000 rpm (450 g) for 5 min at 4 °C in phosphate-buffered saline (PBS; pH 7.4) followed by resuspension in RPMI-1640 culture medium with 10% FBS.

Erythrocytes were isolated from the blood of mice. Blood was taken from CD-1 mice (18–20 g). The mice were anesthetized with diethyl ether, their chests were rapidly opened, and blood was collected in cold (4 °C) 10 mM phosphate-buffered saline pH 7.4 (PBS) without an anticoagulant. Erythrocytes were washed by centrifugation with centrifuge LABOFUGE 400R (Heraeus, Germany) for 5 min 3 times (2000 rpm) in PBS using at least 10 vol. of washing solution. Then, the residue of erythrocytes were resuspended in ice cold phosphate-buffered saline (pH 7.4) to a final optical density of 1.0 at 700 nm and kept on ice.

All experiments were carried out in accordance with the EU Animals (Scientific Procedures) Act, 1986 and associated guidelines, EU Directive 2010/63/EU for animal experiments.

3.3.2. Cell Viability Assay

MTT Assay

Stocks of substances were prepared in DMSO at a concentration of 10 mM. All studied compounds were tested in concentrations from 25 μM using twofold dilution. All tested compounds were added to the wells of the plates in a volume of 20 μL diluted in PBS (final DMSO concentration < 1%). A known cytotoxic agent—triterpene glycoside cucumarioside A_2-2 from holothurian *Cucumaria japonica*—was used as positive control. Cells (3×10^4 cells/well) were incubated with different concentrations of 1,4-NQs in a CO_2 incubator for 24 h at 37 °C. After incubation, the medium with tested substances was replaced with 100 μL of pure medium. Then, 10 μL of MTT (Sigma-Aldrich, St. Louis, MO, USA) (thiazolyl blue tetrazolium bromide) stock solution (5 mg/mL) was added to each well and the microplate was incubated for 4 h. After that, 100 μL of SDS-HCl solution (1 g SDS/10 mL dH_2O/17 μL 6 N HCl) was added to each well followed by incubation for 4–18 h. The absorbance of the converted dye formazan was measured using a Multiskan FC microplate photometer (Thermo Scientific, USA) at a wavelength of 570 nm [32]. The results were presented as percent of control data, and concentration required for 50% inhibition of cell viability (EC_{50}) was calculated. Selectivity index (SI) is defined as the ratio of EC_{50} for normal cells and EC_{50} for tumor cell lines (SI = EC_{50} (normal cells)/EC_{50} (tumor cells)).

Nonspecific Esterase Activity Assay

The test solution (20 μL) and 180 μL of the cell suspension were placed into each well of a 96-well microplate (3.5×10^4 cells/well). The plates were incubated in a CO_2 incubator at 37 °C for 24 h. A stock solution of the probe fluorescein diacetate (FDA) (Sigma-Aldrich, St. Louis, MO, USA) in DMSO (1 mg/mL) was prepared. After incubation, cells were washed with PBS, 10 μL of FDA solution (50 μg/mL) was added to each well and the plate was incubated at 37 °C for 15 min, and fluorescence was measured with a Fluoroskan Ascent plate reader (Thermo Labsystems, Finland) at λ_{ex} = 485 nm and λ_{em} = 518 nm. All experiments were repeated in triplicate. Cytotoxic activity was expressed as the percent of cell viability.

Hemolytic Assay

Erythrocytes were used at a concentration that provided an optical density of 1.0 at 700 nm for a non-hemolyzed sample. In addition, 20 μL of a solution of test substances with a fixed concentration was added to a well of a 96-well plate containing 180 μL of the erythrocyte suspension. The erythrocyte suspension was incubated with substances for 1 h at 37 °C. After that, the optical density of the obtained solutions was measured at 700 nm and EC_{50} for hemolytic activity of each compound was calculated.

3.3.3. Antimicrobial Assay

Microorganism Cultures

Bacteria *Bacillus cereus* ATCC 10702, *Escherichia coli* K-12, *Pseudomonas aeruginosa* ATCC 27853, and *Staphylococcus aureus* ATCC 21027 and fungus *Candida albicans* KMM 453 were obtained from the Collection of Marine Microorganisms (KMM) of the G.B. Elyakov Pacific Institute of Bioorganic Chemistry, Far East Branch, Russian Academy of Sciences.

Agar Diffusion Method

Test substances were dissolved in DMSO at concentrations of 1.0, 0.1, and 0.01 mg/mL. All strains were grown on trypticase-soy agar TSA (BBL) at 37 °C. After overnight incubation of each test organism in Petri dishes with the same medium, wells with a diameter of 7 mm were cut and 100 μL of the test substance solution was added. After incubation of the dishes at 37 °C for 24 h (for bacteria) and 48 h (for the fungus) at 28 °C, the diameters of the growth inhibition zones of pathogenic microorganisms (mm) were measured, including the diameter of the wells. Commercial antibacterial drugs (vancomicin and gentamicin) and antifungal (clotrimazol) drugs were used as positive controls.

Determination of Minimum Inhibitory Concentration (MIC)

Stocks of substances were prepared in DMSO at a concentration of 10 mM. All studied compounds were tested in concentrations from 1.5 μM to 100 μM using twofold dilution. All tested compounds were add to the wells of the plates in a volume of 20 μL diluted in PBS (DMSO concentration < 1%). The bacterial culture of *S. aureus* ATCC 21027 (Collection of Marine Microorganisms PIBOC FEB RAS) was cultured in a Petri dish at 37 °C for 24 h on solid medium with the following composition; pepton—5.0 g/L, K_2HPO_4—0.2 g/L, glucose—1.0 g/L, $MgSO_4$—0.05 g/L, yeast extract—1 g/L, agar—16.0 g/L, and distilled water—1.0 L. The pH of the medium was adjusted to 7.2–7.4 with NaOH solution.

The antimicrobial activity of the compounds was determined by the minimum inhibitory concentration (MIC) according to the method adopted by the Clinical and Laboratory Standards Institute (CLSI). Methods for dilution antimicrobial susceptibility tests for bacteria that grow aerobically followed the approved standard—tenth edition. CLSI document M07-A10, PA: Clinical and Laboratory Standards Institute; 2015, with a slight modification of the medium. The assays were performed in 96-well microplates in appropriate broth (pepton—5.0 g/L, K_2HPO_4—0.2 g/L, glucose—1.0 g/L, $MgSO_4$—0.05 g/L, yeast extract—1 g/L, casein hydrolyzate—2.5 g/L, distilled water—1L). The *S. aureus* bacterial suspension had a concentration of 10×10^9 CFU/mL in the medium, 180 μL was then added to each well of the microplates, followed by incubation at 37 °C for 24 h. The MIC value is defined as the lowest concentration of compounds resulting in the complete inhibition of bacterial growth by measuring the absorbance at 600 nm with a microplate reader (BioTek, Winooski, Vt, USA). All experiments were performed twice in triplicate. Gentamicin was used as a positive control in concentration 1 mg/mL; 1% DMSO solution in PBS as a negative. The selectivity index for S. *aureus* has been defined as the ratio of EC_{50} for normal cells and the MIC value for the bacterial culture (SI = EC_{50}/MIC).

4. Conclusions

In summary, a series of new tetracyclic oxathiine-fused quinone-thioglycoside conjugates based on biologically active 1,4-naphthoquinones (chloro-, hydroxy-, and methoxysubstituted) have been synthesized, characterized, and evaluated for their cytotoxic and antimicrobial activities. It was shown that the most active compounds are tetracyclic conjugates of juglone (5-hydroxy-1,4-naphthoquinone), which showed high cytotoxic activity with EC_{50} in the range of 0.3 to 0.9 μM for all cancer and noncancer cell lines. Furthermore, for the first time the antimicrobial activity for this type of compounds was evaluated by the agar diffusion method. Among the tested conjugates, the activity of juglone derivatives with a D-xylose or L-arabinose moiety and hydroxyl group at C-7 position of naphthoquinone core was comparable with antibiotics vancomicin and gentamicin against Gram-positive bacteria *S. aureus* and *B. cereus*. In liquid media, the juglone-arabinosidic tetracycles showed highest activity with a MIC of 6.25 μM against *S. aureus* strain. Thus, the positive effect of heterocyclization with mercaptosugars on cytotoxic and antimicrobial activity for a group of 1,4-naphthoquinones was shown. The effect of chloro-, hydroxy-, and metoxysubstituents on teracycles activity was also studied and a significant effect of the hydroxy group on activity was shown. It can be assumed that further modification of such tetracycles may lead to new compounds with selectivity for cancer cells and great antimicrobial activity.

Supplementary Materials: ^{1}H and ^{13}C-NMR spectra for synthesized compounds are available online.

Author Contributions: Y.E.S. and S.G.P. performed the chemical experiments; S.G.P. and Y.E.S. conceived and designed the chemical experiments; V.A.D. carried out the NMR analyses and analyzed the data, R.S.P. performed MS measurements; E.S.M., L.S.S., E.A.C. and A.R.C. performed bioactivity testing supervised by D.L.A. and V.V.M.; Y.E.S.—original draft preparation, S.G.P.—review and editing, D.L.A. and V.V.M.—supervision. All authors have read and agreed to the published version of the manuscript.

Funding: This research was funded by the Russian Foundation for Basic Research (Grant № 18-33-00492_mol_a) and particularly by the Russian Science Foundation (Grant № 19-14-00047).

Acknowledgments: We thank V. P. Glazunov for recording IR spectra.

Conflicts of Interest: The authors declare no conflict of interest.

References

1. Thomson, R.H. *Naturally Occurring Quinones IV*; Blackie: London, UK, 1997.
2. Asche, C. Antitumour quinones. *Mini-Rev. Med. Chem.* **2005**, *5*, 449–467. [CrossRef] [PubMed]
3. Constantino, L.; Barlocco, D. Privileged structures as leads in medicinal chemistry. *Curr. Med. Chem.* **2006**, *13*, 65–85. [CrossRef]
4. Wellington, K.W. Understanding cancer and the anticancer activities of naphthoquinones—A review. *RSC Adv.* **2015**, *5*, 20309–20338. [CrossRef]
5. Service, M.; Wardlaw, A.C. Echinochrome-A as a bactericidal substance in the coelomic fluid of *Echinus esculentus* (L.). *Comp. Biochem. Physiol. Part. B Comp. Biochem.* **1984**, *79*, 161–165. [CrossRef]
6. Sanchez-Calvo, J.M.; Barbero, G.R.; Guerrera-Vasquez, G.; Duran, A.G.; Macıas, M.; Rodrıguez-Iglesias, M.A.; Molinillo, J.M.G.; Macıas, F.A. Synthesis, antibacterial and antifungal activities of naphthoquinone derivatives: A structure–activity relationship study. *Med. Chem. Res.* **2016**, *25*, 1274–1285. [CrossRef]
7. Bringmann, G.; Rudenauer, S.; Irmer, A.; Bruhn, T.; Brun, R.; Heimberger, T.; Stühmer, T.; Bargou, R.; Chatterjee, M. Antitumoral and antileishmanial dioncoquinones and ancistroquinones from cell cultures of *Triphyophyllum peltatum* (Dioncophyllaceae) and *Ancistrocladus abbreviatus* (Ancistrocladaceae). *Phytochemistry* **2008**, *69*, 2501–2509. [CrossRef]
8. El Hage, S.; Ane, M.; Stigliani, J.-L.; Marjorie, M.; Vial, H.; Baziard-Mouysset, G.; Payard, M. Synthesis and antimalarial activity of new atovaquone derivatives. *Eur. J. Med. Chem.* **2009**, *44*, 4778–4782. [CrossRef]
9. Mishchenko, N.P.; Fedoreev, S.A.; Bagirova, V.L. Histochrome: A new original domestic drug. *Pharm. Chem. J.* **2003**, *37*, 48–52. [CrossRef]
10. Patai, S. *The Chemistry of Quinonoid Compounds. Part. 1*; John Wiley & Sons: New York, NY, USA, 1974.
11. Jordao, A.K.; Vargas, M.D.; Pinto, A.C.; da Silva, F.C.; Ferreira, V.F. Lawsone in organic synthesis. *RSC Adv.* **2015**, *5*, 67909–67943. [CrossRef]
12. Aminin, D.L.; Polonik, S.G. 1,4-Naphthoquinones: Some biological properties and application. *Chem. Pharm. Bull.* **2020**, *68*, 46–57. [CrossRef]
13. Moreira, C.S.; Silva, A.C.; Novais, J.S.; Sá Figueiredo, A.M.; Ferreira, V.F.; da Rocha, D.R.; Castro, H.C. Searching for a Potential Antibacterial Lead Structure Against Bacterial Biofilms Among New Naphthoquinone Compounds. *J. Appl. Microbiol.* **2017**, *122*, 651–662. [CrossRef] [PubMed]
14. Novais, J.S.; Moreira, C.S.; Silva, A.C.J.; Loureiro, R.S.; Sá Figueiredo, A.M.; Ferreira, V.F.; Castro, H.C.; da Rocha, D.R. Antibacterial Naphthoquinone Derivatives Targeting Resistant Strain Gram-negative Bacteria in Biofilms. *Microb. Pathogen.* **2018**, *118*, 105–114. [CrossRef] [PubMed]
15. Pelageev, D.N.; Dyshlovoy, S.A.; Pokhilo, N.D.; Denisenko, V.A.; Borisova, K.L.; von Amsberg, G.K.; Bokemeyer, C.; Fedorov, S.N.; Honecker, F.; Anufriev, V.P. Quinone-carbohydrate nonglucoside conjugates as a new type of cytotoxic agents: Synthesis and determination of in vitro activity. *Eur. J. Med. Chem.* **2014**, *77*, 139–144. [CrossRef] [PubMed]
16. Dyshlovoy, S.A.; Pelageev, D.N.; Hauschild, J.; Borisova, K.L.; Kaune, M.; Krisp, C.; Venz, S.; Sabutskii, Y.E.; Khmelevskaya, E.A.; Busenbender, T.; et al. Successful targeting of the Warburg effect in prostate cancer by glucose-conjugated 1,4-naphthoquinones. *Cancers* **2019**, *11*, 1690. [CrossRef] [PubMed]
17. Polonik, S.G.; Krylova, N.V.; Kompanets, G.G.; Iunikhina, O.V.; Sabutski, Y.E. Synthesis and screening of anti-HSV-1 activity of thioglucoside derivatives of natural polyhydroxy-1,4-naphthoquinones. *Nat. Prod. Commun.* **2019**, *14*, 1–7. [CrossRef]

18. Polonik, S.G. Synthesis and properties of water-soluble 5,8-dihydroxy-1,4-naphthoquinone thioglucosides structurally related to echinochrome. *Russ. J. Org. Chem.* **2009**, *45*, 1474–1480. [CrossRef]
19. Ottoni, F.M.; Gomes, E.R.; Padua, R.M.; Oliveira, M.C.; Silva, I.T.; Alves, R.J. Synthesis and cytotoxicity evaluation of glycosidic derivatives of lawsone against breast cancer cell lines. *Bioorg. Med. Chem. Lett.* **2020**, *30*, 126817. [CrossRef]
20. Eremenko, E.M.; Antimonova, O.I.; Shekalova, O.G.; Polonik, S.G.; Margulis, B.A.; Guzhova, I.V. Novel compounds that increase expression of Hsp70 and its biological activity. *Cell. Tiss. Biol.* **2010**, *4*, 251–257. [CrossRef]
21. Yurchenko, E.A.; Menchinskaya, E.S.; Polonik, S.G.; Agafonova, I.G.; Guzhova, I.V.; Margulis, B.A.; Aminin, D.L. Hsp70 induction and anticancer activity of U-133, the acetylated trisglucosydic derivative of echinochrome. *Med. Chem.* **2015**, *5*, 263–271. [CrossRef]
22. Polonik, S.G.; Denisenko, V.A. Synthesis and properties of fused tetracyclic derivatives of 1,4-naphthoquinone thioglycosides. *Russ. Chem. Bull.* **2009**, *58*, 1062–1066. [CrossRef]
23. Fedorov, S.N.; Shubina, L.K.; Kuzmich, A.S.; Polonik, S.G. Antileukemic properties and structure-activity relationships of *O*-and *S*-glycosylated derivatives of juglone and related 1,4-naphthoquinones. *Open Glycosci.* **2011**, *4*, 1–5. [CrossRef]
24. Stanek, J.; Sindlerova, M.; Cerny, M. Derivatives of D-thioxylopyranose and of some reducing 1-deoxy-1-thiodisaccharides. *Collect. Czech. Chem. C* **1965**, *30*, 297–303. [CrossRef]
25. Floyd, N.; Vijayakrishnan, B.; Koeppe, J.R.; Davis, B.G. Thiyl glycosylation of olefinic proteins: S-linked glycoconjugate synthesis. *Angew. Chem. Int. Edit.* **2009**, *48*, 7798–7802. [CrossRef] [PubMed]
26. Doyle, L.M.; O'Sullivan, S.; Di Salvo, C.; McKinney, M.; McArdle, P.; Murphy, P.V. Stereoselective epimerizations of glycosyl thiols. *Org. Lett.* **2017**, *19*, 5802–5805. [CrossRef]
27. Sabutskii, Y.E.; Denisenko, V.A.; Popov, R.S.; Polonik, S.G. The synthesis of thioglucosides substituted 1,4-naphthoquinones and their conversion in oxathiane fused quinone-thioglucoside conjugates. *Arkivoc* **2017**, *iii*, 302–315. [CrossRef]
28. Tran, N.-C.; Le, M.-T.; Nguyen, D.N.; Tran, T.D. *Synthesis and Biological Evaluation of Halogen Substituted 1,4-Naphthoquinones as Potent Antifungal Agents*; The 13rd International Electronic Conference on Synthetic Organic Chemistry; MDPI AG: Basel, Switzerland, 2009; pp. 1–7. [CrossRef]
29. Wurm, G.; Gurka, H.-J.; Geres, U. Untersuchungen an 1.4-Naphthochionen, 15. Mitt. Reaktion von 2- und 3-Chlor/Bromjuglonderivaten mit methanolischer Lauge (Teil 1: Monomere Produkte). *Arch. Pharm.* **1986**, *319*, 1106–1113. [CrossRef]
30. Sabutskii, Y.E.; Denisenko, V.A.; Popov, R.S.; Polonik, S.G. Acid-catalyzed heterocyclization of trialkylnaphthazarin thioglucosides in angular quinone-carbohydrate tetracycles. *Russ. J. Org. Chem.* **2019**, *55*, 147–151. [CrossRef]
31. Menchinskaya, E.S.; Pislyagin, E.A.; Kovalchyk, S.N.; Davydova, V.N.; Silchenko, A.S.; Avilov, S.A.; Kalinin, V.I.; Aminin, D.L. Antitumor Activity of Cucumarioside A_2-2. *Chemotherapy* **2013**, *59*, 181–191. [CrossRef]
32. Horiuchi, N.; Nakagava, K.; Sasaki, Y.; Minato, K.; Fujiwara, Y.; Nezu, K.; Ohe, Y.; Saijo, N. In vitro antitumor activity of mitomycin C derivative (RM49) and a new anticancer antibiotics (FK973) against lung cancer lines determined by tetrazolium dye (MTT) assay. *Cancer Chemoth. Pharm.* **1988**, *22*, 246–250. [CrossRef]

Sample Availability: Samples of the compounds are available from the authors.

molecules

MDPI

Communication

In Vitro Anticancer and Radiosensitizing Activities of Phlorethols from the Brown Alga *Costaria costata*

Olesya S. Malyarenko *,†, Tatiana I. Imbs † and Svetlana P. Ermakova

G.B. Elyakov Pacific Institute of Bioorganic Chemistry, Far-Eastern Branch of the Russian Academy of Sciences, 159, 100-let Vladivostoku Prospect, 690022 Vladivostok, Russia; tatyanaimbs@mail.ru (T.I.I.); ermakova@piboc.dvo.ru (S.P.E.)

* Correspondence: malyarenko.os@gmail.com; Tel.: +7-423-231-07-05; Fax: +7-423-231-40-50

† These authors contributed equally to this work.

Academic Editors: Cédric Delattre and Luisa Tesoriere

Received: 18 June 2020; Accepted: 12 July 2020; Published: 14 July 2020

Abstract: The anticancer and radiosensitizing effects of high-molecular-weight phlorethols **CcPh** (Mw = 2520 Da) isolated from the brown algae of *Costaria costata* on human colorectal carcinoma HCT 116 and HT-29 cells were investigated. Phlorethols **CcPh** possessed cytotoxic activity against HT-29 (IC_{50} = 92 µg/mL) and HCT 116 (IC_{50} = 94 µg/mL) cells. **CcPh** at non-toxic concentrations inhibited the colony formation in colon cancer cells and significantly enhanced their sensitivity to low non-toxic X-ray irradiation. The combinatory effect of radiation and **CcPh** was synergistic (Combination index < 0.7). Algal phlorethols might be prospective candidates as radiosensitizers to improve the scheme of radiotherapy.

Keywords: phlorotannins; phlorethols; anticancer activity; colorectal cancer; radiosensitizer; radiotherapy

1. Introduction

Radiation therapy is traditionally considered one of the most effective methods for influencing tumors [1]. The standard methods of radiation therapy are limited because any attempt to increase the total irradiation dose is associated with a high risk of severe post-radiation damage [2]. This necessitates new approaches to realize the main objective of radiation therapy, which is to increase the radiosensitivity of cancer cells. An increase in the effect of ionizing radiation can be achieved using radiosensitizers even without increasing the total irradiation dose [3,4].

Brown algae are an attractive source of biologically active compounds for researchers. They serve as raw materials for various therapeutic and prophylactic preparations, since they contain a number of bioactive substances, such as polysaccharides, mannitol, vitamins, macro- and microelements, polyphenols, iodine-containing organic compounds, and polyunsaturated fatty acids

Phlorotannins are a type of phenolic metabolite found in brown algae. This group of phenolic compounds contains a large number of hydroxyl groups and chelates divalent metal ions; they have polymer structures and are highly soluble in water and firmly bound to proteins, polysaccharides, and other biopolymers. The monomer unit of phlorotannins is phloroglucinol (1,3,5-trihydroxybenzene). Based on the type of monomer bond, phlorotannins can be divided into four classes: fuhalols and phlorethols (ether bond), fucols (phenyl bond), fucophlorethols (ether and phenyl bonds), and eckols and carmalols (dibenzodioxin bond). Within each class, the binding of monomers to each other can occur at different positions of the phloroglucinol ring, which leads to the formation of structural isomers in addition to conformational ones. One species of algae is often known to produce phlorotannins with different structures and different degrees of polymerization [5].

We have previously shown that the aqueous ethanol extract of *Costaria costata* containing abundant phlorotannins possesses antitumor potential [6]. Moreover, phlorotannins from the same species of brown algae were found to be effective inhibitors of fucoidanase (enzyme hydrolyzed fucoidan) [7]. Since phlorotannins from brown algae are considered to have different types of biological activities and up to a certain concentration do not exert any toxic effect [8–10], they are of research interest as perspective bioactive agents for cancer therapy.

In the present work, we checked the hypothesis whether phlorotannins from *C. costata* inhibited cell viability and colony formation of human colorectal cancer cells as well as sensitized HT-29 and HCT 116 cells to X-ray radiation.

2. Results and Discussion

2.1. Characterization of the Phlorethols Isolated from Costaria costata

Phlorotannins constitute an extremely heterogeneous group of molecules, which differ by the types of bonds between the phloroglucinol units. Fucols have only phenyl bonds in the polymer molecule, phlorethols and fuhalols have ether bonds, and fucophlorethols have both of these bond types. The presence of a dibenzodioxin element in the structure is specific for eckols. Within each class, the binding of monomers to each other can take place at different positions of the phloroglucinol ring, resulting in the formation of structural isomers. The separation of longer, condensed phlorotannins is currently a challenge. [11].

The phlorotannins were isolated from an aqueous ethanol extract of the brown alga *C. costata.* Earlier the structural characteristics of phlorethols from *C. costata* were determined [7]. In the present work, NMR analysis showed that the phlorotannins of *C. costata* belong to the class of phlorethols (**CcPh**). The average molecular weight of **CcPh** was 2520 Da, as measured by ESI-MS in negative ionization mode. The degree of polymerization (DP) of the obtained phlorethols ranged from 12 to 25 phloroglucinol units with the most abundant phlorethols containing between 16 and 20 phloroglucinol units. [7] (Figures S1 and S2, and Table S1).

2.2. Bioactivity of the Phlorethols from C. costata

As a first step, the effect of phlorethols (**CcPh)** from brown alga *C. costata* (0–1000 μg/mL) on the viability of human colorectal carcinoma HT-29 and HCT 116 cell lines, breast cancer MCF-7 cells, and melanoma SK-MEL-28 cells was tested by MTS assay (Figure 1). The treatment of cells by **CcPh** was found to induce a concentration-dependent inhibition of cell's viability. The inhibiting concentration of **CcPh** that caused a 50% reduction in cell viability (IC_{50}) of HT-29 was 92 ± 2.8 μg/mL; HCT, 116–94 ± 3.08 μg/mL; MCF-7, 96 ± 3.3 μg/mL; and melanoma cells SK-MEL-28, 102 ± 4.8 μg/mL.

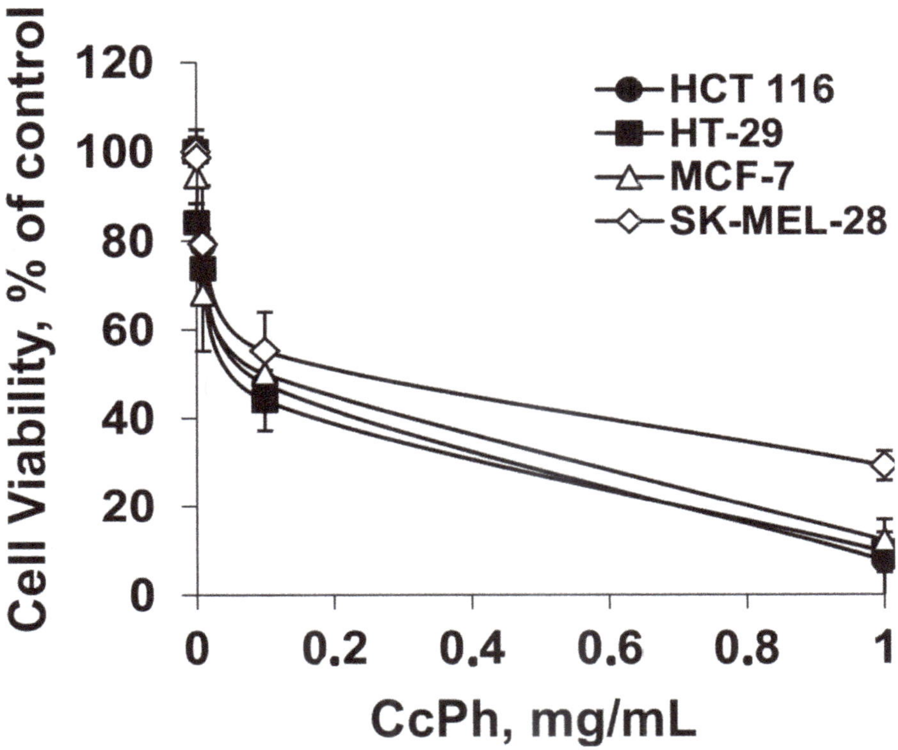

Figure 1. The cytotoxic activity of phlorethols from *C. costata* (**CcPh**) against colorectal carcinoma HT-29 and HCT 116 cells, breast cancer MCF-7 cells, and melanoma SK-MEL-28 cells. The cells were treated with **CcPh** at concentration of 0.001–1 mg/mL for 24 h. Cell viability was estimated using the MTS assay. Data are represented as the mean ± standard deviation (SD), as determined from triplicate experiments.

Recently, the cytotoxic activity of polyphenols' fractions from different species of brown algae was investigated against pancreatic cancer cells Mia-PaCa-2, Panc-1, Panc-3.27, and BxPC-3 [12], human colorectal adenocarcinoma HT-29, Caco-2, breast carcinoma T-47D, MDA-MB-468 cell lines [13], human leukemia HL60 and THP-1, and prostate cancer PC3 cell lines [14]. The IC_{50} values for investigated polyphenols fractions were in a range from 80 to 200 µg/mL. It was suggested that the number of phenolic ring substituents and phenyl ether linkages as well as the reduction in pH values caused by an oxidation of hydroxyl groups in phenol polymer derivatives mainly influence their cytotoxic activity [15]. Since colorectal carcinoma cells HCT 116 and HT-29 gain resistance under radiation exposure [16,17], we chose them as model to check the hypothesis whether of phlorethols from brown alga *C. costata* is able to effectively sensitize these cells to radiation.

Next, the effect of **CcPh** or X-ray radiation on colony formation in human colorectal cancer cells was determined using the soft agar assay. It was found that the **CcPh** alone decreased the number of colonies in HT-29 by 3%, 7%, and 24% at 10, 20, and 40 µg/mL, respectively and in HCT 116, by 4%, 7% and 18% at the same doses. (Figure 2A,D). It should be noted that **CcPh** did not influence the viability of normal mouse epidermal cells JB6 Cl41 at an effective concentration of 40 µg/mL even after 72-h of treatment (Figure S4).

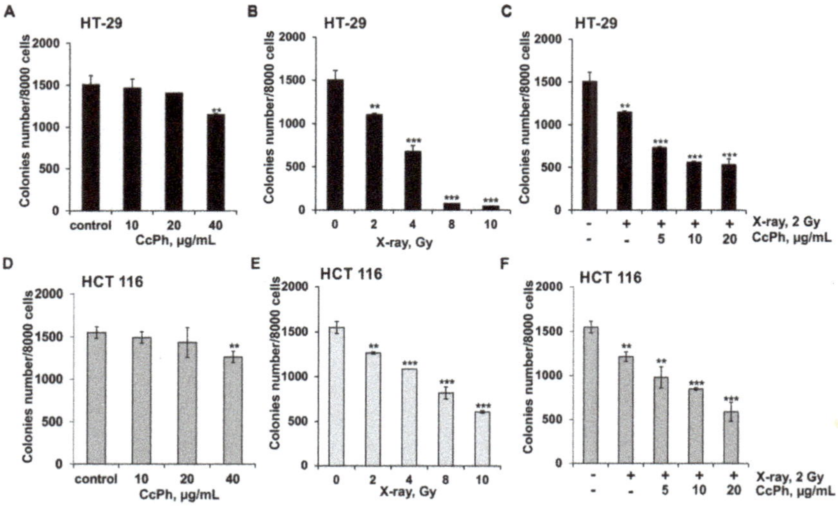

Figure 2. Anticancer and radiosensitizing effects of phlorethols from *C. costata* (**CcPh**) on colony formation in human colorectal carcinoma cells. HT-29 and HCT 116 cells (2.4×10^4) were treated with (**A**,**D**) **CcPh** (10, 20, and 40 µg/mL) or (**B**,**E**) X-ray (2, 4, 8, and 10 Gy) or (**C**,**F**) a combination of X-ray radiation (2 Gy) and **CcPh** (5, 10, and 20 µg/mL) and subcultured onto Basal Medium Eagle BME soft agar and incubated for 2 weeks. The number of colonies was calculated using the ImageJ software bundled with 64-bit Java 1.8.0_112 (NIH, Bethesda, Maryland, USA). All experiments were repeated at least three times in each group. Results are expressed as the mean ± standard deviation (SD). The asterisk (*) indicates a significant decrease in the number of colonies of cancer cells treated with **CcPh** or X-ray compared to PBS-treated cells or **CcPh** in combination with X-ray compared to irradiated cells (** $p < 0.01$, *** $p < 0.001$).

The exposure of HT-29 by X-ray irradiation at doses of 2, 4, 8, and 10 Gy caused the inhibition of the colonies number by 26%, 55%, 95%, and 97%, respectively and HCT 116 by 18%, 30%, 47%, and 61% at the same doses (Figure 2B,E).

To investigate the radiosensitizing effect of phlorethols **CcPh**, HT-29 and HCT116 cells were treated with **CcPh** at non-toxic concentrations of 5, 10, and 20 µg/mL and exposed to a low dose of X-ray of 2 Gy, which alone slightly inhibited the viability or colony formation in cancer cells.

Under X-ray exposure, the number of colonies of HT-29 and HCT 116 cells were decreased by 24% and 21%, respectively, compared to the non-irradiated cells (control) (Figure 2C,F). The combinatory treatment with X-ray (2 Gy) and **CcPh** (5, 10, and 20 µg/mL) caused significant inhibition of colony formation of HT-29 by 28%, 39%, and 41%, respectively and HCT 116 by 15%, 24%, and 40%, respectively, compared to irradiated cells (Figure 2C,F).

To determine the type of combination effect of X-ray radiation with phlorethols **CcPh**, the combination index (CI) was calculated (Table 1). This revealed that the interactions of radiation and **CcPh** were mostly synergistic (CI < 0.7) [18]. **CcPh** strongly sensitized the HCT 116 cells to radiation as compared with HT-29 cells.

Table 1. Combination index (CI) for phlorethols—X-ray irradiation interactions.

Concentration of CcPh, μg/mL	Dose of X-ray, Gy	Combination Effect, % of Control		Combination Index, CI *	
		HT-29	HCT 116	HT-29	HCT 116
2.5	2	40	30	0.67879 ± 0.024	0.58038 ± 0.012
5	2	52	37	0.61559 ± 0.03	0.48424 ± 0.035
10	2	63	45	0 58341 ± 0.027	0.43194 ± 0.042
20	2	65	62	0 66600 ± 0.021	0.33167 ± 0.037

* The Combination Index (CI) is the quantitative measure of the degree of interaction between different treatments. CI values in the range of 0.9–1.1 indicate additive effect; and CI values greater than 1.1, antagonism. CI values in the range from 0.9 to 0.7 indicate slight synergism; and CI values less than 0.7, synergism. CI values were calculated according to the Chou and Talalay mathematical model for drug interactions using the CompuSyn software version 1.0 on the basis of the results of soft agar assay. Data represent mean CI calculated from three independent experiments ± standard deviations.

To the best of our knowledge, this is the first study on the radiosensitizing activity of phlorotannins from brown algae. Since polyphenols from brown algae possessed potent antioxidant activity, their radioprotective effect is a topic of interest of numerous investigations [19]. However there is evidence that several natural compounds are able to exert a dual mode of action after irradiation—radioprotective as well as radiosensitizing—depending on its dose and scheme of treatment [20]. For example, polyphenols of plant origin such as curcumin or resveratrol have been found to protect various systems against ionizing radiation and to sensitize cancer cells to radiation treatment [21].

3. Materials and Methods

3.1. Reagents

Phosphate buffered saline (PBS), L-glutamine, penicillin-streptomycin solution (10,000 U/mL, 10 μg/mL) were purchased from Sigma-Aldrich (St. Louis, MI, USA). The MTS reagent (3-(4,5-dimethylthiazol-2-yl)-5-(3-carboxymethoxyphenyl)-2-(4-sulfophenyl)-2H-tetrazolium) was purchased from Promega (Madison, WI, USA). The Basal Medium Eagle (BME), McCoy's 5A Modified Medium (McCoy's 5A), Minimum Essential Medium Eagle (MEM), trypsin, fetal bovine serum (FBS), and agar were purchased from ThermoFisher Scientific (Waltham, MA, USA). Organic solvents, inorganic acids, and salts were commercially obtained (Laverna, Moscow, Russia)

3.2. Cell Culture

Human colorectal carcinoma HT-29 (ATCC® no. HTB-38™) and HCT 116 (ATCC® CCL-247™) were cultured in McCoy's 5A medium, supplemented with 10% FBS and 1% penicillin-streptomycin solution; breast cancer MCF-7 (ATCC® no. HTB-22™) or melanoma SK-MEL-28 (ATCC® no. HTB-72™) cells were maintained in MEM medium with 10% FBS and 1% penicillin-streptomycin solution. The cell cultures were maintained at 37 °C in humidified atmosphere containing 5% CO_2. Every 3–4 days, the cells were detached with 0.25% trypsin/0.05 M EDTA for 1–3 min and 10–20% of the harvested cells were transferred to a new flask containing fresh complete culture media.

3.3. Isolation of the Phlorethol Fraction (CcPh) from C. costata

The brown alga *C. costata* (Turn) Saund (order Laminariales) was collected in Peter the Great Bay, Sea of Japan in July 2016. The isolation, separation, and structural characterization of phlorethols were conducted as previously reported [7].

3.4. Preparation of the Phlorethol Fraction (CcPh) for Bioassays

CcPh was dissolved in sterile PBS to prepare stock concentrations of 10 mg/mL. Cells were treated with serially diluted phlorethols (culture medium used as diluent) to give the intended final concentrations (5–1000 μg/mL). The vehicle control was the cells treated with an equivalent volume of PBS for all presented experiments.

3.5. MTS Assay

HT-29 and HCT 116 or MCF-7 and SK-MEL-28 cells (1.0×10^4) were seeded in 200 μL of complete McCoy's 5A/10% FBS or MEM/10% FBS media, respectively, and incubated for 24 h at 37 °C in 5% CO_2 incubator. The attached cells were treated by phlorethols **CcPh** at concentrations ranging from 0 to 1000 μg/mL for an additional 24 h. Subsequently, the cells were incubated with 15 μL MTS reagent for 3 h, and the absorbance of each well was measured at 490/630 nm using PowerWave XS Microplate Reader (BioTek, Winooski, VT, USA). All samples were tested in triplicate.

3.6. Soft Agar Assay

To estimate the effect of phlorethols **CcPh** on colony formation (phenotype expression), human colorectal carcinoma cells HT-29 and HCT 116 (2.4×10^4) were treated with PBS (control) or phlorethols **CcPh** (10, 20, and 40 μg/mL) in 1 mL of 0.3% BME agar containing 10% FBS, 2 mM L-glutamine, and 25 μg/mL gentamicin. The cultures were maintained in a 37 °C, 5% CO_2 incubator for 14 days, and the cell colonies were scored using a Motic microscope AE 20 (XiangAn, Xiamen, China) and ImageJ software bundled with 64-bit Java 1.8.0_112 (NIH, Bethesda, MD, USA).

3.7. Cell Irradiation Assay

Human colorectal carcinoma cells HT-29 and HCT 116 were exposed to X-ray radiation using an XPERT 80 X-ray system (KUB Technologies, Inc., Milford, CT, USA). The absorbed dose of radiation was measured by a DRK-1 X-ray radiation clinical dosimeter (Axelbant LLK, Moscow, Russia).

To determine the sensitivity of HT-29 or HCT 116 cells to radiation, the cells (5.0×10^5) were exposed to X-ray at a dose rate from 2 to 10 Gy and recovered at 37 °C in a 5% CO_2 incubator for 3 h. The cells were harvested with 0.25% trypsin/0.05 M EDTA solution and subjected to the "Soft agar assay" as described above.

To determine the radiosensitizing activity of phlorethols, HT-29 or HCT 116 cells (5.0×10^5) were exposed to 2 Gy of X-ray and incubated for 3 h. Then, cells were treated with either PBS (control) or phlorethols **CcPh** (5, 10, and 20 μg/mL) for an additional 24 h. The cells were harvested and used for the "Soft agar assay" as described above.

3.8. Combination Index (CI) Calculation

The calculations of drug concentration/X-ray irradiation dose-effect were performed by CompuSyn software version 1.0 (ComboSyn, Inc., Paramus, NJ, USA) using the median effect method described by Chou and Talalay [22].

3.9. Data Analysis

All assays were performed in at least three independent experiments. Results are expressed as the mean ± standard deviation (SD). The Student's t-test was used to evaluate the data with the following significance levels: * $p < 0.05$, ** $p < 0.01$, *** $p < 0.001$.

4. Conclusions

This study evaluated the in vitro anticancer and radiosensitizing effects of phlorethols (**CcPh**) from *C. costata* on human colorectal carcinoma HCT 116 and HT-29 cell lines. **CcPh** at non-toxic concentrations was found to significantly inhibit the colony formation in HT-29 and HCT 116 alone and in combination with X-ray irradiation. The combinatory effect of radiation and **CcPh** was synergistic (combination index CI < 0.7). The molecular mechanism of this action requires further research. In future studies, we plan to determine the effects of polyphenols on the detoxification of enzymes, the activation of endogenous protective systems, and the repair of DNA strand breaks induced by X-ray exposure.

Supplementary Materials: The following are available online, Figure S1: ^{13}C-NMR spectra of the **CcPh** fraction, Figure S2: HMBC NMR spectra of the **CcPh** fraction, Figure S3: HSQC NMR spectra of the **CcPh** fraction, Table S1: The data of elemental compositions, monoisotopic masses and ions of the phlorethol **CcPh** fraction, Figure S4: The effect of phlorethols from *C. costata* (**CcPh**) on viability of normal mouse epidermal cells JB6 Cl41.

Author Contributions: O.S.M. and T.I.I.: the conceptualization, design, and methodology of the work, data curation, writing-original draft preparation. T.I.I.: the preparation of the phlorotannins. O.S.M. performed the biological assays. S.P.E.: the editing of the manuscript. All authors have read and agreed to the published version of the manuscript.

Funding: The reported study was funded by RSF grant # 16-14-10131.

Conflicts of Interest: The authors declare no conflict of interest.

References

1. Matthias, F.; Debusa, H.; Debusa, J. Radiotherapy for colorectal cancer: Current standards and future perspectives. *Visc. Med.* **2016**, *32*, 172–177.
2. Mishra, J.; Dromund, J.; Quazi, S.H.; Karanki, S.S.; Shaw, J.J.; Chen, B.; Kumar, N. Prospective of colon cancer treatments and scope for combinatorial approach to enhanced cancer cell apoptosis. *Crit. Rev. Oncol. Hematol.* **2013**, *86*, 232–250. [CrossRef] [PubMed]
3. Nambiar, D.; Rajamani, P.; Singh, R.P. Effects of phytochemicals on ionization radiation-mediated carcinogenesis and cancer therapy. *Mutat. Res.* **2011**, *728*, 139–157. [CrossRef] [PubMed]
4. Yue, Q.; Gao, G.; Zou, G.; Yu, H.; Zheng, X. Natural products as adjunctive treatment for pancreatic cancer: Recent trends and advancements. *Biomed. Res. Int.* **2017**, *2017*, 8412508. [CrossRef]
5. Li, Y.X.; Wijeseker, I.; Li, Y.; Kim, S.K. Phlorotannins as bioactive agents from brown algae. *Process. Biochem.* **2011**, *46*, 2219–2224. [CrossRef]
6. Imbs, T.I.; Krasovskaya, N.P.; Ermakova, S.P.; Makarieva, T.N.; Shevchenko, N.M.; Zvyagintseva, T.N. Comparative study of chemical composition and antitumor activity of aqueous–ethanol extracts of brown algae *Laminaria cichorioides, Costaria costata,* and *Fucus evanescens*. *Russ. J. Mar. Biol.* **2009**, *35*, 164–170. [CrossRef]
7. Imbs, T.I.; Silchenko, A.S.; Fedoreev, S.A.; Isakov, V.V.; Ermakova, S.P.; Zvyagintseva, T.N. Fucoidanase inhibitory activity of phlorotannins from brown algae. *Algal. Res.* **2018**, *32*, 54–59. [CrossRef]
8. Javed, A.; Hussain, M.B.; Tahir, A.; Waheed, M.; Anwar, A.; Shariati, M.A.; Plygun, S.; Laishevtcev, A.; Pasalar, M. Pharmacological applications of phlorotannins: A comprehensive review. *Curr. Drug. Discov. Technol.* **2020**. [CrossRef] [PubMed]
9. Murray, M.; Dordevic, A.L.; Bonham, M.P.; Ryan, L. Do marine algal polyphenols have antidiabetic, antihyperlipidemic or anti-inflammatory effects in humans? A systematic review. *Crit. Rev. Food. Sci. Nutr.* **2018**, *58*, 2039–2054. [CrossRef]
10. Pangestuti, R.; Siahaan, E.A.; Ki, S.K. Photoprotective substances derived from marine algae. *Mar. Drugs.* **2018**, *16*, E399. [CrossRef] [PubMed]
11. Imbs, T.I.; Zvyagintseva, T.N. Florotannins are polyphenolic metabolites of brown algae. *Russ. J. Mar. Biol.* **2018**, *44*, 217–227. [CrossRef]
12. Aravindan, S.; Delma, C.R.; Thirugnanasambandan, S.S.; Herman, T.S.; Aravindan, N. Anti-pancreatic cancer deliverables from Sea: First-hand evidence on the efficacy, molecular targets and mode of action for multifarious polyphenols from five different brown-algae. *PLoS ONE.* **2013**, *8*, e61977. [CrossRef] [PubMed]
13. Mahnaz, K.; Maryam, N.; Nargess, S.; Mohammadreza, A.; Jelveh, S.; Seyed, N.; Padideh, G., Seyed, O. Cytotoxic activity of some marine brown algae against cancer cell lines. *Biol. Res.* **2010**, *43*, 31–37.
14. Mansur, A.A.; Brown, M.T.; Billington, R.A. The cytotoxic activity of extracts of the brown alga *Cystoseira tamariscifolia* (Hudson) Papenfuss, against cancer cell lines changes seasonally. *J. Appl. Phycol.* **2020**. [CrossRef]
15. Li, Y.; Qian, Z.J.; Kim, M.; Kim, S.K. Cytotoxic activities of phlorethol and fucophlorethol derivatives isolated from Laminariaceae *Ecklonia cava*. *J. Food. Biochem.* **2011**, *35*, 357–369. [CrossRef]

16. Jin, H.; Gao, S.; Guo, H.; Ren, S.; Ji, F.; Liu, Z.; Chen, X. Re-sensitization of radiation resistant colorectal cancer cells to radiation through inhibition of AMPK pathway. *Oncol. Lett.* **2016**, *11*, 3197–3201. [CrossRef]
17. Yadollahpour, A.; Rezaee, Z.; Bayati, V.; Tahmasebi Birgani, M.J.; Negad Dehbashi, F. Radiotherapy Enhancement with Electroporation in Human Intestinal Colon Cancer HT-29 Cells. *Asian. Pac. J. Cancer Prev.* **2018**, *19*, 1259–1262.
18. Bijnsdorp, I.V.; Giovannetti, E.; Peters, G.J. Analysis of drug interactions. *Methods Mol. Biol.* **2011**, *731*, 421–434.
19. Zhang, R.; Kang, K.A.; Piao, M.J.; Ko, D.O.; Wang, Z.H.; Lee, I.K.; Kim, B.J.; Jeong, I.Y.; Shin, T.; Park, J.W.; et al. Eckol protects V79-4 lung fibroblast cells against γ-ray radiation-induced apoptosis via the scavenging of reactive oxygen species and inhibiting of the c-Jun NH_2-terminal kinase pathway. *Euro. J. Pharmacol.* **2008**, *591*, 114–123. [CrossRef]
20. Nair, C.K.; Parida, D.K.; Nomura, T. Radioprotectors in radiotherapy. *J. Radiat. Res.* **2001**, *42*, 21–37. [CrossRef]
21. Sebastià, N.; Montoro, A.; Hervás, D.; Pantelias, G.; Hatzi, V.I.; Soriano, J.M.; Villaescusa, J.I.; Terzoudi, G.I. Curcumin and trans-resveratrol exert cell cycle-dependent radioprotective or radiosensitizing effects as elucidated by the PCC and G2-assay. *Mutat. Res.* **2014**, *766-767*, 49–55. [CrossRef] [PubMed]
22. Chou, T.C. Drug combination studies and their synergy quantification using the Chou-Talalay method. *Cancer Res.* **2010**, *70*, 440–446. [CrossRef] [PubMed]

Sample Availability: Sample of the phlorethols from *C. costata* (1 mg) is available from the authors.

molecules

MDPI

Article

Sea Anemone *Heteractis crispa* Actinoporin Demonstrates In Vitro Anticancer Activities and Prevents HT-29 Colorectal Cancer Cell Migration

Aleksandra Kvetkina [1,*,†], Olesya Malyarenko [1,†], Aleksandra Pavlenko [1], Sergey Dyshlovoy [2,3,4], Gunhild von Amsberg [2,3], Svetlana Ermakova [1] and Elena Leychenko [1]

1 G.B. Elyakov Pacific Institute of Bioorganic Chemistry, Far Eastern Branch, Russian Academy of Sciences, 159, Pr. 100 let Vladivostoku, Vladivostok 690022, Russia; malyarenko.os@gmail.com (O.M.); apavlenko141@gmail.com (A.P.); swetlana_e@mail.ru (S.E.); leychenko@gmail.com (E.L.)

2 Department of Oncology, Hematology and Bone Marrow Transplantation with Section Pneumology, Hubertus Wald-Tumorzentrum, University Medical Center Hamburg-Eppendorf, 20251 Hamburg, Germany; dyshlovoy@gmail.com (S.D.); g.von-amsberg@uke.de (G.v.A.)

3 Martini-Klinik, Prostate Cancer Center, University Hospital Hamburg-Eppendorf, 20251 Hamburg, Germany

4 School of Natural Sciences, Far Eastern Federal University, Vladivostok 690922, Russia

* Correspondence: kvetkinaan@gmail.com; Tel.: +7-423-231-1168

† These two authors contribute equally to this work.

Academic Editor: Luisa Tesoriere

Received: 3 November 2020; Accepted: 15 December 2020; Published: 17 December 2020

Abstract: Actinoporins are the most abundant group of sea anemone cytolytic toxins. Their membranolytic activity is of high interest for the development of novel anticancer drugs. However, to date the activity of actinoporins in malignant cells has been poorly studied. Here, we report on recombinant analog of Hct-S3 (rHct-S3), belonging to the combinatory library of *Heteractis crispa* actinoporins. rHct-S3 exhibited cytotoxic activity against breast MDA-MB-231 (IC_{50} = 7.3 μM), colorectal HT-29 (IC_{50} = 6.8 μM), and melanoma SK-MEL-28 (IC_{50} = 8.3 μM) cancer cells. The actinoporin effectively prevented epidermal growth factor -induced neoplastic transformation of JB6 Cl41 cells by 34% ± 0.2 and decreased colony formation of HT-29 cells by 47% ± 0.9, MDA-MB-231 cells by 37% ± 1.2, and SK-MEL-28 cells by 34% ± 3.6. Moreover, rHct-S3 decreased proliferation and suppressed migration of colorectal carcinoma cells by 31% ± 5.0 and 99% ± 6.4, respectively. The potent anti-migratory activity was proposed to mediate by decreased matrix metalloproteinases-2 and -9 expression. In addition, rHct-S3 induced programmed cell death by cleavage of caspase-3 and poly (ADP-ribose) polymerase, as well as regulation of Bax and Bcl-2. Our results indicate rHct-S3 to be a promising anticancer drug with a high anti-migratory potential.

Keywords: actinoporin; sea anemone; *Heteractis crispa*; anticancer activity; anti-migratory activity

1. Introduction

Cancer is a major public burden with tens of millions people being diagnosed around the world every year. Eventually, more than half of the patients succumb to their disease despite new developments [1]. In 2018, 9.6 million people died from cancer according to the World Health Organization [2], with increasing numbers in developing countries. Lung, breast, colorectal, and prostate cancer belong to the most frequently diagnosed malignancies worldwide [2]. These diseases claim the lives of more than a million people annually.

Although scientific and technological progress allowed the development of new approaches such as gene-therapy [3], stem cell transplantation [4], immunotherapy [5], and therapy by nanoparticles [6] for cancer treatment, the traditional cancer therapy including a combination of surgery, radio- and chemotherapy is still the most commonly used [7–9]. Standard cancer therapy is accompanied by various drawbacks, such as a lack of tumor-specific drug delivery systems, regular application of toxic anticancer drugs leading to adverse side effects, in particular normal cell death, drug resistance, as well as cancer recurrence after surgical removal of solid tumors [5,10]. Therefore, the search for more effective anticancer compounds is ongoing.

In this context, actinoporins, cytolytic toxins derived from sea anemones (marine venomous cnidarians), were identified as promising candidates for cancer therapy. This unique group of small basic α-pore-forming proteins includes a compact β-fold lacking disulfide bounds formed by 12 β-sheets and two α-helices—one of which, functional and more extended, is located at the N-terminus, and the second one, short, is at the C-terminus [11]. Cytotoxicity of actinoporins relies on the formation of pores within sphingomyelin-containing membranes, which disrupts ion gradients that lead to osmotic swelling and ultimately to cell death [12–14]. Cytolytic activity of actinoporins was observed in different cells including platelets, fibroblasts, parasite cells, lactotrophs and some cancer cells [15,16]. Due to pore-forming ability and selective binding to sphingomyelin on the cell membrane surface, high stability to temperature and proteolytic cleavage, they are currently considered as antibacterial and anticancer agents as well as components of immunotoxins [17,18]. These immunotoxins include StnI from *Stichodactyla helianthus* [17], Gigantoxin-4 from *Stichodactyla gigantea* [19], EqII from *Actinia equina* [20] and its mutant EqTx-II(I18C) [21], as well as FraC from *Actinia fragacea* [22].

In our previous studies, we reported the structure and first functional analyses of actinoporins isolated from *Heteractis crispa* (=*Radianthus macrodactylus*) [23–28]. We were able to demonstrate cytotoxic activity of actinoporin RTX-A in monocytic leukemia (THP-1) cells, cervix carcinoma (HeLa), breast (MDA-MB-231) and colon (SNU-C4) cancer cells. In addition, epidermal growth factor (EGF)-induced tumor transformation of mouse epidermal JB6 P^+ Cl41 cells was observed. Activity was mediated by induction of p53-independent apoptosis as well as inhibition of the oncogenic nuclear factors AP-1 and NF-κB [16]. Moreover, we found the combinatorial library of *H. crispa* actinoporins encoded by the multigene family including at least 47 representatives [26–28]. Here, we report the in vitro anticancer activity of the recombinant analog of Hct-S3, the most abundant isoform of *H. crispa* actinoporins.

2. Results

2.1. Obtaining of Recombinant Analog of Hct-S3 Actinoporin

Hct-S3 (177 amino acid residues) is one of the most represented isoforms belonging to the multigene family of *H. crispa* actinoporins. A recombinant analog of Hct-S3 (rHct-S3) was expressed in *Escherichia coli* strain Rosetta (DE3) as fusion protein containing glutation-S-transpherase, polyhistidine tag, enteropeptidase cleavage site, and actinoporin. In order to avoid the denaturation of the target polypeptide and increase its yield, we applied a high-pressure homogenization approach using a French-press homogenizer for the cell destruction instead of ultrasonication. The fusion protein with a molecular mass of ~50 kDa was isolated using a metal-affinity chromatography and cleaved by enteropeptidase. Next, the targeted actinoporin was purified on a soybean trypsin inhibitor -affinity column and further desalted (Figure 1a). The final yield of rHct-S3 after high-pressure homogenization was 1 mg/L of cell culture in contrast to 0.2 mg/L yield after ultrasonication. The molecular mass of the polypeptide was determined by MALDI-TOF/TOF mass-spectrometry as 19,393 Da (Figure 1b), which is consistent with the predicted molecular mass (19,390 Da). The N-terminal amino acid sequence (15 aa) determined by the automated Edman degradation matched well with the amino acid sequence deduced from cDNA earlier.

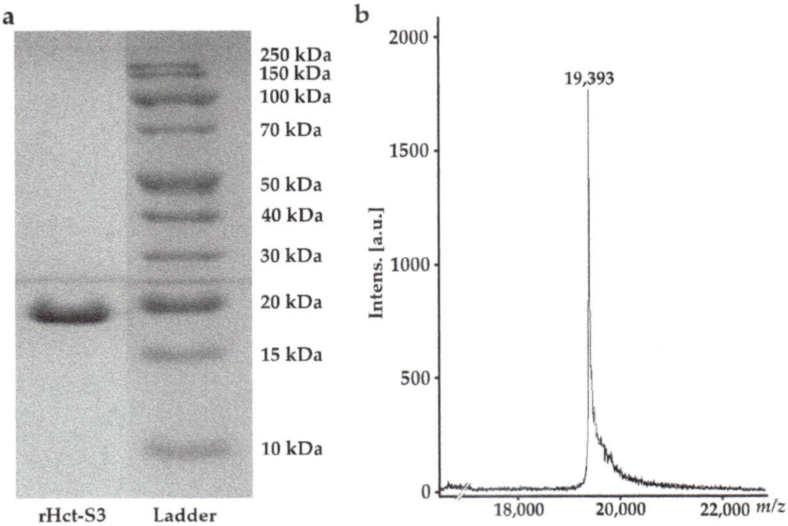

Figure 1. Electrophoregram (**a**) and mass spectrum (**b**) of purified rHct-S3.

2.2. *The Effect of rHct-S3 on Cell Viability*

In order to determine the cytotoxic effect of rHct-S3, the panel of human cancer cell lines HT-29 (colorectal carcinoma), MDA-MB-231 (triple negative breast cancer), SK-MEL-28 (malignant melanoma), as well as normal mouse epidermal JB6 Cl41 cells and human embryonic kidney HEK 293 cells were treated by rHct-S3 at a concentration range 0.01 µM–10 µM for 24 h and cell viability was estimated by MTS assay. rHct-S3 had comparable effects on viability of cell lines with an IC_{50} of 8.6 µM for JB6 Cl41 cells (Figure 2a), 8.5 µM for HEK 293 (Figure 2b), 6.8 µM for HT-29 cells (Figure 2c), 7.3 µM for MDA-MB-231 cells (Figure 2d), and 8.3 µM for SK-MEL-28 cells (Figure 2e).

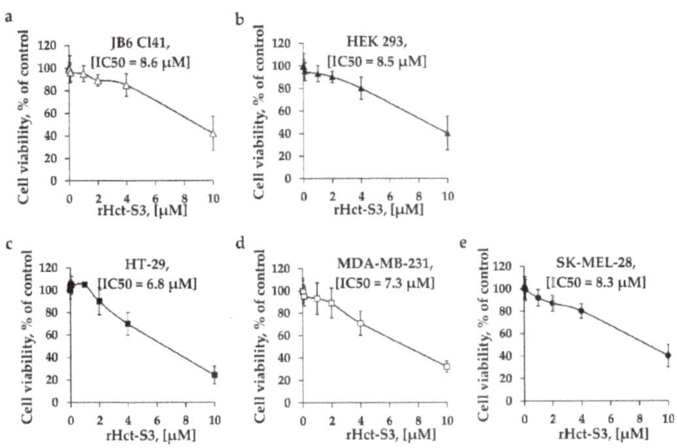

Figure 2. The effect of rHct-S3 on cell viability of (**a**) normal mouse epidermal cells JB6 Cl41, human (**b**) embryonic kidney HEK 293, (**c**) colorectal carcinoma HT-29, (**d**) breast cancer MDA-MB-231, and (**e**) melanoma SK-MEL-28 cell lines. The cytotoxic activity was determined by MTS assay after 24 h of treatment. The results are expressed as the percentage of inhibition that produced a reduction in absorbance by rHct-S3 treatment compared the non-treated cells. Results are expressed as the mean ± standard deviation (SD).

2.3. The Effect of rHct-S3 on EGF-Induced Neoplastic Transformation of Normal Cells and Colony Formation of Cancer Cells

The effects of rHct-S3 on neoplastic transformation of JB6 Cl41 cells induced by EGF, colony formation and growth of cancer cells were studied by soft agar assay, which is considered to be the most accurate type of in vitro test for detecting malignant transformation of cells [29]. The actinoporin was found to inhibit the EGF-induced neoplastic transformation of JB6 Cl41 cells by 10% ± 5.0, 23% ± 2.5, and 34% ± 0.2 at subtoxic concentrations of 1, 2, and 4 μM, respectively (Figure 3a,b). Moreover, rHct-S3 decreased the number of colonies of HT-29 cells by 25% ± 1.8, 33% ± 0.1, and 47% ± 0.9, at concentrations of 1, 2, and 4 μM, respectively, compared to non-treated cells (control) (Figure 3c). At the same doses, rHct-S3 inhibited the colony formation of MDA-MB-231 cells by 17% ± 2.4, 20% ± 2.5, 37% ± 1.2 and SK-MEL-28 cells by 18% ± 1.5, 24% ± 0.5, 34% ± 3.6, respectively (Figure 3d,e). Because activity of rHct-S3 was most pronounced in colorectal carcinoma HT-29 cells, further experiments were carried out with this cell line. It should be noted that chemotherapeutic drug, cisplatin, used as a positive control in this study, inhibited colony formation of HT-29, MDA-MB-231, and SK-MEL-28 cells by 46%, 75%, and 39% at a non-cytotoxic dose of 3 μM, respectively (Figure 3). These results indicate that the actinoporin has a promissing anticancer potential.

2.4. The Effect of rHct-S3 on Migration of Colorectal Carcinoma HT-29 Cells

We investigated the effect of rHct-S3 on the migration of colorectal carcinoma HT-29 cells with high metastatic potential, using a scratch assay. It was demostrated that rHct-S3 supressed migration of HT-29 cells by 33% ± 10.2, 50% ± 7.5, and 99% ± 6.4, respectively, at concentrations of 1, 2, and 4 μM, compared to the control group (Figure 4a,b). In order to reveal the impact of inhibition of proliferation by rHct-S3 on migration, the antiproliferative activity of rHct-S3 against HT-29 cells was checked in 24, 48, 72, and 96 h of treatment (Supplementary Figure S1). It was found that rHct-S3 at concentrations 1, 2, and 4 μM slightly (not more than 10%) decreased the proliferation rate of HT-29 cells after 24 h and 48 h of treatment, while it inhibited cells proliferation by 11% ± 3.0, 26% ± 1.2, and 31% ± 5.0, respectively, after 96 h of treatment. These results indicate that rHct-S3 possess a moderate antiproliferative activity.

To elucidate the potential mechanism of this anti-migratory activity, we evaluated the effect of rHct-S3 on the expression level of the matrix metalloproteinases (MMP)-2 and MMP-9, playing a pivotal role in cancer cell invasion and metastasis. Indeed, actinoporin effectively inhibited the expression of MMP-2 and MMP-9 (Figure 4c) at a concentration of 2 μM. In addition, we estimated whether rHct-S3 affect the activation of caspase-3, a known executor of apoptosis. The upregulation of cleaved caspase-3 was detected in HT-29 cells treated with rHct-S3. Additionally and in line with this, we have detected a degradation of poly (ADP-ribose) polymerase (PARP) as well as Bcl-2 down-regulation and Bax up-regulation (Figure 4d). Thus, rHct-S3 decreases the migratory activity of colorectal carcinoma HT-29 cells by the inhibition of MMP-2 and MMP-9 and induces the apoptosis via the activation of caspase-3.

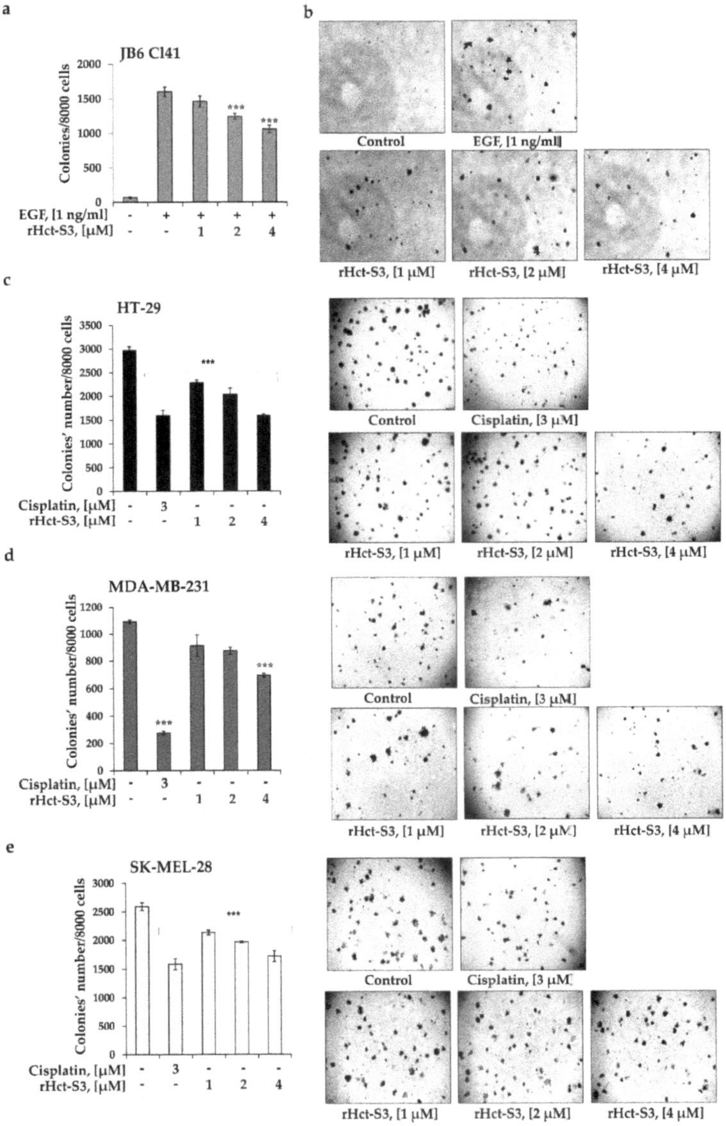

Figure 3. Effect of rHct-S3 on EGF-induced neoplastic cells transformation of JB6 Cl41 cells and colony formation of human colorectal carcinoma HT-29, breast cancer MDA-MB-231, and melanoma SK-MEL-28 cell lines. (**a**,**b**) JB6 Cl41 cells (2.4×10^4 /mL) treated with/without EGF (1 ng/mL) or investigated compound (1, 2, and 4 μM) in 1 mL of 0.3% Basal medium Eagle (BME) agar containing 10% FBS and overlaid with 3.5 mL of 0.5% BME's agar containing 10% FBS. The culture was maintained at 37 °C in a 5% CO_2 atmosphere for 2 weeks. (**c**) HT-29, (**d**) MDA-MB-231, (**e**) SK-MEL-28 cells (2.4×10^4 /mL) treated with/without investigated compound (1, 2, and 4 μM) or cisplatin at 3 μM (positive control) and subjected into a soft agar. The culture was maintained at 37 °C in a 5% CO_2 atmosphere for 2 weeks. The colonies were counted under a microscope with the aid of the ImageJ software program (n = 6 for control and each compound, n—quantity of photos). The magnification of representative photos of colonies is ×10. The asterisks (*** $p < 0.001$) indicate a significant decrease in colony formation in cells treated with compound compared with the non-treated cells (control).

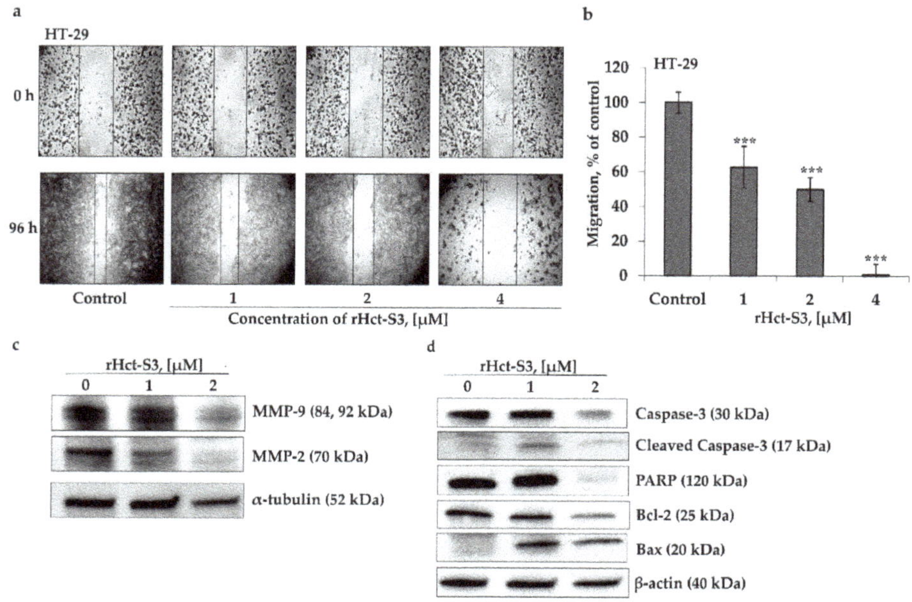

Figure 4. Effects of rHct-S3 on migration of HT-29 cells, MMPs and apoptotic proteins. (**a**,**b**) The HT-29 cells migration distance was measured the width of the wound and expressed as a percentage of each control for the mean of wound closure area. All experiments were repeated at least three times in each group (n = 18 for control and each compound, n—quantity of photos). The magnification of representative photos is ×10. The asterisks (*** $p < 0.001$) indicate a significant decrease in migration of cells treated with rHct-S3 compared with the non-treated cells (control) (**c**,**d**) rHct-S3 inhibited MMP-9, MMP-2 expression and regulated caspase-3, cleaved caspase-3, PARP, Bcl-2 and Bax in HT-29 cells, as determined by Western Blotting with specific antibodies.

3. Discussion

Actinoporins are the major components of sea anemone venom, which disrupt cell membranes by pore formation [30]. *H. crispa* venom contains numerous actinoporin isoforms, encoded by the multigene family [28]. Hct-S3 is one of the isoform of *H. crispa* actinoporins belonging to Hct-S group with Ser at N-terminus (Figure 5). Earlier, the recombinant analog of Hct-S3 was obtained [26,31] and its hemolytic activity was comparable with well-characterized actinoporins such as RTX-A from *H. crispa* [32], EqII from *A. equina* [33] and StnII from *S. helianthus* [34]. Comparative analysis of amino acid sequences of known actinoporins and Hct-S3 revealed that Hct-S3 shared 87–89% identity with Gigantoxin-4 from *S. gigantea*, and RTX-A and StnI from *S. helianthus*, which possess anticancer activity (Figure 5). However, their anticancer mechanism has not been studied in detail. We attempted to elucidate the mechanism of action of actinoporins, in particular, Hct-S3, in different human cancer cells.

The recombinant analog of Hct-S3 was obtained using a previously developed scheme [26] with a changing of cell disruption approach. The lack of carbohydrates and disulfide bridges simplifies the production of recombinant actinoporins by heterologous expression in *E. coli*. However, there are some difficulties with the isolation of soluble actinoporins due to the protein aggregation as inclusion bodies during cells' ultrasonication. It is known that ultrasonic homogenization is a high-energy process of cell disruption. This fact may lead to the samples heating and result in the denaturation of proteins. Therefore, to minimize the protein denaturation we used a high-pressure homogenization of *E. coli* cells that allow us to increase the yield of soluble form of rHct-S3 by five times.

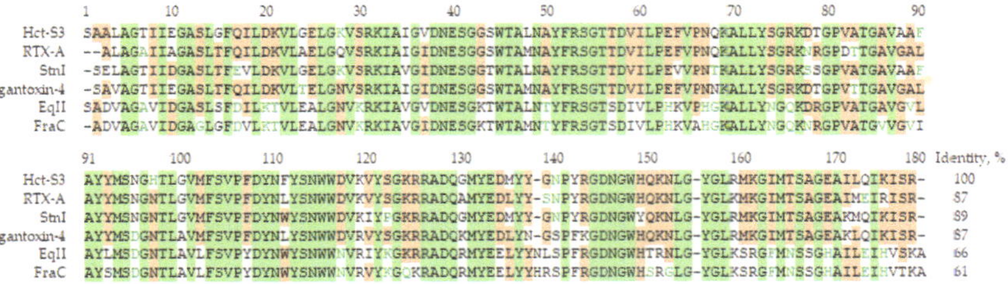

Figure 5. Multiple sequence alignment of actinoporins. EqII (P61914) from *Actinia equina*, FraC (B9W5G6) from *Actinia fragacea*, Gigantoxin-4 (H9CNF5) from *Stichodactyla gigantea*, StnI (P81662) from *Stichodactyla helianthus*, and RTX-A (P58691) from *Heteractis crispa*. The identical and conservative amino acid residues are shown on green and brown background.

Cytotoxic effects of rHct-S3 were studied in normal mouse epidermal, human embryonic kidney cells and human colon carcinoma, breast cancer, and melanoma cells. rHct-S3 exhibited cytotoxic activity against all tested cell lines with comparable IC_{50} values (Figure 2), which were 100–1000-fold higher than those found for other actinoporins [17,18,35]. Previously, StnI and hemolytic fraction of *S. helianthus* were shown to possess cytotoxic activities against colorectal cancer or breast cancer cells, respectively, while RTX-A demonstrated potent cytotoxic activity against both tested cancer cell lines [16].

Carcinogenesis is known to be a multistage process, which includes the initiation (transformation of normal cells into cancer cells), development (formation of colonies of cancer cells) and progression (growth of colonies of cancer cells) of cancer. Cancer prevention is gaining increasing attention because it may be a promising alternative to cancer treatment sparing complications caused by advanced diseases. The involvement of multiple factors and developmental stages and our increased understanding of cancer at the epigenetic, genetic, molecular, and cellular levels is opening up enormous opportunities to interrupt and reverse the initiation and progression of the disease and provide scientists with numerous targets to arrest by physiological and pharmacologic mechanisms, with the goal of preventing end-stage, invasive disease and impeding or delaying the development of cancer [36]. One of the promising strategies for combating carcinogenesis is to search for substances that can prevent the transformation of normal cells into cancer cells induced by various stimulating factors, e.g., epidermal growth factor (EGF), triphorbol ether (TPA), ultraviolet radiation (UV), etc. The promotion-sensitive mouse epidermal JB6 Cl41 cells are known to respond irreversibly to tumor promoters such as epidermal growth factor (EGF) with induction of anchorage-independent growth in soft agar [37]. Therefore, this well-established culture system was used to study the cancer-preventive activity of rHct-S3. Indeed, the actinoporin delayed the EGF-induced neoplastic transformation of JB6 Cl41 cells (Figure 3a,b) and suppressed colony formation of all cancer cell lines (Figure 3c–e), with the inhibition level of HT-29 and SK-MEL-28 cells comparable to cisplatin. Similar activity was previously demonstrated for RTX-A [16]. This polypeptide prevented malignant transformation of JB6 P^+ Cl41 cells and suppressed the growth of HeLa cell colonies at nanomolar concentrations [16].

The most significant cancer-preventive activity of rHct-S3 was found in colon cancer cells. Therefore, we examined the effects of rHct-S3 on the migration of colon cancer cells, as well as their proliferation, in order to incorporate the influence of cell proliferation in the interpretation of the results of migration assays. In fact, tumor cell migration essentially contributes to invasion and metastatic spread, ultimately resulting in progression of disease. More than 30% of patients with colorectal carcinoma have clinically detectable metastases at the time of primary diagnosis [38]. Since the most serious complication and the main cause of death of patients with colorectal carcinoma are distant metastases, the evolution of antimetastatic activity of potential therapeutic agents continues to be an important and urgent task. The mechanism of metastases formation is complex and not fully understood. The migration, intravasation, extravasation of cancer cells and formation of a new vessels (neoangiogenesis) to consolidate a secondary tumor at a distant site are the most important steps of the metastasis process [39]. Remarkably, rHct-S3 almost completely suppressed the migration of HT-29 cells at a concentration of 4 μM (Figure 4a,b). Moreover, the actinoporin possessed a moderate antiproliferative activity (Supplementary Figure S1), but its impact on the anti-migratory activity of rHct-S3 was not significant.

During metastasis, the degradation of extracellular matrix (ECM) and components of the basement membrane by proteases facilitates the detachment of cancer cells, their crossing of tissue boundaries, and invasion into adjacent tissue compartments [40]. In recent years, the importance of cancer-associated proteases such as matrix metalloproteinases MMP-2 and MMP-9 in invasion and metastasis has been reported for a variety of solid malignant tumors [41]. Indeed, the actinoporin was found to effectively inhibit an expression of MMP-2 and MMP-9 (Figure 4c) that apparently resulted in the decrease in HT-29 cell migration. Recently, it was shown that caspase-3 is also able to influence the migration and invasion of colorectal cells [42]. In addition, caspase-3 is a key executioner of programmed cell death. In fact, rHct-S3 treatment cleavage of total caspase-3, followed by PARP cleavage, mediate both anti-migratory activity and induction of apoptosis in HT-29 cells (Figure 4d). In line with the pro-apoptotic activity of rHct-S3, an up-regulation of pro-apoptotic Bax and suppression of anti-apoptotic Bcl-2 were observed.

In conclusion, *H. crispa* actinoporin shows promising anticancer activity with a strong inhibiting effect on the migratory potency of cancer cells. We revealed for the first time that the actinoporin was able to inhibit cancer colony formation and cell migration via suppression of MMP-2 and MMP-9 expression and induce cell apoptosis via activation of caspase-3, cleavage of PARP, activation of Bax and suppression of Blc-2 expressions. The results indicate a high potential of the actinoporin to prevent cancer disease progression. Deep investigations of the underlying mechanism of the effect on apoptotic PI3K/AKT/mTOR, and of cell adhesion signaling pathways, are still to be performed.

4. Materials and Methods

4.1. Expression and Isolation of Recombinant Hct-S3

The recombinant plasmid obtained earlier was transformed into *E. coli* strain Rosetta (DE3) (Novagen, Merck KGaA, Darmstadt, Germany). Transformed cells were cultured at 37 °C in 1 L of Luria-Bertani medium containing 50 μg/mL kanamycin (Gibco, Thermo Fisher Scientific, Gaithersburg, MD, USA) until the optical density (OD_{600}) ~0.5 was reached. After induction with IPTG at a concentration of 0.2 mM, the cells were incubated at 19 °C for 18 h at 180 rpm, centrifuged for 6 min at 6000 rpm at 4 °C, and supernatant was removed. The presence of rHct-S3 was determined in 12% polyacrylamide gel by Laemmli's SDS-PAGE method [43]. Precipitate was resuspended in the start buffer for affinity chromatography (400 mM NaCl, 20 mM Tris-HCl buffer, pH 8.0) and disrupted by French-press homogenizer (Thermo Fisher Scientific, Waltham, MA, USA) using the mini-cell (3.7 mL). Lysed cells were centrifuged for 10 min at 10,000 rpm to remove all insoluble particles. Supernatant was applied to a Ni-NTA agarose (Qiagen, Venlo, Netherlands), the fusion protein was purified with 5 volume of wash buffer (400 mM NaCl, 50 mM imidazole, 20 mM Tris-HCl buffer, pH 8.0) and 5 volume of start buffer. The fusion protein was cleaved by enteropeptidase (1 unit/mg protein) at room temperature for 18 h at 80 rpm. The recombinant actinoporin were purified from enteropeptidase on soybean trypsin inhibitor-agarose (Sigma-Aldrich, St. Louis, MO, USA) and desalted on a centrifugal filter tube (Millipore, Lenexa, KS, USA) with capacity < 3000 Da. The molecular masses of the purified rHct-S3 were analyzed by Ultra Flex III MALDI-TOF/TOF mass spectrometer (Bruker, Bremen, Germany). The amino acid sequence of rHct-S3 were determined on an automated sequencer protein Procise 492 Clc (Applied Biosystems, Foster City, CA, USA).

The purified rHct-S3 was dissolved in PBS, filtered by 0.22 μm "Millipore" membranes (Billerica, MA, USA) and used for the bioactivity experiments.

4.2. Cell Lines

Normal mouse epidermal cells JB6 Cl41 (ATCC® no. CRL-2010™), human colorectal carcinoma HT-29 cells (ATCC® no. HTB-38™), breast cancer MDA-MB-231 cells (ATCC® HTB-26™), and melanoma SK-MEL-28 cells (ATCC® no. HTB-72™) were obtained from the American Type Culture Collection (Manassas, VA, USA).

4.3. Cell Culture

JB6 Cl41, HT-29, MDA-MB-231, and SK-MEL-28 cells were cultured in complete MEM/5% FBS, McCoy's 5A/10% FBS, DMEM/10% FBS, and MEM/10% FBS medium, respectively, containing 1% penicillin-streptomycin solution. The cell cultures were maintained at 37 °C in humidified atmosphere containing 5% CO_2. Every 3–4 days, cells were rinsed with PBS, detached from the tissue culture flask by 0.25% trypsin/0.5 mM EDTA, and 10–20% of the harvested cells were transferred to a new flask containing fresh culture media. The cells after 4–5 passages were used for the experiments. The cells were passaged for a month.

4.4. MTS Assay

To determine cytotoxic activity of rHct-S3, cells (1.0×10^4) were seeded in 96-well plates ("Jet Biofil", Guangzhou, China) and cultured in 200 µL of complete culture medium for 24 h at 37 °C in a 5% CO_2 incubator. The cell monolayer was washed with PBS and treated either with PBS (control) or various concentrations of rHct-S3 (0.01 µM–10 µM) in fresh appropriate culture medium for 24 h. Subsequently, the cells were incubated with 15 µL MTS reagent ("Promega", Madison, WI, USA) for 3 h, and the absorbance of each well was measured at 490/630 nm using Power Wave XS microplate reader ("BioTek", Wynusky, VT, USA).

To determine the antiproliferative activity of rHct-S3, cells (0.7×10^4) were seeded in 96-well plates and cultured in 200 µL of complete culture medium for 24 h at 37 °C in a 5% CO_2 incubator. The cell monolayer was washed with PBS and treated either with PBS (control) or various concentrations of rHct-S3 (1, 2, and 4 µM) in fresh appropriate culture medium for 24, 48, 72, 96 h. Subsequently, the cells were incubated with 15 µL MTS reagent ("Promega", Madison, WI, USA) for 3 h, and the absorbance of each well was measured at 490/630 nm using Power Wave XS microplate reader ("BioTek", Wynusky, VT, USA).

4.5. The Soft Agar Colony Formation Assay

JB6 Cl41 cells (2.4×10^4) were exposed to EGF (1 ng/mL) and treated with rHct-S3 (1, 2, and 4 µM) in 1 mL of 0.3% Basal Medium Eagle (BME) agar containing 10% FBS, 2 mM L-glutamine, and 25 µg/mL gentamicin. The cultures were maintained at 37 °C in a 5% CO_2 incubator for 14 days, and the cell's colonies were scored using a microscope, Motic AE 20 (XiangAn, Xiamen, China) and the ImageJ software.

To estimate the effect of rHct-S3 on colony formation, human cancer cells (2.4×10^4/mL) were seeded into 6-well plate and treated with rHct-S3 (1, 2, and 4 µM) or cisplatin (3 µM) in 1 mL of 0.3% Basal Medium Eagle (BME) agar containing 10% FBS, 2 mM L-glutamine, and 25 µg/mL gentamicin. The cultures were maintained in a 37 °C, 5% CO_2 incubator for 14 days, and the cell colonies were scored as described above.

4.6. Scratch-Wound Assay

Cell migration assay was performed as previously described [44]. Briefly, HT-29 cells (3×10^5 cells/mL) were seeded into 6-well plates and 24 h later the culture medium was removed and straight scratch was created using a 200 µL sterile pipette tip. Cells were washed twice with PBS to remove cellular debris, replaced with appropriate complete culture media containing rHct-S3 (1, 2, and 4 µM) and incubated for 96 h. All experiments were repeated at least three times in each group.

For the image analysis, cell migration into the wound area was photographed at the stages of 0 and 96 h using a microscope, Motic AE 20, and ImageJ software. The cells migration distance was estimated by measuring the width of the wound and expressed as a percentage of each control for the mean of the wound closure area.

4.7. Protein Preparation and Western Blotting

The cells (0.5×10^6 cells/well) were seeded in 6-well plates and incubated overnight. The medium was replaced with fresh medium (1 mL/well) containing the investigated actinoporin at the indicated concentrations, and the cells were incubated for the next 48 h. Cells were harvested by scrapers, pelleted (centrifugation, 5 min, 453× *g*), and washed with PBS (10 mL/samples followed by centrifugation, 3 times). Cells were lysed with 70 µL/sample of lysis buffer (0.88% (*w*/*v*) NaCl, 50 mM Tris-HCl (pH 7.6), 1% NP-40 (*v*/*v*), 0.25% (*w*/*v*) sodium cholate, 1 mM PMSF, 1 mM Na_3VO_4, containing cOmplete™ EASYpacks protease inhibitors cocktail and PhosSTOP™ EASYpacks phosphatase inhibitors cocktail (Roche, Mannheim, Germany)) on ice for 20 min. Lysates were frozen for 1 h, centrifuged (10 min, 11,170× *g*), the protein-containing supernatants were taken, and the protein concentrations were evaluated using the Bradford assay. Protein extracts were diluted with loading buffer according to the manufacture's recommendations, heated for 5 min at 99 °C, and subjected to electrophoresis in 4–12.5% gradient SDS-PAGE (Cat. No. 4568083, Bio-Rad, Hercules, CA, USA) at 120 V. A quantity of 20 µg of total protein was loaded to each slot of the gel. Proteins were then transferred from gel to a 0.2 µm pore PVDF membrane (Millipore, Bedford, MA, USA) using the Bio-Rad transfer system (Cat. No. 10026938, Bio-Rad) and the membranes were blocked with 5% (*w*/*v*) BSA in 0.05% Tween-20/TBS. The membranes were consequently treated with the primary and secondary antibodies, according to the manufacturer's protocol. Signals were detected using the enhanced chemiluminescence system (Thermo Fisher Scientific, Rockford, IL, USA) according to the manufacturer's protocol. The following primary and secondary antibody were used: anti-caspase-3 (mAb, anti-rabbit, #9662, dilution 1:1000, Cell Signaling), anti-MMP-2 (mAb, anti-rabbit, #13132, dilution 1:1000, Cell Signaling), anti-MMP-9 (mAb, anti-rabbit, #13667, dilution 1:1000, Cell Signaling), anti-PARP (pAb, anti-rabbit, #9542, dilution 1:1000, Cell Signaling), anti-Bcl-2 (pAb, anti-rabbit, #2876, dilution 1:1000, Cell Signaling), anti-α-tubulin (mAb, anti-mouse, T5168, dilution 1:5000, Sigma-Aldrich), anti-Bax (mAb, anti-rabbit, #5023, dilution 1:1000, Cell Signaling),anti-β-actin (mAb, anti-mouse, #CP01, dilution 1:10,000, Calbiochem), secondary anti-mouse IgG-HRP (sheep, #NXA931, dilution 1:10,000, GE Healthcare), secondary anti-rabbit IgG-HRP (goat, #7074, dilution 1:5000, Cell Signaling).

4.8. Statistical Analysis

All assays were performed in at least three independent experiments. Results are expressed as the mean ±standard deviation (SD). Student's t test was used to evaluate the data with the following significance levels: * $p < 0.05$, ** $p < 0.01$, *** $p < 0.001$.

Supplementary Materials: The following are available online, Figure S1: The effect of rHct-S3 on proliferation of human colorectal carcinoma HT-29 cell lines.

Author Contributions: E.L., A.K., and O.M. designed the experiment and wrote the manuscript; A.K. and A.P. obtained recombinant analog of actinoporin; O.M., S.D. provided experiments on anticancer activity; E.L., S.E. and G.v.A. critically reviewed the manuscript. All authors have read and agreed to the published version of the manuscript.

Funding: This research was funded by the Russian Scientific Foundation, grant number 18-74-10028.

Acknowledgments: The authors are grateful to Pavel Dmitrenok for determination of molecular mass. The authors are grateful to Jessica Hauschild (University Medical Center Hamburg-Eppendorf) for the technical assistance of biological experiments.

Conflicts of Interest: The authors declare no conflict of interest.

References

1. Ma, X.; Yu, H. Global burden of cancer. *Yale J. Biol. Med.* **2006**, *79*, 85–94. [PubMed]
2. World Health Organization. *Who Report on Cancer: Setting Priorities, Investing Wisely and Providing Care for All*; WHO: Geneva, Switzerland, 2020; ISBN 9789240001299.
3. Naldini, L. Gene therapy returns to centre stage. *Nature* **2015**, *526*, 351–360. [CrossRef] [PubMed]
4. Hawsawi, Y.M.; Al-Zahrani, F.; Mavromatis, C.H.; Baghdadi, M.A.; Saggu, S.; Oyouni, A.A.A. Stem cell applications for treatment of cancer and autoimmune diseases: Its promises, obstacles, and future perspectives. *Technol. Cancer Res. Treat.* **2018**, *17*, 1–12. [CrossRef] [PubMed]
5. Allahyari, H.; Heidari, S.; Ghamgosha, M.; Saffarian, P.; Amani, J. Immunotoxin: A new tool for cancer therapy. *Tumor Biol.* **2017**, *39*, 1–11. [CrossRef] [PubMed]
6. Sahakyan, N.; Haddad, A.; Richardson, S.; Forcha-Etieundem, V.; Christopher, L.; Alharbi, H.; Campbell, R. Personalized nanoparticles for cancer therapy: A call for greater precision. *Anticancer. Agents Med. Chem.* **2017**, *17*, 1033–1039. [CrossRef]
7. Peer, D.; Karp, J.M.; Hong, S.; Farokhzad, O.C.; Margalit, R.; Langer, R. Nanocarriers as an emerging platform for cancer therapy. *Nat. Nanotechnol.* **2007**, *2*, 751–760. [CrossRef]
8. Amit, D.; Hochberg, A. Development of targeted therapy for bladder cancer mediated by a double promoter plasmid expressing diphtheria toxin under the control of H19 and IGF2-P4 regulatory sequences. *J. Transl. Med.* **2010**, *8*, 134. [CrossRef]
9. Kang, T.H.; Mao, C.P.; He, L.; Tsai, Y.C.; Liu, K.; La, V.; Wu, T.C.; Hung, C.F. Tumor-targeted delivery of IL-2 by NKG2D leads to accumulation of antigen-specific CD8+ T cells in the tumor loci and enhanced anti-tumor effects. *PLoS ONE* **2012**, *7*, 1–9. [CrossRef]
10. Wang, L.; Dong, C.; Li, X.; Han, W.; Su, X. Anticancer potential of bioactive peptides from animal sources (Review). *Oncol. Rep.* **2017**, *38*, 637–651. [CrossRef]
11. Athanasiadis, A.; Anderluh, G.; Maček, P.; Turk, D. Crystal structure of the soluble form of equinatoxin II, a pore-forming toxin from the sea anemone *Actinia equina*. *Structure* **2001**, *9*, 341–346. [CrossRef]
12. MacGregor, R.D.; Tobias, C.A. Molecular sieving of red cell membranes during gradual osmotic hemolysis. *J. Membr. Biol.* **1972**, *10*, 345–356. [CrossRef] [PubMed]
13. Anderluh, G.; Maček, P. Cytolytic peptide and protein toxins from sea anemones (Anthozoa: Actinaria). *Toxicon* **2002**, *40*, 111–124. [CrossRef]
14. Yap, W.Y.; Hwang, J.S. Response of cellular innate immunity to cnidarian pore-forming toxins. *Molecules* **2018**, *23*, 2537. [CrossRef] [PubMed]
15. Tejuca, M.; Anderluh, G.; Dalla Serra, M. Sea anemone cytolysins as toxic components of immunotoxins. *Toxicon* **2009**, *54*, 1206–1214. [CrossRef]
16. Fedorov, S.; Dyshlovoy, S.; Monastyrnaya, M.; Shubina, L.; Leychenko, E.; Jin, J.O.; Kwak, J.Y.; Bode, A.M.; Dong, Z.; Stonik, V. The anticancer effects of actinoporin RTX-A from the sea anemone *Heteractis crispa* (=*Radianthus macrodactylus*). *Toxicon* **2010**, *5*, 811–817. [CrossRef]
17. Avila, A.D.; Mateo de Acosta, C.; Lage, A. A carcinoembryonic antigen-directed immunotoxin built by linking a monoclonal antibody to a hemolytic toxin. *Int. J. Cancer* **1989**, *43*, 926–929. [CrossRef]
18. Álvarez, C.; Mancheño, J.M.; Martínez, D.; Tejuca, M.; Pazos, F.; Lanio, M.E. Sticholysins, two pore-forming toxins produced by the Caribbean sea anemone *Stichodactyla helianthus*: Their interaction with membranes. *Toxicon* **2009**, *54*, 1135–1147. [CrossRef]
19. Lv, X.; Zhang, J.; Xu, R.; Dong, Y.; Sun, A.; Shen, Y.; Wei, D. Gigantoxin-4-4D5 scFv is a novel recombinant immunotoxin with specific toxicity against HER2/neu-positive ovarian carcinoma cells. *Appl. Microbiol. Biotechnol.* **2016**, *100*, 6403–6413. [CrossRef]
20. Pederzolli, C.; Belmonte, G.; Serra, M.D.; Menestrina, G.; Maček, P. Biochemical and cytotoxic properties of conjugates of transferrin with equinatoxin II, a cytolysin from a sea anemone. *Bioconjug. Chem.* **1995**, *6*, 166–173. [CrossRef]
21. Potrich, C.; Anderluh, G.; Maˇcek, P. Construction of new immunotoxins by linking equinatoxin II to monoclonal antibodies via the biotin-avidin interaction. Cytotoxic effects on human tumor cells. *Acta Biol. Slov.* **2000**, *43*, 47–51.

22. Mutter, N.L.; Soskine, M.; Huang, G.; Albuquerque, I.S.; Bernardes, G.J.L.; Maglia, G. Modular pore-forming immunotoxins with caged cytotoxicity tailored by directed evolution. *ACS Chem. Biol.* **2018**, *13*, 3153–3160. [CrossRef] [PubMed]
23. Klyshko, E.V.; Issaeva, M.P.; Monastyrnaya, M.M.; Il'yna, A.P.; Guzev, K.V.; Vakorina, T.I.; Dmitrenok, P.S.; Zykova, T.A.; Kozlovskaya, E.P. Isolation, properties and partial amino acid sequence of a new actinoporin from the sea anemone *Radianthus macrodactylus*. *Toxicon* **2004**, *44*, 315–324. [CrossRef] [PubMed]
24. Il'ina, A.; Lipkin, A.; Barsova, E.; Issaeva, M.; Leychenko, E.; Guzev, K.; Monastyrnaya, M.; Lukyanov, S.; Kozlovskaya, E. Amino acid sequence of RTX-A's isoform actinoporin from the sea anemone, *Radianthus macrodactylus*. *Toxicon* **2006**, *47*, 517–520. [CrossRef] [PubMed]
25. Monastyrnaya, M.; Leychenko, E.; Isaeva, M.; Likhatskaya, G.; Zelepuga, E.; Kostina, E.; Trifonov, E.; Nurminski, E.; Kozlovskaya, E. Actinoporins from the sea anemones, tropical *Radianthus macrodactylus* and northern *Oulactis orientalis*: Comparative analysis of structure-function relationships. *Toxicon* **2010**, *56*, 1299–1314. [CrossRef] [PubMed]
26. Tkacheva, E.S.; Leychenko, E.V.; Monastyrnaya, M.M.; Issaeva, M.P.; Zelepuga, E.A.; Anastuk, S.D.; Dmitrenok, P.S.; Kozlovskaya, E.P. New Actinoporins from sea anemone *Heteractis crispa*: Cloning and functional expression. *Biochem. Biokhimii͡a* **2011**, *76*, 1131–1139. [CrossRef] [PubMed]
27. Leichenko, E.V.; Monastirnaya, M.M.; Zelepuga, E.A.; Tkacheva, E.S.; Isaeva, M.P.; Likhatskaya, G.N.; Anastyuk, S.D.; Kozlovskaya, E.P. Hct-A is a new actinoporin family from the *Heteractis crispa* sea anemone. *Acta Naturae* **2014**, *6*, 89–98. [CrossRef]
28. Leychenko, E.; Isaeva, M.; Tkacheva, E.; Zelepuga, E.; Kvetkina, A.; Guzev, K.; Monastyrnaya, M.; Kozlovskaya, E. Multigene family of pore-forming toxins from sea anemone *Heteractis crispa*. *Mar. Drugs* **2018**, *16*, 183. [CrossRef]
29. Borowicz, S.; Van Scoyk, M.; Avasarala, S.; Karuppusamy Rathinam, M.K.; Tauler, J.; Bikkavilli, R.K.; Winn, R.A. The soft agar colony formation assay. *J. Vis. Exp.* **2014**, 1–6. [CrossRef]
30. Rojko, N.; Dalla Serra, M.; MačEk, P.; Anderluh, G. Pore formation by actinoporins, cytolysins from sea anemones. *Biochim. Biophys. Acta Biomembr.* **2016**, *1858*, 446–456. [CrossRef]
31. Tkacheva, E.S. Functional expression of actinoporin RTX-S3, a new cytolytic toxin of the sea anemone *Heteractis crispa* (=*Radianthus macrodactylus*). *Vestn. FEB RAS* **2010**, *4*, 134–137.
32. Monastyrnaya, M.M.; Zykova, T.A.; Apalikova, O.V.; Shwets, T.V.; Kozlovskaya, E.P. Biologically active polypeptides from the tropical sea anemone *Radianthus macrodactylus*. *Toxicon* **2002**, *40*, 1197–1217. [CrossRef]
33. Maček, P.; Lebez, D. Kinetics of hemolysis induced by equinatoxin, a cytolytic toxin from the sea anemone *Actinia equina*. Effect of some ions and pH. *Toxicon* **1981**, *19*, 233–240. [CrossRef]
34. Lanio, M.E.; Morera, V.; Alvarez, C.; Tejuca, M.; Gómez, T.; Pazos, F.; Besada, V.; Martínez, D.; Huerta, V.; Padrón, G.; et al. Purification and characterization of two hemolysins from *Stichodactyla helianthus*. *Toxicon* **2001**, *39*, 187–194. [CrossRef]
35. Tejuca, M.; Diaz, I.; Figueredo, R.; Roque, L.; Pazos, F.; Martinez, D.; Iznaga-Escobar, N.; Perez, R.; Alvarez, C.; Lanio, M.E. Construction of an immunotoxin with the pore forming protein StI and ior C5, a monoclonal antibody against a colon cancer cell line. *Int. Immunopharmacol.* **2004**, *4*, 731–744. [CrossRef] [PubMed]
36. Amin, A.R.M.R.; Kucuk, O.; Khuri, F.R.; Shin, D.M. Perspectives for cancer prevention with natural compounds. *J. Clin. Oncol.* **2009**, *27*, 2712–2725. [CrossRef]
37. Colburn, N.H.; Former, B.F.; Nelson, K.A.; Yuspa, S.H. Tumour promoter induces anchorage independence irreversibly. **1979**, *281*, 589–591. [CrossRef] [PubMed]
38. Davies, R.J.; Miller, R.; Coleman, N. Colorectal cancer screening: Prospects for molecular stool analysis. *Nat. Rev. Cancer* **2005**, *5*, 199–209. [CrossRef]
39. Hunter, K.W.; Crawford, N.P.S.; Alsarraj, J. Mechanisms of metastasis. *Breast Cancer Res.* **2008**, *10*, 1–10. [CrossRef]
40. Fingleton, B. Matrix metalloproteinases: Roles in cancer and metastasis. *Front. Biosci.* **2006**, *11*, 479–491. [CrossRef]
41. Mysliwiec, A.G.; Ornstein, D.L. Matrix metalloproteinases in colorectal cancer. *Clin. Colorectal Cancer* **2002**, *1*, 208–219. [CrossRef]
42. Zhou, M.; Liu, X.; Li, Z.; Huang, Q.; Li, F.; Li, C.Y. Caspase-3 regulates the migration, invasion and metastasis of colon cancer cells. *Int. J. Cancer* **2018**, *143*, 921–930. [CrossRef] [PubMed]

43. Laemmli, U.K. Cleavage of structural proteins during the assembly of the head of bacteriophage T4. *Nature* **1970**, *227*, 680–685. [CrossRef] [PubMed]
44. Malyarenko, O.S.; Usoltseva, R.V.; Zvyagintseva, T.N.; Ermakova, S.P. Laminaran from brown alga *Dictyota dichotoma* and its sulfated derivative as radioprotectors and radiosensitizers in melanoma therapy. *Carbohydr. Polym.* **2019**, *206*, 539–547. [CrossRef] [PubMed]

Sample Availability: Sample of the compound rHct-S3 are available from the authors.

molecules

MDPI

Article

Human Lanosterol 14-Alpha Demethylase (CYP51A1) Is a Putative Target for Natural Flavonoid Luteolin 7,3′-Disulfate

Leonid Kaluzhskiy [1,*], Pavel Ershov [1], Evgeniy Yablokov [1], Tatsiana Shkel [2], Irina Grabovec [2], Yuri Mezentsev [1], Oksana Gnedenko [1], Sergey Usanov [2], Polina Shabunya [2], Sviatlana Fatykhava [2], Alexander Popov [3], Aleksandr Artyukov [3], Olga Styshova [3], Andrei Gilep [2], Natallia Strushkevich [4,*] and Alexis Ivanov [1]

[1] Institute of Biomedical Chemistry, 10 Building 8, Pogodinskaya Street, 119121 Moscow, Russia; pavel.ershov@ibmc.msk.ru (P.E.); evgeniy.yablokov@ibmc.msk.ru (E.Y.); yuri.mezentsev@ibmc.msk.ru (Y.M.); oksana@ibmh.msk.su (O.G.); alexei.ivanov@ibmc.msk.ru (A.I.)

[2] Institute of Bioorganic Chemistry NASB, 5 Building 2, V.F. Kuprevich Street, 220141 Minsk, Belarus; tvshkel@gmail.com (T.S.); grabovec-irina@mail.ru (I.G.); usanov@iboch.by (S.U.); iboh_lfhi@rambler.ru (P.S.); fsa1981@tut.by (S.F.); agilep@iboch.by (A.G.)

[3] G.B. Elyakov Pacific Institute of Bioorganic Chemistry, Far Eastern Branch of the Russian Academy of Science, 159 Prospect 100-letiya Vladivostoka, 690022 Vladivostok, Russia; popovam@piboc.dvo.ru (A.P.); artyukova@mail.ru (A.A.); krivoshapkoon@mail.ru (O.S.)

[4] Skolkovo Institute of Science and Technology, Bolshoy Boulevard 30, bld. 1, 121205 Moscow, Russia

* Correspondence: leonid.kaluzhskiy@ibmc.msk.ru (L.K.); n.strushkevich@skoltech.ru (N.S.)

Citation: Kaluzhskiy, L.; Ershov, P.; Yablokov, E.; Shkel, T.; Grabovec, I.; Mezentsev, Y.; Gnedenko, O.; Usanov, S.; Shabunya, P.; Fatykhava, S.; et al. Human Lanosterol 14-Alpha Demethylase (CYP51A1) Is a Putative Target for Natural Flavonoid Luteolin 7,3′-Disulfate. *Molecules* **2021**, *26*, 2237. https://doi.org/10.3390/molecules26082237

Academic Editor: Celestino Santos-Buelga

Received: 10 March 2021
Accepted: 12 April 2021
Published: 13 April 2021

Publisher's Note: MDPI stays neutral with regard to jurisdictional claims in published maps and institutional affiliations.

Abstract: Widespread pathologies such as atherosclerosis, metabolic syndrome and cancer are associated with dysregulation of sterol biosynthesis and metabolism. Cholesterol modulates the signaling pathways of neoplastic transformation and tumor progression. Lanosterol 14-alpha demethylase (cytochrome P450(51), CYP51A1) catalyzes one of the key steps in cholesterol biosynthesis. The fairly low somatic mutation frequency of CYP51A1, its druggability, as well as the possibility of interfering with cholesterol metabolism in cancer cells collectively suggest the clinical importance of CYP51A1. Here, we show that the natural flavonoid, luteolin 7,3′-disulfate, inhibits CYP51A1 activity. We also screened baicalein and luteolin, known to have antitumor activities and low toxicity, for their ability to interact with CYP51A1. The K_d values were estimated using both a surface plasmon resonance optical biosensor and spectral titration assays. Unexpectedly, in the enzymatic activity assays, only the water-soluble form of luteolin—luteolin 7,3′-disulfate—showed the ability to potently inhibit CYP51A1. Based on molecular docking, luteolin 7,3′-disulfate binding suggests blocking of the substrate access channel. However, an alternative site on the proximal surface where the redox partner binds cannot be excluded. Overall, flavonoids have the potential to inhibit the activity of human CYP51A1 and should be further explored for their cholesterol-lowering and anti-cancer activity.

Keywords: lanosterol 14-alpha demethylase; flavonoids; enzyme inhibition; surface plasmon resonance; spectral titration; molecular docking

1. Introduction

Cholesterol is a major source of bioactive sterols. Cholesterol modulates the signaling pathways of neoplastic transformation and tumor progression by covalently modifying hedgehog and smoothened family proteins [1,2] and it is also involved in atherosclerosis and metabolic syndrome progression [3]. Tumor growth is accompanied by a significant increase in the expression level of cholesterol biosynthetic enzymes, including lanosterol 14-alpha demethylase (cytochrome P450(51), CYP51A1) [4]. CYP51A1 belongs to the evolutionarily conserved family of cytochrome P450 and catalyzes the oxidative removal of the alpha-methyl group at the C14-position of the steroid substrate in three steps [5]. According to the COSMIC (http://cancer.sanger.ac.uk/cosmic accessed on 29 June 2020) resource, the CYP51A1 gene has a fairly low somatic mutation frequency (<0.1%) in various cancers. It has been shown that CYP51A1 gene expression correlates with the estrogen and progesterone receptor status of breast cancer [6] and could be one of the factors in assessing the survival rate of patients with gastric adenocarcinoma [7]. CYP51A1 catalyzes the production of 4,4-dimethyl-5α-cholesta-8,14,24-triene-3β-ol (follicular fluid meiosis-activating sterol, FF-MAS), one of the modulators of meiosis [8]. The ploidy disturbance characteristic for cancer cells is caused by processes similar to meiosis [9]. Thus, FF-MAS might be linked to the ploidy balance of tumor cells. CYP51A1 gene knockout blocked de novo cholesterol synthesis [10], while CYP51A1 inhibition led to the induction of apoptosis in cancer cells [11], indicating the clinical significance of this protein.

We analyzed the potential of natural flavonoids (Figure 1) as modulators of CYP51A1 function using purified human protein. Baicalein and luteolin are flavonoids that were originally isolated from plants of the *Scutellaria* and *Reseda* genus. The inhibitory activity of baicalein and luteolin was demonstrated for some cytochrome P450 isozymes [12–14]. Baicalein exhibited broad antifungal activity [15] and demonstrated strong synergy with fluconazole [16], a known inhibitor of fungal CYP51A1. Luteolin possesses an antibacterial effect in vivo, increasing membrane permeability [17], but does not directly perturb the model membranes in vitro [18,19]. Both baicalein and luteolin possess antitumor activity [20,21] and, as well as their derivatives, have been used in preclinical studies and in experimental oncology [22–27]. Epidemiological studies showed that foods rich in polyphenolic compounds (flavonoids, phenolic acids, lignans and stilbenes) included in the diet reduced the total risk of cancer by up to 50% [28]. Overall, 14 clinical trials were initiated to study baicalein and luteolin as dietary supplements (https://clinicaltrials.gov/, accessed on 1 March 2021). The G.B. Elyakov Pacific Institute of Bioorganic Chemistry has a broad collection of natural compounds from unique Far-Eastern plants and marine species with a wide range of biological activities. Luteolin 7,3′-disulfate, a water-soluble luteolin derivative originally obtained from the seagrass *Zostera marina* [29], also exhibits antitumor activity [30,31]. It was shown that sulfation at the 7-position of the luteolin molecule decreases cytotoxicity [32]. Moreover, the activity of luteolin 7,3′-disulfate in some cases is stronger than that of luteolin [30,33,34], possibly due to bypassing the stage of conjugation by intestinal and liver cells. Cholesterol is a crucial component of membranes, maintaining their permeability and fluidity. We hypothesized that flavonoids might target its synthesis via CYP51A1 and selected baicalein, luteolin and luteolin 7,3′-disulfate for testing. Using surface plasmon resonance (SPR), we showed that only luteolin 7,3′-disulfate interacted with CYP51A1 with high affinity. However, in the spectral binding experiments luteolin 7,3′-disulfate does not induce spectral changes. In contrast, baicalein and luteolin induce a reverse type I response in the difference absorption spectra of CYP51A1, indicating changes around heme iron. In the reconstituted enzymatic assay, among the three tested flavonoids, only luteolin 7,3′-disulfate inhibited the lanosterol 14α-demethylase activity of human CYP51A1 with significant potency. The binding mode distant from the heme was predicted for luteolin 7,3′-disulfate by the performed molecular docking, showing the binding not in the hydrophobic active site but rather in the access channel. The inhibitory effect of the most hydrophilic form of tested flavonoids—luteolin 7,3′-disulfate—is quite unusual. We suggest that, besides the predicted binding, luteolin 7,3′-disulfate could also bind to the proximal surface of CYP51A1, interfering with the interaction to the redox partner. The obtained data open up a new valuable source of flavonoid modulators of CYP51A1 activity as an alternative to the classic inhibition by azole compounds.

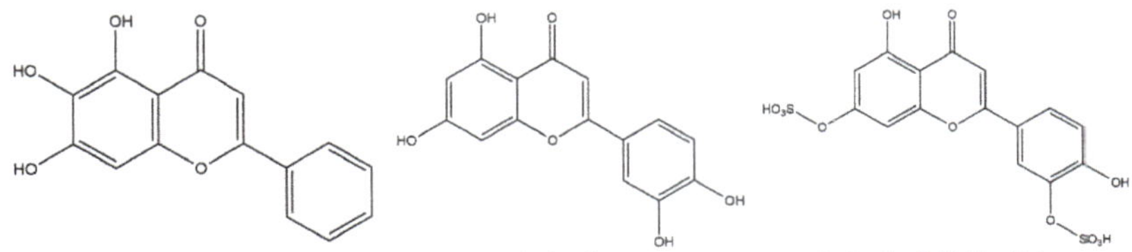

Baicalein **Luteolin** **Luteolin 7,3'-disulfate**

Figure 1. Flavonoids used in this work.

2. Results

2.1. Surface Plasmon Resonance

The CYP51A1 complex formation with flavonoids was detected using a SPR-biosensor. Lanosterol was used as a positive control to confirm the ability of immobilized CYP51A1 to bind ligands. With the CYP51A1 immobilized on the biosensor chip surface, we were able to detect the interaction with baicalein, luteolin and luteolin 7,3′-disulfate (Figure 2).

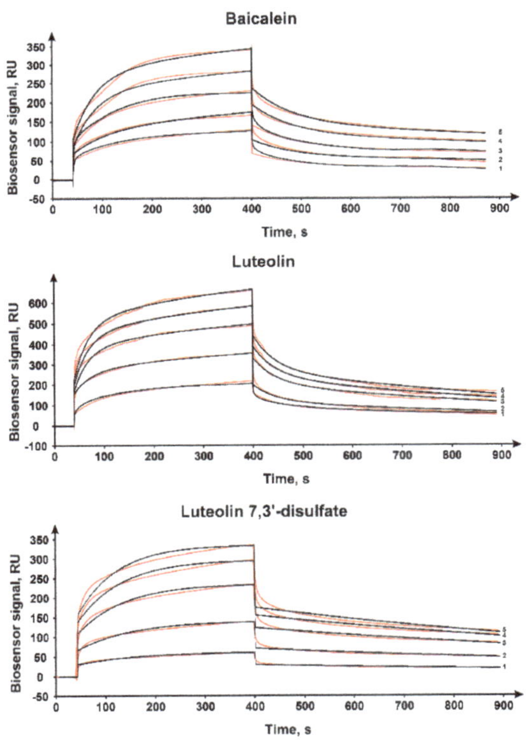

Figure 2. Typical surface plasmon resonance sensorgrams of binding between immobilized CYP51A1 on the optical chip and baicalein, luteolin and luteolin 7.3′-disulfate at different concentrations: 10 (1), 25 (2), 50 (3), 75 (4) and 100 μM (5). Fitting curves (theoretical models) are highlighted in black; Chi^2 = 25.3 (baicalein), 68.2 (luteolin), 10.2 (luteolin 7,3′-disulfate).

The equilibrium dissociation constant (K_d) values of CYP51A1/flavonoid complexes were in the range of 2.9–20 μM, calculated association and dissociation rate constants are shown in Table 1. The obtained Kd value of the CYP51A1/lanosterol complex was 2.4 μM, which is comparable with the previously published data [35]. The association rate of the CYP51A1 complex with luteolin 7,3′-disulfate is seven times faster compared to the complex formation with lanosterol, while the dissociation rate is about eight times higher. The resulting K_d value for both complexes is similar. CYP51A1 complex formation with baicalein and luteolin is characterized by the increased association rate compared to lanosterol, but the main differences in the resulting K_d value are due to the great increase in dissociation rates of the complexes. Overall, the binding of flavonoids is faster compared to the natural substrate, but the dissociation of the complexes is faster as well. The highest affinity was detected for luteolin 7,3′-disulfate, which is more soluble.

Table 1. Kinetic and equilibrium parameters of cytochrome P450(51) (CYP51A1) complex formation with lanosterol, baicalein, luteolin and luteolin 7,3′-disulfate.

Compound	k_{on}	k_{off}	K_d, μM	Evaluation Model
lanosterol	k_{on} (1/Ms) = 41.4 ± 5.0	k_{off} (1/s × 10^{-4}) = 1.0 ± 0.2	2.4	Langmuir 1:1
baicalein	k_{on1} (1/Ms) = 146 ± 20 k_{on2} (1/s × 10^{-4}) = 27 ± 3	k_{off1} (1/s × 10^{-4}) = 100 ± 20 k_{off2} (1/s × 10^{-4}) = 6 ± 1	12.5	Two state reaction
luteolin	k_{on1} (1/Ms) = 282 ± 40 k_{on2} (1/s × 10^{-4}) = 33 ± 4	k_{off1} (1/s × 10^{-4}) = 190 ± 30 k_{off2} (1/s × 10^{-4}) = 14 ± 2	20.0	Two state reaction
luteolin 7,3′-disulfate	k_{on} (1/Ms) = 294.0 ± 32.3	k_{off} (1/s × 10^{-4}) = 8.4 ± 2.0	2.9	Langmuir 1:1

The table shows the average values of the parameters ± standard deviation, n = 3.

2.2. Spectral Titration Analysis

The difference spectra of CYP51A1 were obtained by titration with baicalein, luteolin and luteolin 7,3′-disulfate in the presence of lanosterol. Baicalein and luteolin induced a reverse type I spectral response with absorbance minimum at 390 nm and maximum at 420 nm for luteolin and 436 nm for baicalein (Figure 3). These spectral changes are consistent with the previously detected interaction of cytochrome P450 1B1 with compounds of flavonoid class [36]. Titration with luteolin 7,3′-disulfate (up to 30 μM) does not cause changes in the difference spectrum of CYP51A1. The apparent dissociation constant (K_{dapp}) values of the complexes of CYP51A1 with baicalein and luteolin were 8.2 ± 0.4 and 5.1 ± 0.5 μM, respectively. It should be noted that the K_d values from spectrophotometric titration experiments differ from those obtained using SPR. These differences can be attributed to the different affinities of the complexes in solution and immobilized on the surface of the optical chip. Interaction with the different sites of the enzyme cannot be excluded during SPR measurements and the measured Kd reflects all possible interactions between the ligand and enzyme, while spectral assays detect interactions of ligand only within close vicinity of the heme cofactor buried in the CYP active site.

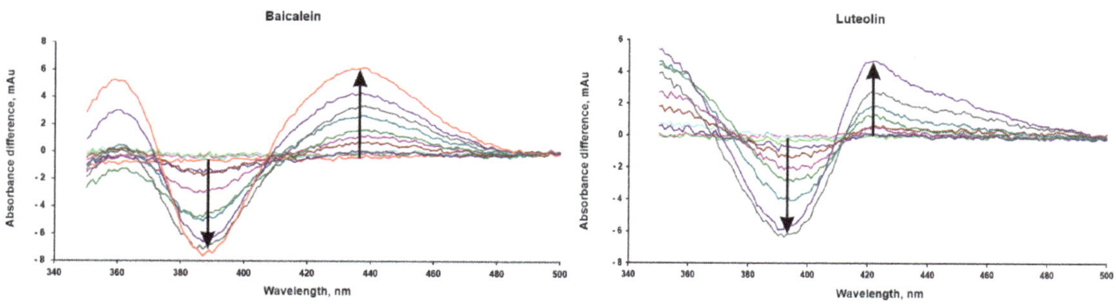

Figure 3. Difference spectra of CYP51A1 in the presence of lanosterol after addition of baicalein (up to 30 μM) and luteolin (up to 15 μM). The arrows indicate the direction of the spectral changes with increasing ligand concentration.

2.3. Enzyme Activity Assay

Lanosterol 14α-demethylase activity of human CYP51A1 in the presence of flavonoids was determined in the reconstituted system. Only luteolin 7,3′-disulfate can inhibit the activity of the CYP51A1 (Table 2). Surprisingly, luteolin, being a more hydrophobic molecule compared to its sulfated derivative, does not have a similar effect. The apparent IC50 for luteolin 7,3′-disulfate is greater than 25 μM. At the same time, the level of inhibition by ketoconazole (94.6% at a concentration of compound of 5 μM) significantly exceeds the effect of luteolin 7,3′-disulfate (50.1% at a concentration of compound of 25 μM). Overall, the inhibition of CYP51A1 utilizing highly hydrophobic substrate by the water-soluble luteolin 7,3′-disulfate could not be predicted. This observation suggests a different mode of binding in the active site. To visualize the binding of luteolin and its disulfate in the active site we performed molecular docking.

Table 2. Effect of compounds on catalytic activity of human CYP51A1 (lanosterol 14α-demethylase) in the reconstituted system in vitro.

Compound	Relative Activity, %
No compound	100.0
Baicalein (25 μM)	89.4
Luteolin (25 μM)	92.6
Luteolin 7,3′-disulfate (25 μM)	49.9
Ketoconazole (5 μM)	5.4

The final concentrations of CYP51A1 and cytochrome P450 reductase (CPR) were 0.5 and 2.0 μM, respectively. The final concentration of lanosterol was 50 μM.

2.4. Molecular Docking

We used a CYP51A1 crystal structure Protein Data Bank (PDB) ID: 3LD6 for molecular docking. The resulting models were selected based on the higher values of scoring function. The obtained docking poses are shown in Figure 4. Based on the docking results, luteolin binds very close to the heme coordinating iron (less than 3 Å) by the 3-OH-group of the phenyl ring. In contrast, luteolin 7,3′-disulfate binds at >8.5 Å from the heme. The docking results are consistent with the spectral titration data—luteolin induces reverse type I spectra, while luteolin 7,3′-disulfate does not change the spectral response.

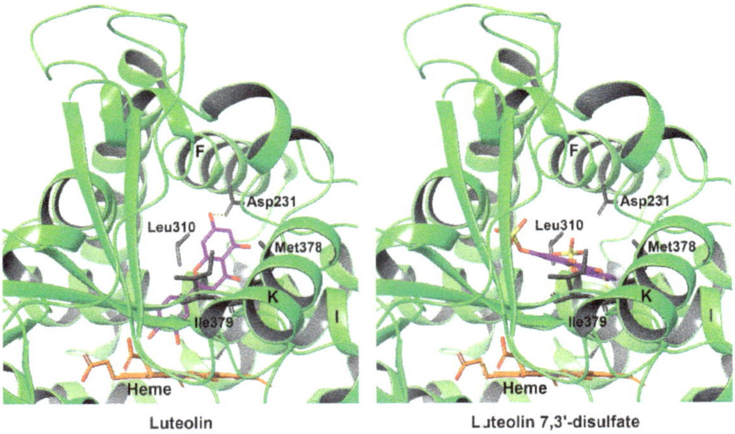

Figure 4. Luteolin and luteolin 7,3′-disulfate docked to the active site of human CYP51A1. The secondary structure of the protein is depicted as a ribbon and colored green. The amino acid side chains are shown as sticks and are colored in grey. The flavonoids and heme are shown as sticks and are colored in magenta and orange, respectively.

Asp231 (C-terminal part of the F-helix) H-bonded to luteolin and is important for the enzymatic activity of CYP51 [37]. The negative charge in this position is highly conserved in Prokaryotes and Eukaryotes [38]. Residues Leu310 (part of I-helix), Met378 and Ile379 (both K-helix–β1-4 loop) are involved in the interaction with luteolin 7,3′-disulfate. Residues Leu310 and Met378 are conservative among *Chordata*, and Ile379 is conservative among primates [38]. Notably, these structural elements were shown to interact with the elongated azole inhibitors (PDB ID: 3LD6, 4UHI and 6Q2T), suggesting that several residues of the active site are utilized for the distant binding of luteolin 7,3′-disulfate.

The docking pose obtained for luteolin 7,3′-disulfate showed binding in the access channel (Figure 4). Thus, the inhibition effect could be the result of blocking of the substrate access channel. However, the inhibition of CYP51A1 by luteolin 7,3′-disulfate does not exclude the modulation of interaction with its redox partner. The proximal surface of CYP51A1—where the redox partner, cytochrome P450 reductase, is binding—contains positively charged amino acids which can interact with the negatively charged sulfate groups of luteolin 7,3′-disulfate.

3. Discussion

Plant flavonoids have a variety of biological activities in animals. However, despite numerous studies in this field, the mechanism/s of action of flavonoids remain poorly understood. Using animal models, it was shown that some flavonoids, luteolin in particular, may mitigate the toxicity of drugs [13,39]. However, the protective effect of flavonoids in humans has not been reliably ascertained [13]. There have been a number of studies reporting the effect of flavonoids, mostly on xenobiotic transformation by CYP enzymes and drug–drug interactions. In particular, baicalein showed an inhibition activity to CYP1A, CYP2B and CYP3A4, with IC50 values in the range of 0.5–36 μM [12,40]. Both baicalein and luteolin inhibit diclofenac 4′-hydroxylase activity in the CYP2C9 RECO system, with baicalein acting as a competitive inhibitor of CYP2C9 [41]. The most effective luteolin inhibition was shown for CYP2C8, while its close homologs, CYP2C9 and CYP2C19, were less effectively inhibited [13]. The activity of CYP1A2, CYP3A, CYP2B6, CYP2E1 and CYP2D6 was also inhibited by luteolin, with an IC50 in the range of 1.6–132.6 μM [13,42]. Notably, luteolin selectively inhibits CYP2D6-mediated metabolism with different substrates. For example, O-demethylation of 3-[2-(N,N-diethyl-N-methyl-ammonium)ethyl]-7-methoxy-4-methylcoumarin was inhibited to 40% by the administration of 20 μM luteolin, while the same concentration of luteolin showed less than a 5% inhibition in reaction with dextromethorphan [14]. Overall, the baicalein and luteolin inhibitory concentration on drug metabolizing CYPs is in the micromolar range.

The inhibition effect of flavonoids was also shown for CYPs involved in the biosynthesis of steroid hormones, neurosteroids, prostaglandins, as well as other regulatory metabolites. The effect of different flavonoids was evaluated on cortisol production in human adrenocortical H295R cells, and the competitive mechanism of inhibition was established for CYP21B1 [43]. The inhibition effect of luteolin was shown for human aromatase CYP19 [44]. A synthetic analogue of dihydrodaidzein, NV-52, inhibited the thromboxane A2 synthase CYP5A1 [45], while isoflavonoids inhibited the oxidation of vitamin D3 by CYP24A1 [46].

Sulfation, methylation and glucuronidation, occurring in the enterocytes and liver, are major factors affecting flavonoid bioavailability and are crucial for their transport via the blood [47]. Non-conjugated flavonoids are generally not present in plasma, however, there is an indication that a small amount of non-conjugated flavonoids can be transported through the blood system [48]. To the best of our knowledge, no studies have been conducted on the inhibition of CYP activity by sulfated forms of baicalein and luteolin. The inhibitory potential of other sulfated derivatives was probed with drug-metabolizing CYPs. It was shown that sulfated derivatives of quercetin and chrysin can inhibit several CYPs in vitro. In particular, quercetin 3′-sulfate has a selective concentration-dependent inhibition activity to CYP2C19 and CYP3A4 up to 30 μM, but the inhibition effect did not exceed 50% and overall was less than that of ticlopidine and ketoconazole (used as positive controls) [49]. Chrysin 7-sulfate has an IC50 value of 2.7 μM to CYP2C9, which is comparable to that of the positive control sulfaphenazole [50]. Additionally, chrysin 7-sulfate showed a slight inhibition effect on CYP2C19 and CYP3A4 [50].

CYP51A1 is considered as a potent target for cholesterol-lowering drugs [51]. There is an indication that the regulation of CYP51A1 function could be important to the treatment of oncological pathologies [11]. It was shown that anticancer drugs, abiraterone and galeterone, which are steroidal inhibitors of CYP17A1, can interact with human CYP51A1. However, their inhibition potential was not estimated. The Kdapp values determined for the abiraterone and galeterone, were 22 and 16 μM, respectively [52], and are significantly higher than those for baicalein and luteolin obtained in this work (8.2 and 5.1 μM, respectively). In contrast, non-steroidal pyridine derivative LK-935 [53] and azole inhibitors, ketoconazole and econazole [38], have an affinity in the submicromolar range due to direct coordination with heme iron. However, azole derivatives have a poor bioavailability and relatively low selectivity which might cause adverse reactions.

We showed the inhibition of CYP51A1 activity by the sulfated derivative of luteolin isolated from seagrass within the family *Zosteraceae* (*Zostera marina* and *Zostera asiatica*). It was previously demonstrated that luteolin 7,3′-disulfate has a wide range of biological activities that might be linked to its higher bioavailability [33,54]. Considering that natural flavonoids and their biological activities are currently a subject of great interest and in light of our data, it is plausible to suggest that CYP51A1 activity could also be modulated by this group of compounds. Obtained data on inhibition by luteolin 7,3′-disulfate could be further explored for the development of a new class of CYP51A1 inhibitors.

4. Materials and Methods

4.1. Samples

Highly purified (>95% by SDS-PAGE) recombinant human CYP51A1 protein was expressed and purified as previously described [38]. Low molecular weight compounds: lanosterol (PubChem CID 246983, CAS Number 79-63-0), natural substrate of CYP51A1, and ketoconazole (PubChem CID 456201, CAS Number 65277-42-1), azole inhibitor of CYP51A1, were obtained from Cayman Chemicals (Ann Arbor, MI, USA), baicalein (PubChem CID 5281605, CAS Number 491-67-8) was obtained from Sigma Aldrich (St. Louis, MO, USA), luteolin (PubChem CID 5280445, CAS Number 491-70-3) and luteolin 7,3′-disulfate (PubChem CID 44258153) were purified in the G.B. Elyakov Pacific Institute of Bioorganic Chemistry (Vladivostok, Russia) by water–alcohol extraction, followed by chromatographic purification from the sea plants of *Zosteraceae* genus [55,56].

4.2. Surface Plasmon Resonance

SPR analyses were carried out at 25 °C using the optical biosensors Biacore T200 and Biacore 8K (GE Healthcare, Chicago, IL, USA) and sensor chips of CM5 series S type (Cytiva, Marlborough, MA, USA). HBS-N (10 mM HEPES, 150 mM NaCl, pH 7.4) (Cytiva) was used as a running buffer for CYP51A1 immobilization. Carboxyl groups of biosensor chip dextran were activated for 5 min by injection of the 1:1 mixture of 0.2 M 1-ethyl-3-(3-dimethylaminopropyl)carbodiimide hydrochloride (EDC) and 0.05 M N-hydroxysuccinimide (NHS) at a flow rate of 5 μL/min, followed by 1 min wash with HBS-N buffer at the same flow rate. Next, CYP51A1 (25 μg/mL) in 10 mM sodium acetate (pH 5.0) was injected into the working channel of the biosensor for 5 min at a flow rate of 5 μL/min. The final level of immobilization was 13,500 RU (13.5 ng of protein). Reference channel without immobilized CYP51A1 was used to correct the effects of the non-specific binding of analytes to the chip surface.

Baicalein and luteolin were prepared as 10 mM stock solutions in 100% dimethyl sulfoxide (DMSO). Experimental samples of baicalein and luteolin were prepared in an HBS-N buffer at the concentration range 10–100 μM and 1% DMSO. The same amount of solvent was added to the HBS-N running buffer to minimize bulk-effects introduced by the difference between the refractive indexes of the running buffer and the experimental samples. Refractive indexes of running buffer and experimental samples were matched with a precision refractometer RX-5000 (Atago, Saitama, Japan). If needed, the concentration of solvent in the running buffer was corrected according to the equation:

$$C(DMSO)_{running\ buffer} = C(DMSO)_{sample} \times \frac{\eta_1 - \eta_2}{\eta_3 - \eta_2},$$

where $C(DMSO)_{running\ buffer}$—DMSO final concentration in running buffer, $C(DMSO)_{sample}$—DMSO concentration in experimental sample, η_1—analyzed sample refractive index, η_2—HBS-N buffer refractive index, η_3—HBS-N buffer containing the DMSO of the same concentration as experimental sample refractive index.

Luteolin 7,3′-disulfate 10 mM stock solution and experimental samples at final concentrations of 10–100 µM were prepared with HBS-N buffer without organic solvent. The same buffer was used as a running buffer with luteolin 7,3′-disulfate to minimize the bulk-effects on the obtained experimental data. A total of 10 mM stock solution of lanosterol was prepared in ethanol. Lanosterol experimental samples at the final concentrations of 10–100 µM, as well as the running buffer, were prepared by the same protocol as for baicalein and luteolin but in ethanol instead of DMSO.

Low molecular weight compounds were injected through biosensor channels (working and reference) at a flow rate of 10 µL/min (luteolin 7,3′-disulfate) and 50 µL/min (baicalein and luteolin) for 6 min. Dissociation of the formed CYP51A1/compound complexes were registered at the same flow rate for no less than 6 min after the sample injection. After each biosensor cycle, a bound analyte was removed with two-times injection of regenerating solution (2 M NaCl, 1% CHAPS) at a flow rate of 30 µL/min for 30 s.

SPR sensorgrams were processed in Biacore T200 Evaluation Software v.1.0 (GE Healthcare) and BIAevaluation Software v 4.1.1 (GE Healthcare) using "1:1 (Langmuir) binding" and "Two-state (conformational change) binding" data processing models. The 1:1 (Langmuir) binding model is a model for the 1:1 interaction between compound (C) with immobilized protein (P), and is equivalent to the Langmuir isotherm for adsorption to a surface: C + P ↔ CP. Two-state (conformational change) binding model describes a 1:1 binding of compound (C) to immobilized protein (P) followed by a conformational change in the complex (CP ↔ CP*). It is assumed that the conformationally changed complex can dissociate only through the reverse of the conformational change: C + P ↔ CP ↔ CP*. The final kinetic parameters were obtained from the models with best fit of the experimental curves according to the minimum of the obtained chi^2 value. The equations describing used models are as follows:

(1) 1:1 (Langmuir) binding [57]: $K_d = \frac{k_{off}}{k_{on}}$, where K_d—equilibrium dissociation constant, k_{off}—dissociation rate constant, k_{on}—association rate constant.

(2) Two-state (conformational change) binding [58]: $K_d = \frac{k_{off1}}{k_{on1}} \times \left(1 + \frac{k_{on2}}{k_{off2}}\right)^{-1}$, where K_d—equilibrium dissociation constant, k_{off1}—dissociation rate constant, k_{on1}—association rate constant, k_{on2}—forward rate constant for CP ↔ CP* transition, k_{off2}—backward rate constant for CP ↔ CP* transition.

4.3. Spectral Titration Analysis

Spectrophotometric titration was used to determine the apparent dissociation constants (K_{dapp}) for the enzyme–ligand complexes. The spectral measurements were performed on Cary Series UV-Vis-NIR (Agilent Technologies, Santa Clara, CA, USA) spectrophotometer using tandem quartz cuvettes (1 cm optical path) to exclude the absorption of ligands. Natural substrate, lanosterol, at final concentration 5 µM was added before the titration to the CYP51A1 (final concentration 4 µM) in 50 mM potassium phosphate buffer, pH 7.4. For titration, the ligand solution was added to the experimental cuvette (baicalein and luteolin 7,3′-disulfate were added up to a final concentration of 30 µM, luteolin was added up to a final concentration of 15 µM) and an equal volume of solvent was added to the control cuvette. The difference spectra were recorded after each addition of ligand at room temperature in the range of 350–500 nm. The apparent dissociation constants were determined by plotting the absorbance changes in the difference spectra versus the concentration of free ligand and evaluated by using the Hill equation (OriginPro 8.1 statistical data analysis package):

$$A_{obs} = A_{max} \times \left(\frac{S \times n}{K_{dapp} \times n + S \times n} \right),$$

where A_{obs}—the observed change in the absorption, A_{max}—the absorbance change at ligand saturation, K_{dapp}—the apparent dissociation constant for the ligand–enzyme complex, S—the ligand concentration, n—a Hill coefficient.

4.4. Enzyme Assay

Lanosterol 14-alpha-demethylase activity of human CYP51A1 was determined at 37 °C in 50 mM KPB, 4 mM $MgCl_2$, 0.1 mM DTT in presence of lipids (0.15 mg/mL mixture 1:1:1 of L-α-dilauroyl-sn-glycero-3-phosphocholine, L-α-dioleoyl-sn-glycero-3-phosphocholine and L-α-phosphatidyl-L-serine). The final concentrations of CYP51A1 and CPR were 0.5 and 2.0 μM, respectively. Aliquots of concentrated recombinant proteins were mixed and preincubated for 5 min at room temperature. Lanosterol (10 mM stock solution in ethanol) was added to the reaction mixture at a final concentration of 50 μM. Tested compounds were added to the reaction mixture at a final concentration of 25 μM. To estimate the apparent IC50, the following concentrations of luteolin 7,3′-disulfate were used: 5, 10, 25, 50 and 100 μM. Ketoconazole at a concentration of 5 μM was used as a positive control. After 10 min of preincubation at 37 °C, the reaction was started by adding NADPH at a final concentration of 0.25 mM. Aliquots (0.5 mL) were taken from the incubation mixture at chosen time intervals. Steroids were extracted with 5 mL of ethyl acetate. The mixture was vigorously mixed, the water and organic phases were separated by centrifugation at 3000 rpm for 10 min. The organic layer was carefully removed and dried under argon flow. A total of 50 μL of methanol was added to the pellet and steroids were analyzed on a computerized Agilent 1200 series HPLC instrument (Agilent Technologies, USA) equipped with Agilent Triple Quad 6410 mass spectrometer (Agilent Technologies). Samples were analyzed by gradient elution on Zorbax Eclipse XDB C18 column (4.6 × 150 mm; 5 μm) (Agilent Technologies). A total of 0.1% (*v/v*) FA in water was used as the mobile phase A and 0.1% (*v/v*) FA in methanol:1-propanol mix (75:25, *v/v*) as mobile phase B. The gradient was 75–100% B in 0–5 min. The flow rate was 500 μL per min. The column temperature was maintained at 40 ± 1 °C. Mass spectrometry experiments were performed with atmospheric pressure chemical ionization source (APCI) at positive ion mode. The following APCI settings were used: gas temperature 200 °C, vaporizer 250 °C, gas flow 7 L/min, nebulizer pressure 40 psig, Vcap 4000 V, corona 4 μA, fragmentor 100 V. The data acquisition mode was MS2Scan from 200 to 550 Da.

4.5. Molecular Docking

Crystal structure of CYP51A1 PDB ID 3LD6 was used for docking. 3D structures of luteolin (CID 5280445) and luteolin 7,3′-disulfate (CID 44258153) were obtained from the PubChem database (https://pubchem.ncbi.nlm.nih.gov/, accessed on 7 November 2020). Removal of water and ligand molecules from the original protein PDB files and molecular docking over the entire surface of CYP51A1 were performed automatically in the Flare software package (Cresset, Litlington, UK) with default settings [59]. Docking hypotheses were arranged according to score functions values: Lead Finder (LF) Rank Score, LF dG, LF VSscore. The lower is the LF Rank Score, the higher is the likelihood that the docked pose reproduces the crystallographic pose. LF dG has been designed to perform accurate estimation of the free energy of protein–ligand binding for a given protein–ligand complex. LF VSscore has been designed to produce maximum efficiency in virtual screening experiments, i.e., to assign higher scores to active ligands (true binders) and lower scores to inactive ligands. Molecular graphics visualization tool Maestro version 12.5.139 (Schrödinger, New York, NY, USA) was used to analyze the selected docking hypotheses.

5. Conclusions

In this work, we identified a new ligand of human CYP51A1 among natural flavonoid—luteolin 7,3′-disulfate—that inhibits 14α-demethylase activity. Potential inhibitory mechanisms include blocking of either a substrate access channel or the interaction with a redox partner. Obtained results suggest further exploration of polyphenols for the cholesterol lowering ability and anti-cancer potential via CYP51A1.

Author Contributions: Conceptualization, L.K., P.E. and E.Y.; methodology, L.K., P.E. and E.Y.; formal analysis, L.K., P.E., T.S., P.S., S.F., I.G. and A.G.; investigation, L.K., P.E., E.Y., Y.M. and O.G.; resources, T.S., S.U., A.P., A.A., O.S. and N.S.; data curation, L.K., P.E. and E.Y.; writing—original draft preparation, P.E.; writing—review and editing, L.K., P.E., E.Y., Y.M., O.G., A.G. and N.S.; visualization, L.K., P.E. and E.Y.; supervision, N.S.; project administration, A.I. All authors have read and agreed to the published version of the manuscript.

Funding: Surface plasmon resonance analysis was funded by the Russian Foundation for Basic Research (grant № 20-04-00014) and was performed using the equipment of "Human Proteome" Core Facility of the Institute of Biomedical Chemistry supported by MINOBRNAUKI, Agreement № 075-15-2019-1502 from 5 September 2019. Catalytic activity analysis was funded by SPSR of Belarus "Chemical processes, reagents and technologies, bioregulators and bioorganic chemistry", (project 2.3.1.1). Spectral titration analysis and molecular docking simulations were done in the framework of the Russian Federation fundamental research program for the long-term period for 2021–2030.

Institutional Review Board Statement: Not applicable.

Informed Consent Statement: Not applicable.

Data Availability Statement: Datasets are available from the authors.

Conflicts of Interest: The authors declare no conflict of interest.

Sample Availability: Samples of the compounds are available from the authors.

References

1. Porter, J.A.; Young, K.E.; Beachy, P.A. Cholesterol Modification of Hedgehog Signaling Proteins in Animal Development. *Science* **1996**, *274*, 255–259. [CrossRef] [PubMed]
2. Xiao, X.; Tang, J.-J.; Peng, C.; Wang, Y.; Fu, L.; Qiu, Z.-P.; Xiong, Y.; Yang, L.-F.; Cui, H.-W.; He, X.-L.; et al. Cholesterol Modification of Smoothened Is Required for Hedgehog Signaling. *Mol. Cell* **2017**, *66*, 154–162.e10. [CrossRef] [PubMed]
3. McNeill, A.M.; Rosamond, W.D.; Girman, C.J.; Golden, S.H.; Schmidt, M.I.; East, H.E.; Ballantyne, C.M.; Heiss, G. The Metabolic Syndrome and 11-Year Risk of Incident Cardiovascular Disease in the Atherosclerosis Risk in Communities Study. *Diabetes Care* **2005**, *28*, 385–390. [CrossRef]
4. Kumarakulasingham, M.; Rooney, P.H.; Dundas, S.R.; Telfer, C.; Melvin, W.T.; Curran, S.; Murray, G.I. Cytochrome P450 Profile of Colorectal Cancer: Identification of Markers of Prognosis. *Clin. Cancer Res.* **2005**, *11*, 3758–3765. [CrossRef] [PubMed]
5. Aoyama, Y. Recent Progress in the CYP51 Research Focusing on Its Unique Evolutionary and Functional Characteristics as a Diversozyme P450. *Front. Biosci.* **2005**, *10*, 1546–1557. [CrossRef] [PubMed]
6. Murray, G.I.; Patimalla, S.; Stewart, K.N.; Miller, I.D.; Heys, S.D. Profiling the Expression of Cytochrome P450 in Breast Cancer. *Histopathology* **2010**, *57*, 202–211. [CrossRef]
7. Sun, M.; Qiu, J.; Zhai, H.; Wang, Y.; Ma, P.; Li, M.; Chen, B. Prognostic Implications of Novel Gene Signatures in Gastric Cancer Microenvironment. *Med. Sci. Monit.* **2020**, *26*, e924604-1–e924604-17. [CrossRef] [PubMed]
8. Gabitova, L.; Restifo, D.; Gorin, A.; Manocha, K.; Handorf, E.; Yang, D.-H.; Cai, K.Q.; Klein-Szanto, A.J.; Cunningham, D.; Kratz, L.E.; et al. Endogenous Sterol Metabolites Regulate Growth of EGFR/KRAS-Dependent Tumors via LXR. *Cell Rep.* **2015**, *12*, 1927–1938. [CrossRef]
9. Salmina, K.; Huna, A.; Kalejs, M.; Pjanova, D.; Scherthan, H.; Cragg, M.S.; Erenpreisa, J. The Cancer Aneuploidy Paradox: In the Light of Evolution. *Genes* **2019**, *10*, 83. [CrossRef]
10. Keber, R.; Motaln, H.; Wagner, K.D.; Debeljak, N.; Rassoulzadegan, M.; Ačimovič, J.; Rozman, D.; Horvat, S. Mouse Knockout of the Cholesterogenic Cytochrome P450 Lanosterol 14alpha-Demethylase (Cyp51) Resembles Antley-Bixler Syndrome. *J. Biol. Chem.* **2011**, *286*, 29086–29097. [CrossRef]

11. Martínez-Botas, J.; Ferruelo, A.J.; Suárez, Y.; Gómez-Coronado, D.; Lasunción, M.A. Induction of Apoptosis in P53-Null HL-60 Cells by Inhibition of Lanosterol 14-Alpha Demethylase. *Biochimie* **1998**, *80*, 887–894. [CrossRef]
12. Noh, K.; Nepal, M.R.; Jeong, K.S.; Kim, S.-A.; Um, Y.J.; Seo, C.S.; Kang, M.J.; Park, P.-H.; Kang, W.; Jeong, H.G.; et al. Effects of Baicalin on Oral Pharmacokinetics of Caffeine in Rats. *Biomol. Ther.* **2015**, *23*, 201–206. [CrossRef] [PubMed]
13. Cao, L.; Kwara, A.; Greenblatt, D.J. Metabolic Interactions between Acetaminophen (Paracetamol) and Two Flavonoids, Luteolin and Quercetin, through in-Vitro Inhibition Studies. *J. Pharm. Pharmacol.* **2017**, *69*, 1762–1772. [CrossRef] [PubMed]
14. Fliszár-Nyúl, E.; Mohos, V.; Csepregi, R.; Mladěnka, P.; Poór, M. Inhibitory Effects of Polyphenols and Their Colonic Metabolites on CYP2D6 Enzyme Using Two Different Substrates. *Biomed. Pharmacother.* **2020**, *131*, 110732. [CrossRef] [PubMed]
15. Da, X.; Nishiyama, Y.; Tie, D.; Hein, K.Z.; Yamamoto, O.; Morita, E. Antifungal Activity and Mechanism of Action of Ou-Gon (Scutellaria Root Extract) Components against Pathogenic Fungi. *Sci. Rep.* **2019**, *9*, 1683. [CrossRef] [PubMed]
16. Huang, S.; Cao, Y.Y.; Dai, B.D.; Sun, X.R.; Zhu, Z.Y.; Cao, Y.B.; Wang, Y.; Gao, P.H.; Jiang, Y.Y. In Vitro Synergism of Fluconazole and Baicalein against Clinical Isolates of Candida Albicans Resistant to Fluconazole. *Biol. Pharm. Bull.* **2008**, *31*, 2234–2236. [CrossRef]
17. Joung, D.-K.; Lee, Y.-S.; Han, S.-H.; Lee, S.-W.; Cha, S.-W.; Mun, S.-H.; Kong, R.; Kang, O.-H.; Song, H.-J.; Shin, D.-W.; et al. Potentiating Activity of Luteolin on Membrane Permeabilizing Agent and ATPase Inhibitor against Methicillin-Resistant Staphylococcus Aureus. *Asian Pac. J. Trop. Med.* **2016**, *9*, 19–22. [CrossRef]
18. Popov, A.M.; Osipov, A.N.; Korepanova, E.A.; Krivoshapko, O.N.; Artyukov, A.A.; Klimovich, A.A. A Study of the Antioxidant and Membranotropic Activities of Luteolin Using Different Model Systems. *BIOPHYSICS* **2016**, *61*, 843–850. [CrossRef]
19. Ravishankar, D.; Watson, K.A.; Boateng, S.Y.; Green, R.J.; Greco, F.; Osborn, H.M.I. Exploring Quercetin and Luteolin Derivatives as Antiangiogenic Agents. *Eur. J. Med. Chem.* **2015**, *97*, 259–274. [CrossRef]
20. Gao, Y.; Snyder, S.A.; Smith, J.N.; Chen, Y.C. Anticancer Properties of Baicalein: A Review. *Med. Chem. Res.* **2016**, *25*, 1515–1523. [CrossRef]
21. Imran, M.; Rauf, A.; Abu-Izneid, T.; Nadeem, M.; Shariati, M.A.; Khan, I.A.; Imran, A.; Orhan, I.E.; Rizwan, M.; Atif, M.; et al. Luteolin, a Flavonoid, as an Anticancer Agent: A Review. *Biomed. Pharmacother.* **2019**, *112*, 108612. [CrossRef] [PubMed]
22. Tuli, H.S.; Aggarwal, V.; Kaur, J.; Aggarwal, D.; Parashar, G.; Parashar, N.C.; Tuorkey, M.; Kaur, G.; Savla, R.; Sak, K.; et al. Baicalein: A Metabolite with Promising Antineoplastic Activity. *Life Sci.* **2020**, *259*, 118183. [CrossRef] [PubMed]
23. Shen, J.; Li, P.; Liu, S.; Liu, Q.; Li, Y.; Sun, Y.; He, C.; Xiao, P. Traditional Uses, Ten-Years Research Progress on Phytochemistry and Pharmacology, and Clinical Studies of the Genus Scutellaria. *J. Ethnopharmacol.* **2021**, *265*, 113198. [CrossRef] [PubMed]
24. Nik Salleh, N.N.H.; Othman, F.A.; Kamarudin, N.A.; Tan, S.C. The Biological Activities and Therapeutic Potentials of Baicalein Extracted from Oroxylum Indicum: A Systematic Review. *Molecules* **2020**, *25*, 5677. [CrossRef]
25. Farooqi, A.A.; Butt, G.; El-Zahaby, S.A.; Attar, R.; Sabitaliyevich, U.Y.; Jovic, J.J.; Tang, K.-F.; Naureen, H.; Xu, B. Luteolin Mediated Targeting of Protein Network and MicroRNAs in Different Cancers: Focus on JAK-STAT, NOTCH, MTOR and TRAIL-Mediated Signaling Pathways. *Pharmacol. Res.* **2020**, *160*, 105188. [CrossRef] [PubMed]
26. Tuorkey, M.J. Molecular Targets of Luteolin in Cancer. *Eur. J. Cancer Prev.* **2016**, *25*, 65–76. [CrossRef] [PubMed]
27. Barreca, D.; Mandalari, G.; Calderaro, A.; Smeriglio, A.; Trombetta, D.; Felice, M.R.; Gattuso, G. Citrus Flavones: An Update on Sources, Biological Functions, and Health Promoting Properties. *Plants* **2020**, *9*, 288. [CrossRef]
28. Zhou, Y.; Zheng, J.; Li, Y.; Xu, D.-P.; Li, S.; Chen, Y.-M.; Li, H.-B. Natural Polyphenols for Prevention and Treatment of Cancer. *Nutrients* **2016**, *8*. [CrossRef]
29. Harborne, J.B.; Williams, C.A. Occurrence of Sulphated Flavones and Caffeic Acid Esters in Members of the Fluviales. *Biochem. Syst. Ecol.* **1976**, *4*, 37–41. [CrossRef]
30. Popov, A.M.; Krivoshapko, O.N.; Artyukov, A.A. Comparative estimation of pharmacological activities of luteolin and luteolin 7,3′-disulphate at modelling of different pathologies. *Russ. J. Biopharm.* **2011**, *3*, 27–33.
31. Klimovich, A.A.; Popov, A.M.; Styshova, O.N.; Artyukov, A.A.; Tsybulsky, A.V. A Comparative Evaluation of the Actions of Different Secondary Metabolites of Marine Hydrobionts on the Redox Status of Tumor and Immune Cells. *Biophysics* **2018**, *63*, 763–768. [CrossRef]
32. Kwak, J.Y.; Seok, J.K.; Suh, H.-J.; Choi, Y.-H.; Hong, S.S.; Kim, D.S.; Boo, Y.C. Antimelanogenic Effects of Luteolin 7-Sulfate Isolated from Phyllospadix Iwatensis Makino. *Br. J. Dermatol.* **2016**, *175*, 501–511. [CrossRef] [PubMed]
33. Popov, A.M.; Krivoshapko, O.N. Protective Effects of Polar Lipids and Redox-Active Compounds from Marine Organisms at Modeling of Hyperlipidemia and Diabetes. *J. Biomed. Sci. Eng.* **2013**, *6*, 543–550. [CrossRef]
34. Popov, A.M.; Krivoshapko, O.N.; Klimovich, A.A.; Artyukov, A.A. Biological activity and mechanisms of therapeutic action of rosmarinic acid, luteolin and its sulphated derivatives. *Biomed. Khim.* **2016**, *62*, 22–30. [CrossRef]
35. Kaluzhskiy, L.A.; Gnedenko, O.V.; Gilep, A.A.; Strushkevich, N.V.; Shkel, T.V.; Chernovetsky, M.A.; Ivanov, A.S.; Lisitsa, A.V.; Usanov, A.S.; Stonik, V.A.; et al. Screening of Human Cytochrome P450(51) (CYP51A1) Inhibitors: Structural Lanosterol Analogues of Plant and Animal Origin. *Biochem. (Moscow) Suppl. B Biomed. Chem.* **2014**, *8*, 349–360. [CrossRef]
36. Shimada, T.; Tanaka, K.; Takenaka, S.; Foroozesh, M.K.; Murayama, N.; Yamazaki, H.; Guengerich, F.P.; Komori, M. Reverse Type I Binding Spectra of Human Cytochrome P450 1B1 Induced by Flavonoid, Stilbene, Pyrene, Naphthalene, Phenanthrene, and Biphenyl Derivatives That Inhibit Catalytic Activity: A Structure-Function Relationship Study. *Chem. Res. Toxicol.* **2009**, *22*, 1325–1333. [CrossRef] [PubMed]

37. Nitahara, Y.; Kishimoto, K.; Yabusaki, Y.; Gotoh, O.; Yoshida, Y.; Horiuchi, T.; Aoyama, Y. The Amino Acid Residues Affecting the Activity and Azole Susceptibility of Rat CYP51 (Sterol 14-Demethylase P450). *J. Biochem.* **2001**, *129*, 761–768. [CrossRef]
38. Strushkevich, N.; Usanov, S.A.; Park, H.-W. Structural Basis of Human CYP51 Inhibition by Antifungal Azoles. *J. Mol. Biol.* **2010**, *397*, 1067–1078. [CrossRef]
39. Yousef, M.I.; Omar, S.A.M.; El-Guendi, M.I.; Abdelmegid, L.A. Potential Protective Effects of Quercetin and Curcumin on Paracetamol-Induced Histological Changes, Oxidative Stress, Impaired Liver and Kidney Functions and Haematotoxicity in Rat. *Food Chem. Toxicol.* **2010**, *48*, 3246–3261. [CrossRef]
40. Cho, Y.-A.; Choi, J.-S.; Burm, J.-P. Effects of the Antioxidant Baicalein on the Pharmacokinetics of Nimodipine in Rats: A Possible Role of P-Glycoprotein and CYP3A4 Inhibition by Baicalein. *Pharmacol. Rep.* **2011**, *63*, 1066–1073. [CrossRef]
41. Si, D.; Wang, Y.; Zhou, Y.-H.; Guo, Y.; Wang, J.; Zhou, H.; Li, Z.-S.; Fawcett, J.P. Mechanism of CYP2C9 Inhibition by Flavones and Flavonols. *Drug Metab. Dispos.* **2009**, *37*, 629–634. [CrossRef]
42. Quintieri, L.; Palatini, P.; Nassi, A.; Ruzza, P.; Floreani, M. Flavonoids Diosmetin and Luteolin Inhibit Midazolam Metabolism by Human Liver Microsomes and Recombinant CYP 3A4 and CYP3A5 Enzymes. *Biochem. Pharmacol.* **2008**, *75*, 1426–1437. [CrossRef] [PubMed]
43. Ohno, S.; Shinoda, S.; Toyoshima, S.; Nakazawa, H.; Makino, T.; Nakajin, S. Effects of Flavonoid Phytochemicals on Cortisol Production and on Activities of Steroidogenic Enzymes in Human Adrenocortical H295R Cells. *J. Steroid Biochem. Mol. Biol.* **2002**, *80*, 355–363. [CrossRef]
44. Hodek, P.; Trefil, P.; Stiborová, M. Flavonoids-Potent and Versatile Biologically Active Compounds Interacting with Cytochromes P450. *Chem.-Biol. Interact.* **2002**, *139*, 1–21. [CrossRef]
45. Howes, L.G.; James, M.J.; Florin, T.; Walker, C. Nv-52: A Novel Thromboxane Synthase Inhibitor for the Treatment of Inflammatory Bowel Disease. *Expert Opin. Investig. Drugs* **2007**, *16*, 1255–1266. [CrossRef] [PubMed]
46. Swami, S.; Krishnan, A.V.; Peehl, D.M.; Feldman, D. Genistein Potentiates the Growth Inhibitory Effects of 1,25-Dihydroxyvitamin D3 in DU145 Human Prostate Cancer Cells: Role of the Direct Inhibition of CYP24 Enzyme Activity. *Mol. Cell. Endocrinol.* **2005**, *241*, 49–61. [CrossRef] [PubMed]
47. Thilakarathna, S.H.; Rupasinghe, H.P.V. Flavonoid Bioavailability and Attempts for Bioavailability Enhancement. *Nutrients* **2013**, *5*, 3367–3387. [CrossRef]
48. Williamson, G.; Manach, C. Bioavailability and Bioefficacy of Polyphenols in Humans. II. Review of 93 Intervention Studies. *Am. J. Clin. Nutr.* **2005**, *81*, 243S–255S. [CrossRef]
49. Mohos, V.; Fliszár-Nyúl, E.; Ungvári, O.; Kuffa, K.; Needs, P.W.; Kroon, P.A.; Telbisz, Á.; Özvegy-Laczka, C.; Poór, M. Inhibitory Effects of Quercetin and Its Main Methyl, Sulfate, and Glucuronic Acid Conjugates on Cytochrome P450 Enzymes, and on OATP, BCRP and MRP2 Transporters. *Nutrients* **2020**, *12*, 2306. [CrossRef]
50. Mohos, V.; Fliszár-Nyúl, E.; Ungvári, O.; Bakos, É.; Kuffa, K.; Bencsik, T.; Zsidó, B.Z.; Hetényi, C.; Telbisz, Á.; Özvegy-Laczka, C.; et al. Effects of Chrysin and Its Major Conjugated Metabolites Chrysin-7-Sulfate and Chrysin-7-Glucuronide on Cytochrome P450 Enzymes and on OATP, P-Gp, BCRP, and MRP2 Transporters. *Drug Metab. Dispos.* **2020**, *48*, 1064–1073. [CrossRef]
51. Monostory, K.; Pascussi, J.-M.; Szabó, P.; Temesvári, M.; Köhalmy, K.; Acimovic, J.; Kocjan, D.; Kuzman, D.; Wilzewski, B.; Bernhardt, R.; et al. Drug Interaction Potential of 2-((3,4-Dichlorophenethyl)(Propyl)Amino)-1-(Pyridin-3-Yl)Ethanol (LK-935), the Novel Nonstatin-Type Cholesterol-Lowering Agent. *Drug Metab. Dispos.* **2009**, *37*, 375–385. [CrossRef]
52. Masamrekh, R.; Kuzikov, A.; Veselovsky, A.; Toropygin, I.; Shkel, T.; Strushkevich, N.; Gilep, A.; Usanov, S.; Archakov, A.; Shumyantseva, V. Interaction of 17α-Hydroxylase, 17(20)-Lyase (CYP17A1) Inhibitors—Abiraterone and Galeterone—with Human Sterol 14α-Demethylase (CYP51A1). *J. Inorg. Biochem.* **2018**, *186*, 24–33. [CrossRef]
53. Korošec, T.; Ačimovič, J.; Seliškar, M.; Kocjan, D.; Tacer, K.F.; Rozman, D.; Urleb, U. Novel Cholesterol Biosynthesis Inhibitors Targeting Human Lanosterol 14α-Demethylase (CYP51). *Bioorg. Med. Chem.* **2008**, *16*, 209–221. [CrossRef] [PubMed]
54. Artjukov, A.A.; Kozlovskaja, E.P.; Krivoshapko, O.N.; Krylova, N.V.; Leonova, G.N.; Popov, A.M. Medication, Possessing Antioxidant, Cardioprotective, Antidiabetic, Anti-Inflammatory, Hepatoprotective, Antitumoral and Antiviral Action 2011. Russian Federation Patent № RU2432959C1, 10 November 2011.
55. Kim, J.H.; Cho, Y.H.; Park, S.M.; Lee, K.E.; Lee, J.J.; Lee, B.C.; Pyo, H.B.; Song, K.S.; Park, H.D.; Yun, Y.P. Antioxidants and Inhibitor of Matrix Metalloproteinase-1 Expression from Leaves of *Zostera Marina* L. *Arch. Pharm. Res.* **2004**, *27*, 177–183. [CrossRef] [PubMed]
56. Artjukov, A.A.; Glazunov, V.P.; Kochergina, T.J.; Kozlovskaja, E.P.; Kupera, E.V.; Makhan'kov, V.V.; Novikov, V.L.; Rutskova, T.A. Method of Obtaining Luteolyn 7,3′-Disulphate 2011. Russian Federation Patent № RU2432960C1, 10 November 2011.
57. Müller, K.M.; Arndt, K.M.; Plückthun, A. Model and Simulation of Multivalent Binding to Fixed Ligands. *Anal. Biochem* **1998**, *261*, 149–158. [CrossRef] [PubMed]
58. Lipschultz, C.A.; Li, Y.; Smith-Gill, S. Experimental Design for Analysis of Complex Kinetics Using Surface Plasmon Resonance. *Methods* **2000**, *20*, 310–318. [CrossRef] [PubMed]
59. Cheeseright, T.; Mackey, M.; Rose, S.; Vinter, A. Molecular Field Extrema as Descriptors of Biological Activity: Definition and Validation. *J. Chem. Inf. Model.* **2006**, *46*, 665–676, . [CrossRef]

molecules

MDPI

Article

Porin from Marine Bacterium *Marinomonas primoryensis* KMM 3633^T: Isolation, Physico-Chemical Properties, and Functional Activity

Olga D. Novikova [1,*], Valentina A. Khomenko [1], Natalia Yu. Kim [1], Galina N. Likhatskaya [1], Lyudmila A. Romanenko [1], Ekaterina I. Aksenova [2], Marina S. Kunda [2], Natalia N. Ryzhova [2], Olga Yu. Portnyagina [1], Tamara F. Solov'eva [1] and Olga L. Voronina [2,*]

1 G.B. Elyakov Pacific Institute of Bioorganic Chemistry, Far Eastern Branch, Russian Academy of Sciences, pr. 100 let Vladivostoku, 159, Vladivostok 690022, Russia; homenko.40@mail.ru (V.A.K.); natalya_kim@mail.ru (N.Y.K.); galin@piboc.dvo.ru (G.N.L.); lro@piboc.dvo.ru (L.A.R.); odd64@mail.ru (O.Y.P.); soltaf@mail.ru (T.F.S.)

2 N. F. Gamaleya National Research Center for Epidemiology and Microbiology, Ministry of Health of Russia, Gamaleya Str., 18, Moscow 123098, Russia; aksenova16@gmail.com (E.I.A.); markunda99@gmail.com (M.S.K.); rynatalia@yandex.ru (N.N.R.)

* Correspondence: novolga_05@mail.ru (O.D.N.); olv550@gmail.com (O.L.V.)

Academic Editor: Maksymilian Chruszcz
Received: 26 May 2020; Accepted: 3 July 2020; Published: 8 July 2020

Abstract: *Marinomonas primoryensis* KMM 3633^T, extreme living marine bacterium was isolated from a sample of coastal sea ice in the Amursky Bay near Vladivostok, Russia. The goal of our investigation is to study outer membrane channels determining cell permeability. Porin from *M. primoryensis* KMM 3633^T (MpOmp) has been isolated and characterized. Amino acid analysis and whole genome sequencing were the sources of amino acid data of porin, identified as Porin_4 according to the conservative domain searching. The amino acid composition of MpOmp distinguished by high content of acidic amino acids and low content of sulfur-containing amino acids, but there are no tryptophan residues in its molecule. The native MpOmp existed as a trimer. The reconstitution of MpOmp into black lipid membranes demonstrated its ability to form ion channels whose conductivity depends on the electrolyte concentration. The spatial structure of MpOmp had features typical for the classical gram-negative porins. However, the oligomeric structure of isolated MpOmp was distinguished by very low stability: heat-modified monomer was already observed at 30 °C The data obtained suggest the stabilizing role of lipids in the natural membrane of marine bacteria in the formation of the oligomeric structure of porin.

Keywords: marine bacteria; whole genome sequence; porin; amino acids composition; bilayer lipid membrane; pore-forming activity; spatial structure

1. Introduction

Gram-negative bacteria attract attention because of its extraordinary adaptability to natural environment as well as to anthropogenic, including nosocomial conditions. Marine bacteria studied since 1960s were found to be the most psychro-, piezo-, and halo-tolerant [1]. Moreover marine bacteria inhabit in oligotrophic environment with deficiency or lack of sunlight. Investigation of these bacteria revealed pressure regulation of the proteins expression [2], multiplication of genes important for metabolic pathways, and cell motility and other molecular mechanisms of the adaptation to extreme conditions [3].

Detection of the many carbon and energy utilization pathways in marine bacteria prompted the researchers to study on the properties of these bacteria as a tool for degradation of environmental pollution. Every year more attention is paid to this problem. Oil and petroleum products are the most common pollutants that disrupt and inhibit all life processes. These products accumulate as difficult-oxidized forms of substances that change the direction of metabolism and the natural ratio of the number of microorganisms in ecosystems. This leads to a slowdown in the process of microbial purification of ocean waters.

One of the most promising areas for investigating this problem is the Arctic Ocean, the smallest of the oceans which is almost completely surrounded by earth. For this reason, it has the most extensive shelf areas in comparison with any ocean basin and a significant proportion of terrigenous organic carbon flowing through rivers into the Arctic Ocean [4]. In order to determine the diversity of bacteria involved in bacterial consortia that are capable of degrading hydrocarbons, some studies were conducted in the Chukchi Plateau region [5]. Authors showed that a number of species are involved in this process. So, the potential degraders including *Cycloclasticus, Pseudomonas, Halomonas, Pseudoalteromonas, Marinomonas, Bacillus, Colwellia, Acinetobacter, Alcanivorax, Salinisphaera,* and *Shewanella*, with *Dietzia* as the most abundant, occurred in all sediment samples.

Marine sediment and water samples are the source for peptide-based drug discovery. Non-ribosomal peptides from Proteobacteria have recently attracted much attention as a source of new drugs [6]. Currently, there is considerable research interest in determining the chemical features of major components of cell envelope of marine bacteria, first of all, lipopolysaccharide (LPS) [7]. Lipid A, which is a structural component of the LPS of some marine bacteria, is considered an antagonist to endotoxins of gram-negative bacteria [8].

This is due to the fact that, in comparison with terrestrial bacteria, biologically active compounds of which are successfully used in the pharmaceutical industry, the biological potential of marine bacteria in this regard has not been practically realized [9].

However, the penetration process of the marine bacteria cells is the most intriguing. Gram-negative bacteria are covered by two distinct biological membranes. A well-established response of poikilothermic organisms to low temperature and high pressure exposure is a change in membrane fluidity [10]. There are two major classes of membrane transport proteins: transporters and channels, which are influenced by this change. Outer membrane (OM) channels include porin superfamily. These outer membrane channels share a beta-barrel structure that differ in strand and shear number. Porin superfamily comprises classical (gram-negative) porins, maltoporin-like channels, and ligand-gated protein channels cooperating with a TonB-associated inner membrane complex [11].

Classical (gram-negative) porins are known as OmpF, OmpC, and PhoE in *Escherichia coli*. To date, significant progress has been achieved in the study of the structure and function of porins of the terrestrial gram-negative bacteria, mainly members of the family Enterobacteriaceae [12,13]. However, classical porins are of the greatest interest in the context of the mechanism that mediates uptake of small molecules, including antibiotics [13].

OM porins of marine bacteria play an important role in the adaptation to the extreme environment. Few studies shed light on the role and functioning of porins of aquatic bacteria, including marine bacteria. The consequence of structural changes in the porins may be reduced permeability of the bacterial membrane for solutes. For example, in aquatic inhabitant of *Pseudomonas aeruginosa*, the permeability of the outer membrane is lower in comparison of *E. coli* [14]. Porins of some marine bacteria are sensitive to osmoregulation such as *Vibrio parahaemolyticus* and *V. alginolyticus* [15,16]. In this regard, the porins of marine bacteria are an interesting object of study of the adaptability of microorganisms.

The present study is a continuation of a series of our investigations that are dedicated to the isolation and characterization of functional activities of porins from the OM of marine microorganisms [17,18]. In this study, namely, we have isolated and characterized functional activity and spatial structure of porin from OM of *Marinomonas primoryensis* KMM 3633^{T}. Our interest in this species is due to the fact that the genus *Marinomonas* includes gram-negative bacterial strains common in various marine environments. [19]. In addition, the most recognized species of the genus were isolated from sea water samples collected from different geographical locations [20]. Some of them were isolated from cold media, such as the sub Antarctic regions, for example *Marinomonas polaris* [21] and *M. ushuaiensis* [22].

During the study of microbial communities associated with marine environments in the Sea of Japan, two bacterial strains were isolated from a coastal sea-ice sample, obtained from a sea-ice column at a depth of 0.8 m in Amursky Bay near Vladivostok, Russia, in March 2001 [23]. The isolates were aerobic, gram-negative, heterotrophic microorganisms with a respiratory metabolism, and had phenotypic characteristics similar to those of the genus *Marinomonas*. They were classified as *Marinomonas primoryensis* sp. nov. [23]. *M. primoryensis* KMM 3633^T (= CIP 108051 = IAM 15010) is the type strain of the species. This strain was used in the present study for further investigation.

2. Results

2.1. Isolation Procedure of OM Porin from M. primoryensis KMM 3633^T

2.1.1. Cultivation Condition of Microbial Cells

Bacterial cells of *M. primoryensis* KMM 3633^T were grown at a temperature of 0, 6–8, and 24 °C using the growth medium described in [23]. A comparative analysis of the protein profiles of cell lysates obtained by ultrasonic disintegration was performed using SDS-PAGE (data not shown). Significant differences in the polypeptide composition of cell lysates from bacteria grown at a reduced temperature (0 and 6–8 °C) was not revealed. However, cultivation at room temperature led to some decrease in the intensity of polypeptide bands in the region of high molecular weight proteins. In accordance with the results obtained, for further studies, the microbial cells of M. primoryensis KMM 3633^T grown at a temperature of 6–8 °C were used.

2.1.2. Isolation of Cell Envelope of *M. primoryensis* KMM 3633^T

To remove surface proteins, microbial cells of *M. primoryensis* KMM 3633^T were pre-treated with 5% solution of NaCl. Then, to obtain the cell envelope from the harvested microbial cells ultrasonic disintegration was used. The protein profile of the resulted total crude membrane fraction consisted of several major components: high molecular weight (MW) proteins in the form of a diffuse zone with apparent MW in the range of 95–130 kDa and a number of polypeptides with a MW in the range of 34–72 kDa (Figure 1A, lane 2). Judging by the intensity of the polypeptide bands their content in this area of molecular masses varies greatly. To detect heat-modifiable proteins in the envelope fraction, boiling in denaturing conditions was used (Figure 1A, lane 3). Several polypeptide zones in the region of 34–55 kDa became more clearly visible in MW ladder, while the protein bands in the region of high MW proteins disappeared. In addition, the intensity of the polypeptide band with MW between 43 and 35 kDa increased significantly. Obviously, the oligomeric polypeptide region of cell lysate of *M. primoryensis* KMM 3633^T includes several heat-modifiable proteins which are supposed to be porins. Such behavior in the denaturing conditions of SDS-PAGE is typical for OM porins of gram-negative terrestrial bacteria, in which the oligomers dissociate into monomers after heating above the critical temperature of an irreversible conformational transition. To obtain isolated proteins from the crude cell envelope fraction of *M. primoryensis* KMM 3633^T two methods were used: Sarcosyl treatment and octyl-POE extraction.

2.1.3. Sarcosyl Treatment

Extraction of the crude cell membranes of *M. primoryensis* KMM 3633^T with 1% solution of Sarcosyl [24] resulted in an insoluble fraction of OM proteins. The precipitate obtained after washing with HEPES was a pure sample of the protein oligomer, whose electrophoretic mobility corresponded to MW about 95 kDa (Figure 1B, lane 1). The electrophoretic mobility of the monomer of the target protein under denaturing conditions corresponded to a polypeptide with a molecular weight of about 35 kDa (MpOmp) (Figure 1B, lane 2).

2.1.4. Octyl-POE Extraction

To obtain the fraction with the highest content of heat-modifiable OM proteins, a stepwise extraction of the cell walls of *M. primoryensis* KMM 3633^T with a POE solution was carried out, as is known, selectively extracting OM proteins of gram-negative bacteria [25]. It was found that polypeptides changing their molecular weight when heated under denaturing conditions are extracted at different concentrations of POE in the range from 0.5 to 3.0% by weight [25]. In the case of *M. primoryensis* KMM 3633^T, the greatest amount of the heat-modifiable proteins was obtained using 0.5% detergent solution, so the extraction was repeated three times. According to the SDS-PAGE data an electrophoretically homogeneous protein was isolated in the oligomeric form as a result of purification of the obtained sum fraction by gel permeation chromatography on Sephacryl S-300 (Figure 1B, lane 4). After boiling in SDS solution the protein band observed migrated to the same position as MpOmp monomer isolated with Sarcosyl extraction (Figure 1B, lane 3).

It should be noted that in this case, the unheated protein migrated to a position corresponding to MW of about 72 kDa. This apparent MW value was lower than that of the MpOmp oligomer isolated by Sarcosyl extraction. The cause of this phenomenon could be the presence of residual amount of POE in the protein sample. A similar shift of the protein MW was observed for VhOmp of *Vibrio harveyi* in [26].

Thus, according to the SDS-PAGE data, MW the protein isolated with Sarcosyl extraction was identical to that of the protein obtained by the Garavito method [25]. Taking into account the data obtained, in subsequent experiments a sample obtained as a result of the removal of cytoplasmic proteins with Sarcosyl was used. We chose this method of the target protein isolation because of the simplicity and at the same time the high selectivity of the isolation procedure.

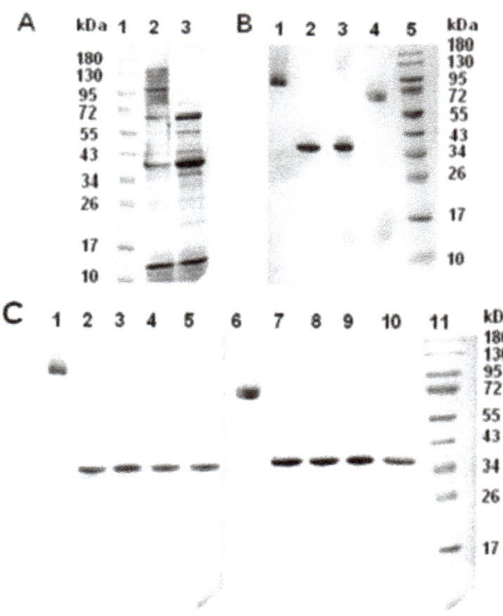

Figure 1. Purification and temperature denaturation of *M. primoryensis* KMM 3633^T porin (MpOmp): (**A**): Crude cell envelope fraction (crude membrane) of *M. primoryensis* KMM 3633^T (2, 3); marker proteins (1). Sample 3 is heated at 99 °C. (**B**): MpOmp isolated from the crude membrane fraction of *M. primoryensis KMM* 3633^T after treatment with 1% Sarcosyl (1, 2); MpOmp isolated from combined OM protein fraction obtained as a result of three time extraction with 0.5% POE of *M. primoryensis* KMM 3633^T cell envelope and gel permeation chromatography on Sephacryl S-300 (3, 4); marker proteins (5). Samples 2 and 3 are heated at 99 °C. (**C**): MpOmp isolated with Sarcosyl extraction (1–5); samples 2, 3, 4, and 5 were heated at 30, 40, 50, and 99 °C, respectively; MpOmp isolated with POE extraction (6–10); samples 7, 8, 9, and 10 were heated at 30, 40, 50, and 99 °C, respectively; marker proteins (11).

2.2. Pore-Forming Activity of OM Porin from M. primoryensis KMM 3633^T

The observed pore-forming activity of the total fraction of OM proteins of *M. primoryensis* KMM 3633^T was instable; nevertheless, this confirmed the presence of porin among the OM proteins of the bacteria (data not shown). When purified MpOmp was added in small quantities (10–100 ng/mL) to the aqueous solution bathing an artificial lipid bilayer, the membrane conductance increased by several orders of magnitude (Figure 2A,B). The current-voltage characteristic of the porin channel is linear in the range up to 180 mV (Figure 2C). The pore-forming activity of MpOmp was found to depend on the salt concentration in the aqueous phase. We failed to record channel formation at NaCl concentration below 0.2 M. The linear dependence of the current through the porin channel is observed when the salt concentration varied in the range 0.5–2.0 M (Figure 2D).

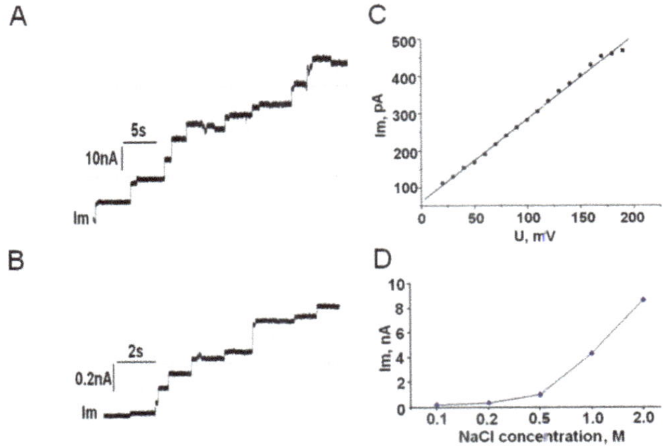

Figure 2. Current traces recorded after the addition of *M. primoryensis* KMM 3633^T porin to bilayer lipid membrane (BLM) in the presence of 1.0 M (**A**) and 0.2 M (**B**) NaCl; current-voltage characteristic of MpOmp channel (**C**); the dependence of the current through the MpOmp channel on NaCl concentration (**D**).

BLM formed from a 1% solution of 1-monooleoylglycerol in *n*-heptane. Aqueous phase: 20 mM Tris-HCl, pH 7.4 (buffer A), 0.5–2.0 M NaCl, protein concentration 10–100 ng/mL. Membrane potential is 50 mV.

2.3. Amino Acid Composition and Amino Acid Sequence of OM Porin from M. primoryensis KMM 3633^T

M. primoryensis KMM 3633^T whole genome was sequenced, assembled, and annotated as described in Methods. The porin of *M. primoryensis* KMM 3633^T was classified as Porin_4 according to the conservative domain searching and analyzed in detail. The characteristics are represented in Table 1. By protein BLAST we revealed proteins with high sequences similarity and proteins with similar domain architecture. Porins of *M. primoryensis* strain AceL and *Marinomonas* sp. strain IMCC 4694 had the highest similarity with porin of strain KMM 3633^T-83.59% and 77.43%, respectively. Porins of *Polaromonas sp.* JS666, *Magnetospirillum magneticum* AMB-1 and *Basfia succiniciproducens* MBEL55E had the most closely related domain architecture with analyzed porin. Porins of *Photobacterium damselae* [27] and *Escherichia coli* [28] were chosen as well-known porins: the first one is from marine Gammaproteobacteria, the fish pathogen, and the second one is from terrestrial Gammaproteobacteria, the classical object of biological and molecular research. Amino acid composition of the isolated proteins [27,28] was supplemented by translation of sequenced genes.

Most of the bacteria represented in Table 1 are aquatic, isolated from sea-, ground-, or bog water. *E. coli* and *B. succiniciproducens* are known as habitants of gastrointestinal tract (GIT). All eight analyzed bacteria were Proteobacteria, six of them were from class Gammaproteobacteria. According to the Pmaf classification porins of aquatic bacteria belong to one superfamily, and GIT bacteria—to another. According to TBCD (Transporter Classification DataBase) all porins belonged to the same superfamily and all but one were assigned to the same family. As follows from the data shown in Table 2, eight porins were not much different in number of amino acids, protein molecular weight, and aliphatic index, but significant differences were revealed in theoretical pI, total number of negatively and positively charged residues, and differences in the content of tryptophan residues. This amino acid was absent in the porins of the three *Marinomonas* sp. strains and *Photobacterium damselae subsp. damselae* and was in low amount in porin of *Polaromonas* sp. JS666 in comparison with porins of the GIT bacteria. All porins analyzed are enriched in polar amino acids and have a significantly higher level of acidic amino acids (15.9–28.8 mol %) compared with the basic amino acids (5.1–16.4 mol %), that is typical for classical gram-negative porins. Despite noted differences all porins had close polarity values calculated by two methods.

The polarity of MpOmp was 44.25 (from amino acid composition) and 50.7 (from DNA translation) calculated as reported previously [29]. The polarity index of the protein studied was 1.81 (from amino acid composition) and 1.95 (from DNA translation) calculated by [30]. The value is close to the average that of similar values for other proteins belonging to the Protein_4 superfamily represented in Table 1. This result and GRAVY index (Table 1) indicate that porin from *M. primoryensis* KMM 3633^T is relatively hydrophilic protein.

ProtParam tool allows the computation of the instability index on the base of dipeptide composition. Since the proteins with instability index value below 40 are stable proteins [31], all porins from our analysis are stable, and porins of *M. primoryensis* KMM 3633^T and *M. primoryensis* AceL was predicted to be the most stable (Table 1).

Table 1. Comparison of porin characteristics of some Proteobacteria.

Bacteria	Marinomonas primoryensis KMM 3633^T DNA Translation GenBank:QES04118.1	Marinomonas primoryensis KMM 3633^T aa Composition (Methods)	Marinomonas primoryensis strain AceL DNA Translation GenBank: WP_112140846.1	Marinomonas sp. IMCC 4694 DNA Translation GenBank: WP_148833815.1	Polaromonas sp. JS666 DNA Translation GenBank: WP_011485260	Magnetospirillum magneticum AMB-1 DNA Translation GenBank: BAE48829.1	Photobacterium damselae subsp. damselae DNA translation GenBank: KJF82573.1	Photobacterium damselae (Vibrio damsela) aa Composition [27]	Escherichia coli DNA Translation GenBank: WP_000768382.1	Escherichia coli aa Sequencing [28]	Basfia succiniciproducens MBEL55E DNA Translation GenBank: AAU37168.1	Average
Class of Proteobacteria	Gamma	Gamma	Gamma	Gamma	Beta	Alpha	Gamma	Gamma	Gamma	Gamma	Gamma	
Superfamily pfam, clan	Porin_4, cl28788		Porin_4, cl28788	Porin_4, cl28788	Porin_4, cl28788	Porin_4, cl28788	Porin_4, cl28788		OM_channels, cl21487		OM_channels, cl21487	
Superfamily TCBD	Outer membrane pore-forming protein (OMPP) superfamily I											
Family TCBD	The General Bacterial Porin (GBP) Family					The Rhodobacter PorCa Porin (RPP) Family	The General Bacterial Porin (GBP) Family					
The best match in TCBD, TCID	1.B.1.1.9 Major outer membrane protein OmpU		1.B.1.1.1 Outer membrane protein F precursor	1.B.1.1.5 Outer membrane porin protein LC PR.	1.B.1.6.1 Outer membrane porin protein 32 pr	1.B.7.1.5 Porin 41 (Por41) OS = Rhodospirillum	1.B.1.1.9 Major outer membrane protein OmpU		1.B.1.1.4 Outer membrane porin protein NMPC		1.B.1.1.14 Major outer membrane protein OS = Pa	
Query: GenBank:QES04118.1 BLAST NCBI, Query Cover	100%		100%	100%	94%	34%	18%		42%		53%	
Query: GenBank:QES04118.1 BLAST NCBI, Per. Ident	100%		83.59%	77.43%	22.42%	24.05%	40.62%		33.90%		33.33%	
Number of amino acids without signal peptide	290		299	296	336	412	317		345		345	330
Molecular weight, kDa	31.23		31.79	31.73	34.70	42.30	34.94		38.93		38.93	35.57
Theoretical pI	3.99		4.17	3.83	9.49	5.73	4.32		4.71		4.71	
Total number of negatively charged residues (Asp + Glu)	48		44	50	21	30	54		48		40	
Total number of positively charged residues (Arg + Lys)	18		21	15	27	26	31		35		35	
Aliphatic index:	63.17		68.49	62.94	60.51	69.88	64.45		61.16		61.16	
Grand average of hydropathicity (GRAVY)	−0.491		−0.383	−0.461	−0.362	−0.201	−0.61		−0.624		−0.624	
Instability index	−0.01		1.40	10.21	24.19	7.74	15.04		16.18		24.22	

* not determined.

Table 2. The amino acid composition of some Proteobacteria porins.

Bacteria	Marinomonas primoryensis KMM 3633^T DNA Translation GenBank:QES04118.1	Marinomonas primoryensis KMM 3633^T aa Composition (Methods)	Marinomonas primoryensis strain AceL DNA Translation GenBank: WP_112140846.1	Marinomonas sp. IMCC 4694 DNA Translation GenBank: WP_148833815.1	Polaromonas sp. JS666 DNA Translation GenBank: WP_011485260	Magnetospirillum magneticum AMB-1 DNA Translation GenBank: BAE48829.1	Photobacterium damselae Subsp. Damselae DNA Translation GenBank: KJF82573.1	Photobacterium damselae (Vibrio damsela) aa Composition [27]	Escherichia coli DNA Translation GenBank: WP_000768382.1	Escherichia coli aa Sequencing [28]	Basfia succiniciproducens MBEL55E DNA Translation GenBank: AAU37168.1	Average
Neutral aliphatic, mol %	**52.1**	**59.8**	**55.5**	**53.7**	**62.6**	**62**	**46.7**	**51.9**	**46**	**50.3**	**46.6**	**53.4**
Gly, mol %	10.3	13	11	10.5	17.6	18	9.8	10.3	11	14.4	11.5	
Ala, mol %	8.3	12.3	10	8.8	10.1	10.2	11.4	10.6	8.4	8.5	7.5	
Val, mol %	9.7	8.3	9.4	7.8	4.2	7.5	4.7	6.9	3.8	6.8	5.6	
Leu, mol %	4.5	5.9	5.7	4.7	5.7	6.6	6	7.9	5.5	6.2	7.1	
Ile, mol %	2.4	3.1	2.3	3.4	4.2	3.2	4.1	4.8	5.2	3.5	4	
Ser, mol %	6.2	8.6	7.4	8.4	11.3	8.7	5	6.2	4.6	4.7	5	
Thr, mol %	10.7	8.6	9.7	10.1	9.5	7.8	5.7	5.2	7.5	6.2	5.9	
Aromatic, mol %	**11.4**	**7.3**	**11.3**	**11.5**	**10.2**	**10.0**	**13.6**	**9.7**	**16.3**	**14.2**	**13**	**11.4**
Phe, mol %	4.5	3.4	4.3	4.4	4.5	2.7	6.3	5	5.8	5.6	7.1	
Tyr, mol %	6.9	3.9	7	7.1	4.8	5.1	7.3	4.7	9.3	8.5	5.9	
Trp, mol %	0	- *	0	0	0.9	2.2	0	-	0.6	1.2	1.2	
Sulphur-containing, mol %	**1.7**	**0**	**0.7**	**1.7**	**0.9**	**1.9**	**0.9**	**2.8**	**1.7**	**0.9**	**1.2**	**1.3**
Met, mol %	1.7	0	0.7	1.7	0.9	1.9	0.9	2.8	1.7	0.9	1.2	
Cys, mol %	0	0	0	0	0	0	0	0	0	0	0	
Imino, mol %	**1.0**	**3.5**	**1.0**	**1.4**	**1.5**	**1.7**	**0.3**	**2.2**	**0.9**	**1.2**	**0.3**	**1.6**
Pro, mol %	1.0	3.5	1.0	1.4	1.5	1.7	0.3	2.2	0.9	1.2	0.3	
Dicarboxylic, mol %	**27.3**	**20.3**	**24.4**	**26.9**	**15.9**	**17**	**28.8**	**21.8**	**24.4**	**24**	**21.2**	**22.9**
Asx, mol %	18.3	12	15.4	16.2	9.9	11.2	18.7	14.4	17.1	16.7	12.5	
Glx, mol %	9	8.3	9	10.7	6	5.8	10.1	7.4	7.3	7.3	8.7	
Basic, mol %	**6.5**	**7.4**	7	**5.1**	**7.7**	**7.5**	**9.8**	**11.5**	**10.7**	**8.8**	**16.4**	**8.9**
His, mol %	0.3	1	0	0	1.2	1.2	0	1.6	0.6	0.3	0.9	
Arg, mol %	0.7	2.1	1	1	6.5	2.4	2.5	4.2	4.6	3.2	5.3	
Lys, mol %	5.5	4.3	6	4.1	1.5	3.9	7.3	5.7	5.5	5.3	10.2	
Polarity index (Hatch)	**1.95**	**1.81**	**2.07**	**2.16**	**2.13**	**1.69**	**2.21**	**1.46**	**2.03**	**1.79**	**1.88**	**1.93**
Polarity (Nitzan)	**50.7**	**44.25**	**48.95**	**50.05**	**46.1**	**43.7**	**52.5**	**45.7**	**51**	**49.55**	**51.3**	**48.5**

2.4. Spatial Structure of OM Porin from M. primoryensis KMM 3633^T

2.4.1. CD and Intrinsic Protein Fluorescence Spectra

One of the important physicochemical characteristics of proteins is their resistance to various denaturing agents and temperature. Therefore, we paid special attention to conformational changes in MpOmp molecule under various conditions: (1) upon dilution of the protein solution, (2) in the presence of non-ionic (POE) and ionic (SDS) detergents, and (3) upon heating. For this purpose, CD analysis and intrinsic protein fluorescence were used. CD spectra were recorded in the far and near UV regions. Intrinsic protein fluorescence spectra were recorded at an excitation wavelength of 280 nm.

It was found that conformational changes were not observed either at the level of the secondary or at the level of the tertiary structure of MpOmp upon dilution of the protein solution without any detergents (Figure 3A,B).

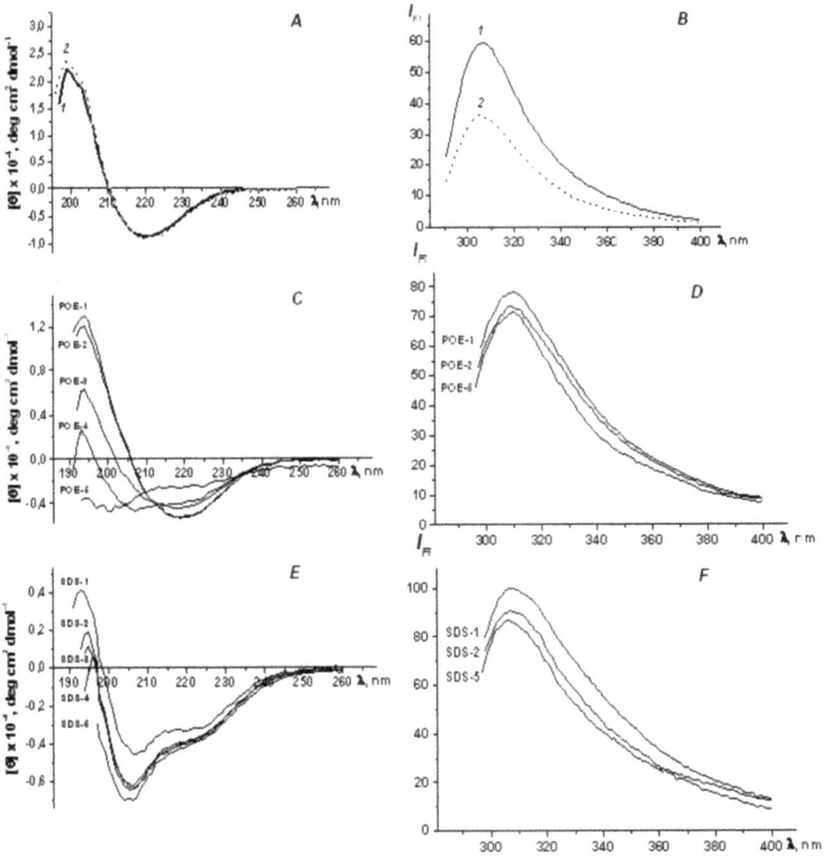

Figure 3. CD spectra in peptide regions (**A**,**C**,**E**) and intrinsic fluorescence spectra at λex 280 nm (**B**,**D**,**F**) of *M. primoryensis* porin (MpOmp) in Tris-HCl buffer(**A**,**B**) and in the presence of POE (**C**,**D**) and SDS (**E**,**F**). The protein samples of MpOmp (1.0 μM and 6.0 μM for peptide and aromatic regions, respectively) were dissolved in 10 mM Tris-HCl buffer, pH 7.5, containing 0.5% POE (1) or 0.25% SDS (2). Samples 2, 3, 4, and 5 were heated at 30, 40, 50, and 80 °C, respectively (**C**,**D**,**E**,**F**).

CD spectra were recorded in the far and near UV regions. CD spectra of MpOmp in the aromatic region (240–320 nm) have a low amplitude and low resolution, which indicates a loosened tertiary structure because of the apparently weak interaction between the protein monomers (data not shown).

CD spectra of the porin studied between 180–240 nm (the region of peptide bonds) in the presence of non-ionic detergent POE are characterized by positive band at 195 nm and only one negative minimum that was centered at 220 nm. The spectra crossed the zero line at 209 nm (Figure 3C). It is typical for proteins with β-pleated sheet structure of (α + β) type [32]. In the presence of SDS a new negative band at 207 nm was observed (Figure 3E), which indicated α-helical structure formation typical of denatured proteins in SDS [33].

Using the CDPro software [34] the content of secondary structure elements of MpOmp was determined. As we can see in Table 3, the secondary structure of the protein in solution of various detergents differed. In POE solution the total β-structure of MpOmp accounted 70% while alpha α-helices make up only 5%. Such ratio of the elements of the regular secondary structure of the protein is similar to that of the porins in the native environment in the bacterial membrane [35].

Table 3. Content of secondary structural elements in porin from *M. primoryensis* KMM 3633^{T} (%).

№	Sample	α-Helix			β-Structure			β-Turn	Random Coil
		I	II	III	I	II	III		
1	MpOmp	**2.8**	1.5	**4.3**	32.5	17.4	**49.9**	**20.3**	**25.5**
2	MpOmp, 0.5% POE	0.8	4.1	**4.9**	34.7	14.4	**49.1**	**20.9**	**25.1**
3	MpOmp, 0.25% SDS	7.6	10.4	**18.0**	18.1	10.2	**28.3**	**22.7**	**31.0**

However, in the presence of SDS, the polypeptide chain of the studied protein contains a significant amount of α-helical regions (3.6 times more than in a solution of non-ionic detergent) and a smaller amount of total β-structure. These data indicate that ionic detergent SDS has a denaturing effect on the secondary structure of MpOmp. Similar denaturing effect of this ionic detergent we observed earlier for *Yersinia* porins [36].

In Figure 3 the intrinsic fluorescence spectra of MpOmp in solutions of POE (Figure 3D) and SDS (Figure 3F) are presented. A significant shift of the maximum of the spectrum to the short- wavelength region in the presence of ionic and nonionic detergents indicates that the radiation of the protein occurs because of tyrosine residues. These results are consistent with the data of amino acid sequence (Table 2), according to which the tryptophan residues are absent in the protein studied.

Thus, according to data of optical spectroscopy the secondary structure of MpOmp depends on the detergent nature present in its solution.

2.4.2. Temperature Stability of OM Porin from *M. primoryensis* KMM 3633^{T}

The instability of MpOmp trimer was demonstrated by heating a protein sample at different temperatures. As follows from the data of SDS-PAGE (Figure 1C), protein samples of MpOmp, both isolated by POE extraction and obtained by treatment of the cell membrane with Sarcosyl, are equally temperature sensitive. Dissociation of the oligomeric form of the protein into monomers was observed already at a temperature of 30 °C. In order to analyze the changes in the spatial structure of porin upon heating and to differentiate the dissociation process from the process of protein denaturation, CD spectra in the peptide region and emission of MpOmp samples heated at different temperatures from 30 to 80 °C were recorded.

As follows from the data in Figure 3C–F, in MpOmp molecule in POE solution, changes at the level of the secondary structure of the protein are observed starting at a temperature of 40 °C. Further heating of the sample leads to more significant changes, manifested in a decrease in the ellipticity of characteristic peaks.

In the SDS solution in the spatial structure of the porin molecule, more serious rearrangements even occur at 30 °C, a change in the peak ratio at 195 and 207 nm is observed, which may indicate a change in the content of α-helices. With a further increase of temperature of the sample treatment, the observed changes were minimal.

The heating of MpOmp sample in solutions of both detergents in the studied temperature range did not lead to any major noticeable changes detected by fluorescence spectroscopy. The data obtained allow us to conclude that only an ionic detergent is capable to effect the conformational changes in the structure of the porin studied.

3. Discussion

3.1. Isolation Procedure of OM Protein from M. primoryensis KMM 3633^T

To date several advances have been made in the development of isolation methods of OM pore-forming proteins. Since porins as integral proteins exist in an intrinsically anisotropic environment within the bacterial membrane the main condition for maximizing protein extraction is the selection of a suitable detergent [25]. Unlike terrestrial, marine bacteria cell envelope is not tightly associated with rigid peptidoglycan layer and therefore the OM may be easily separated [37,38]. Thus, the standard isolation and purification procedure used for OM porins of gram-negative bacteria is not exactly applicable for marine OM proteins. In the case of marine bacteria, the OM proteins may be isolated with the sequential extraction of the microbial cells with solutions of high ionic strength, even without previous cell destruction [39]. In light of the above special attention has been given by us to the choice of cell envelope isolation condition and the selection of a suitable detergent for the extraction of *M. primoryensis* KMM 3633^T OM proteins.

To determine the influence of the cultivation temperature on the protein composition of the cell envelope of the psychrophilic bacterium *M. primoryensis* KMM 3633^T, the bacterial cells were grown at a temperature of 0, 6–8, and 24 °C using the growth medium given in [23]. When growing microorganisms at low temperatures, the most intense polypeptide band was observed in the region corresponding to the molecular masses of the porins of terrestrial bacteria. Taking these results into account, for the further study we selected the *M. primoryensis* KMM 3633^T cells, cultivated at 6–8 °C under aerobic conditions.

In the electrophoretic profile of the crude membrane fraction obtained by ultrasonic disintegration, proteins that changed their electrophoretic mobility after heat pre-treatment of the protein sample were detected. This allows us to assume that heat-modifiable proteins are present in the cell lysate of *M. primoryensis* KMM 3633^T.

To isolate the OM protein fraction from the cell envelope of *M. primoryensis* KMM 3633^T, two methods were tested, commonly used to obtain OM proteins from the terrestrial bacteria. One of them consists in the step-by-step extraction of cell membranes obtained after ultrasound treatment with a nonionic detergent POE [25]. By the author data subsequent extraction with POE allows to yield several fractions highly enriched in *E. coli* porins.

The second method is based on the removal of cytoplasmic proteins from the cell envelope fraction with nonionic detergent Sarcosyl extraction. As known, Sarcosyl is commonly used in the purification procedure of OM proteins of gram-negative terrestrial bacteria, and this method produces samples free of cytoplasmic proteins. OM proteins in this case remain in the sediment. As we have seen earlier, after Sarcosyl pre-treatment the fraction of OM proteins of *Yersinia ruckeri* practically did not contain low molecular weight cytoplasmic proteins [40]. However, in the case of OM proteins of *M. primoryensis KMM* 3633^T in the Sarcosyl-soluble fraction, in addition to cytoplasmic proteins, we found the presence of a certain amount of OM proteins with a molecular weight of 30–50 kDa. This result indicates that the OM proteins of *M. primoryensis* KMM 3633^T are partially soluble in Sarcosyl.

3.2. *Pore-Forming Activity of Porin M. primoryensis KMM 3633*^{T}

It was found that the channels of MpOmp are characterized by a linear current-voltage characteristic and the dependence of conductivity on salt concentration. Purified porin reconstituted in BLM at a concentration 10–100 ng/mL induced stepwise changes in membrane conductance typical for porins. It is noteworthy that the activity of marine porins [17,18] as a rule, is lower (at 100–200 ng/mL) as compared to the activity of Enterobacteriaceae porins (at 1–10 ng/mL) [41].

Within the transmembrane potential of ±180 mV, MpOmp pores acted as ohmic channels, and their conductivity scaled linearly with voltage. Thus, the porin from *M. primoryensis* KMM 3633^T behaves in a voltage-dependent manner, like the other marine porins [26,42].

It was shown that the pore-forming activity of the protein studied depends significantly on the ionic strength of the electrolyte. This is characteristic of the porins of marine bacteria. It is generally accepted that non-linear graphs of the dependence of the electrical conductivity of pores on the symmetric salt concentrations on both sides of the protein-containing membrane indicate the influence of the functionality of the internal channel on the passage of charged particles [26]. In the future we plan to build a theoretical protein model of MpOmp that will allow us to conduct a detailed analysis of charge distribution inside the barrel channel.

3.3. *Physico-Chemical Properties of Porin from M. primoryensis KMM 3633*^{T}

The amino acid composition of porin-like protein from *M. primoryensis* KMM 3633^T has the same characteristic features as those of the classical gram-negative porins. For example, the protein composition is distinguished by a high content of acidic amino acids and low content of sulfur-containing amino acids. Cysteine is completely absent. However, there is a slight difference in the content of individual amino acids in composition of the protein characterized: it contains 1.4 times less aromatic acids and almost 1.4 times less tyrosine residues compared to the OmpF porin *E. coli*. In addition, the composition and properties of MpOmp are quite remarkable, primarily because of the fact that, in contrast to the porins of the other Proteobacteria under consideration (except for the porins from bacteria of the genus *Marinomonas* and *Photobacterium damselae subsp. damselae* presented in Table 1), there are no tryptophan residues in its molecule.

Considering the monomer stability of the analyzed porins, we noticed that predicted instability index had the lowest value for porins of *M. primoryensis* KMM 3633^T and *M. primoryensis* AceL. When characterizing the monomers of these porins as the most stable, we take into account that in the bacterial cell the stability of the protein may be dependent not only on the intrinsic nature of the protein but also on the conditions of the protein milieu, as noted Gamage, et al. in their study [43]. In connection with the foregoing, we should especially dwell on the instability of MpOmp oligomeric structure that we observe.

Porin *M. primoryensis* KMM 3633^T was isolated in the oligomeric form, however porin trimers were extremely unstable. The dissociation of MpOmp trimer into monomers, accompanied by an irreversible conformational transition was already observed at 30 °C. In a number of cases, on the electropherogram of the fraction of OM proteins obtained under denaturing conditions of SDS-PAGE (0.1% SDS), the protein monomer of MpOmp appeared without prior boiling of the sample. In addition, MpOmp trimer dissociated into monomers also after protein precipitation with ethanol in the presence of EDTA, the method used for purification of membrane proteins [25]. So, a trimeric form of the protein was clearly observed only under mild condition of solubilization. Therefore we cannot exclude that isolation conditions of target protein chosen by us do not lead to a partial dissociation of its trimeric form during obtaining of cell envelope and purification from cytoplasmic proteins. Thus, in order to obtain and preserve OM protein from *M. primoryensis* KMM 3633^T as a trimer with a relatively stable structure, it is necessary to choose only a mild non-ionic detergent and strictly monitor the temperature at which the experiment is conducted. Similar instability of native subunit structure of some bacterial pore-forming proteins was described in literature [44,45]. Authors of these articles should have used cross-linking in order to reveal the native oligomers of the proteins studied.

Elucidation of the conformational stability of the porin studied under various conditions using the methods of optical spectroscopy and electrophoresis made it possible to draw the following conclusion. It was found that no conformational changes were observed either at the level of the secondary or at the level of the tertiary structure of the protein upon dilution. However, the presence of ionic and nonionic detergents influenced the spatial organization of MpOmp differently and introduced characteristic features into the conformation of the protein molecule. It has been shown that the SDS ionic detergent has a denaturing effect on the conformation of the porin trimer from *M. primoryensis* KMM 3633^T at the secondary protein structure, significantly increasing the content of the α-helix and decreasing the content of the total β-structure. This is consistent with the widespread literature hypothesis that SDS micelles promote the formation of α-helices in the case of β-structured proteins, containing intrinsically disordered regions (IDRs) [46]. In contrast, in the presence of a non-ionic detergent, minimal changes in protein structure were observed. Thus, the data obtained show, that non-ionic detergent POE is most suitable for solubilization of MpOmp, since in its presence the conformation of the studied protein is as close as possible to the native one.

According to modern concepts, porins are unusual membrane proteins containing a significant amount of hydrophilic amino acid residues. They are able to form β-barrels in contrast to the α-helices of almost all other membrane proteins. The association of monomers within native trimer is stabilized through the hydrophobic and hydrophilic interaction at the subunit interface and involves 35% of the barrel surface area [47].

In addition, porins are amyloidogenic proteins capable of forming amyloid-like structures under certain conditions. This may be facilitated by the availability of hydrophobic sites on the surface of the monomer, buried in the trimer, but released during its dissociation. At the same time, according to the recently published theory of the formation of multimeric proteins, the formation of their quaternary structure occurs because of the balance between the energy of intramolecular interaction and the energy of adhesion between subunits [48].

Given these facts, it can be assumed that the instability of the trimeric form of MpOmp can be caused both by the difference in the values of the indicated energies in favor of the predominant existence of a protein in the form of a monomer, and by differences in the primary structure of porins from marine and terrestrial bacteria, the trimer form of the last exhibit the high thermal stability. So, for example, mutation of residues involved in ionic interactions between the two subunits have been shown to reduce the thermal stability of the trimer significantly [49].

At present, computer models are widely used in investigation of the stabilizing forces in quaternary structure formation. Currently, there is the possibility of a detailed study of protein–protein interaction in the formation of the oligomeric structure of beta-barrel proteins [50]. Transmembrane domains of β-barrel membrane proteins have shown the presence of so called "weakly stable regions" despite an extensive network of hydrogen bonds, as well as ionic and hydrophobic interactions that give high strength to the molecule as a whole [51]. In addition, it was found that one of the ways to stabilize these areas can be by interacting with surrounding lipids [52]. On the other hand, it is known that oligomerization of the proteins, including membrane proteins, can bring various functionally important advantages to a particular protein.

Taking into account the foregoing, we can assume the following. Since the MpOmp monomer has very high stability (Table 1), and the oligomeric structure of the isolated protein, on the contrary, is extremely unstable under the heating, it is possible that the existence of the MpOmp trimer is crucial for the manifestation of the functional activity of this porin in the native membrane. This assumption is based on the fact that isolated MpOmp in monomeric form had a very low efficiency of reconstitution into an artificial bilayer. Given the fact that lipid–protein interaction can be a factor stabilizing the trimeric structure of MpOmp, the study of the pore-forming activity of this novel marine porin depending on the composition of the lipid membrane will undoubtedly be of fundamental interest.

4. Materials and Methods

4.1. Microorganisms

The bacteria of *M. primoryensis* KMM 3633^T were grown at a low (0 and 6–8 °C) and room temperature (24–26 °C) in liquid medium containing sea water as described in [23].

4.2. Isolation of Cell Envelope Fraction

Microbial cells were harvested in the late exponential growth phase (after 48 h) that was determined by the turbidity of the cell suspension. Cells were collected by centrifugation at 7000 rpm for 20 min, washed once with a solution of 5% NaCl, and suspended in buffer containing 50 mM Na_2HPO_4, 100 mM NaCl, sucrose (5%, *w/v*), 1.5 mg of DNAase, and sodium azide (0.02%, *w/v*).The cells were disrupted by sonication (UZDN-2T insonator, Sumy, Ukraine) at 44 kHz (10 × 1 min cycles on ice). The unbroken cells were centrifuged at 10,000× *g* for 15 min at 4 °C, and the supernatant was centrifuged at 25,000× *g* (Heraeus Biofuge stratus, Thermo Fisher Scientific, Waltham, MA, USA) for 1 h at 4 °C. The pellet was then washed with distilled water and used as cell envelope preparation (crude membrane fraction). It contained the total fraction of cytoplasmic and OM proteins.

4.3. Isolation and Purification of Porin from M. primoryensis KMM 3633^T

4.3.1. Extraction with Nonionic Detergent *n*-Octylpolyoxyethylene (POE)

Cell envelope pellet (6.4 mg) was resuspended in 5 mL 20 mM Tris-HCl, pH 7.4 (buffer A), containing 0.5% (*v/v*) POE incubated subsequently at 37 °C for 1 h and at 4 °C overnight, and then centrifuged at 25,000× *g* for 1 h. This procedure was repeated three times. The pellet was washed once with buffer A by centrifugation. The residual pellet was extracted consistently three times with 3 mL of buffer A, containing 0.5, 1, and 2% (*v/v*) POE. According to the SDS-PAGE data the abundant quantity of 35 kDa protein (MPOmp) was found in supernatant after extraction with 0.5% POE. Three 0.5% POE extracts were combined, concentrated to the protein content of 0.5 mg/mL, and subjected to dialysis against a 10-fold volume of buffer A, containing 0.2% POE and then purified by gel permeation chromatography on Sephacryl S-300 (Serva, Heidelberg, Germany).

4.3.2. Extraction with N-Lauroylsarcosine (Sarcosyl)

Cell envelope pellet (6.4 mg) was resuspended in 5 mL 10 mM HEPES, pH 7.5 (buffer B), containing 1% (*w/v*) Sarcosyl (Sigma-Aldrich, St. Louis, MO, USA) and incubated at room temperature for 30 min with shaking. The pellet obtained by centrifugation at 25,000× *g* was washed with 10 mL of buffer B without suspending. According to SDS-PAGE data, this fraction was electrophoretically pure 35 kDa protein (MpOmp) in oligomeric form (approximately 100 kDa).

4.4. SDS-PAGE

Analysis of protein fractions obtained was carried out using SDS-PAGE according to the method described in [49,50]. SDS-PAGE procedure was carried out under native (solubilization temperature 25 °C) and denaturing (after heating at 99 °C) conditions. Apparent MW of proteins in different fraction was determined according to relative mobilities of standard proteins using marker proteins (Thermo Fisher Scientific, Waltham, MA, USA) with MW from 10 to 170 kDa. The proteins separated in the gel were stained with Coomassie R-250 in 3.5% perchloric acid [53]. The protein content of the samples was determined spectrophotometrically at 280 nm, assuming an optical density of 1.0 for a protein concentration of 1 mg/mL.

4.5. Spectroscopic Methods

All absorption, circular dichroism (CD), and fluorescence measurements were performed using the protein samples in buffer A, containing 0.25% SDS at 25 °C. UV absorption spectra were recorded on a UV-visible spectrophotometer UV-1601PC (Shimadzu, Japan) in quartz cuvettes with the path-length of 1 cm. The correction for light scattering of the protein solution was carried out as described in [54]. Specific absorption factor A 0.1%/1 cm of porin was taken equal to 1.00.

Circular dichroism (CD) spectra were recorded on Chirascan Plus CD spectropolarimeter (Applied Photophysics Limited, Leatherhead, UK) in quartz cuvettes with the optical path length of 0.1 and 1 cm for the far-UV or peptide (180–250 nm) and the near-UV or aromatic (250–350 nm) regions, respectively. Samples with the protein concentrations of 1.0 μM and 6.0 μM were used for the CD measurements in the peptide and aromatic regions, respectively.

In the aromatic spectral region, ellipticity [θ]M was calculated as molar ellipticity taking the molecular weights of the protein monomer and trimer as 35 and 105 kDa, respectively. In the peptide region, ellipticity was calculated as the mean residue ellipticity taking the molecular weight of amino acid residue as 110 Da, in standard units (deg cm^2 $dmol^{-1}$) by the equation:

$$[\theta] = [\theta]_{obs} \cdot S \cdot 110/10 \cdot C \cdot l \quad (1)$$

where S is the sensitivity; C is the protein concentration, mg/mL; and l is the optical path length, dm.

The content of the secondary structure elements of the protein was calculated with CD Pro software [34].

Intrinsic protein fluorescence spectra of the porin were measured in quartz cuvettes with optical path length of 1 cm using a RF-5301PC spectrofluorophotometer (Shimadzu, Kyoto, Japan). Porin samples with protein concentration of 0.5 μM were prepared in buffer A, containing 0.25% SDS. Fluorescence was excited by the light with wavelength of 280, 296, and 305 nm. The excitation and emission slits were set at 5 nm. Fluorescence spectra corrected by rhodamine B (Wako Pure Chemical Industries, Osaka, Japan) were recorded by subtracking the Raman band of the buffer solution.

4.6. Lipid Bilayer Experiments

The experiments on reconstitution of protein samples into the BLM were performed under the symmetric conditions, so that the investigated proteins were present on both sides of the BLM. Fluctuations of the current through the BLM in the presence of the porin were measured in the clamp mode on the membrane potential. The current from the membrane was conducted by Ag/AgCl electrodes and registered at a membrane potential of 10–50 mV. The electrical parameters of the BLM were measured with bilayer clamp amplifier BC-535, pCLAMP 10 and Clampfit 10 software suites.

The BLM was formed from a 1% solution of 1-monooleoylglycerol (monoolein) (Sigma- Aldrich, St. Louis, MO, USA) in *n*-heptane with a pipette (a hole of 0.25 mm in diameter) on a Teflon cup placed in a thermostated optically transparent cuvette filled with an electrolyte. The experiments of protein reconstitution in the BLM were performed at room temperature (22 °C). The aqueous phase contained 0.2–2 M NaCl in buffer A and the studied protein in the range of concentrations of 10–100 ng/mL.

4.7. Amino Acid Composition and N-Terminal Sequencing

Amino acid analysis was performed by using a Biochrom 30 amino acid analyzer (Biochrom Ltd. CB4 OFJ, Cambridge, UK), after hydrolysis of purified protein with 6 N HCl according to the method descried in [55]. The amino acid content of the protein sample was calculated in Mol% of each amino acid.

DNA isolation. Preparation of genomic DNA for whole genome sequencing (WGS) was performed according to the protocols [56].

Genome sequencing and assembly. Two kits were used for library preparation: KAPA HyperPlus (Roche, Basel, Switzerland) and Nextera XT DNA Library Prep Kit (Illumina, San Diego, CA, USA).

Libraries were checked on a Bioanalyzer (Agilent, Santa Clara, CA, US) and sequenced on an Illumina MiSeq instrument with paired-end protocol. Genome assembly was performed with CLC Genomic Workbench v.12.

Genome annotation. Genome was annotated by the Rapid Annotations using Subsystems Technology (RAST) server [57,58].

GenBank NCBI submission. The obtained *Marinomonas primoryensis* KMM 3633^{T} genome sequence was submitted in NCBI SRA database under the BioProject number: PRJNA622482. DNA of Porin_4 was translated, complementary annotated in NCBI by conserved domains detection and submitted in GenBank with Accession Number MK820371 (Protein Accession-QES04118).

Sequences, used for analysis. *Marinomonas primoryensis* strain AceL (WP_112140846), *Marinomonas* sp. IMCC 4694 (WP_148833815), *Polaromonas* sp. JS666 (Q123M1), *Magnetospirillum magneticum* AMB-1 (Q2WBE6), *Photobacterium damselae subsp. damselae* (KJF82573.1), *Escherichia coli* (WP_000768382.1), *Basfia succiniciproducens* MBEL55E (Q65V42).

Bioinformatic analysis. Alignment of protein sequences was performed by NCBI BLAST, signaling peptide was predicted by Phobius server (http://phobius.sbc.su.se/) [59]. ProtParam tool was used for the computation of various physical and chemical parameters for the proteins (https://web.expasy.org/protparam/) [60].

Polarity index was calculated according to following formula [30]:

(Asx + Glx + Lys + Arg + Ser + Thr)/(Val + Leu + Ile + Met + Pro + Phe). For the polarity calculation was used formula recommended by [29]:

(Asx + Glx + Lys + Arg) + (Ser + Thr + Tyr + His + Gly)/2

PSORTb program was used for bacterial protein subcellular localization prediction (https://www.psort.org/psortb/) [61,62].

Pfam server was used for protein classification (http://pfam.xfam.org/) [63].

TCBD (Transporter Classification DataBase) and bioinformatics resources of this base were used for porin classification [63].

Author Contributions: O.D.N. Conceptualization, project administration, literature analysis, data analysis and interpretation, writing and editing of the manuscript. V.A.K. Performed experiments (*M. primoryensis KMM* 3633^{T} porin isolation), data analysis and interpretation. N.Y.K. Performed experiments (optical properties of *M. primoryensis KMM* 3633^{T} porin), data analysis. G.N.L. Performed BLM experiments and interpretation of data obtained. L.A.R. isolation and identification of novel *Marinomonas* strain. Cultivation of the microbial cells. O.Y.P. Data analysis and interpretation, writing original draft. T.F.S. Literature analysis, discussion of manuscript structure. E.I.A., M.S.K., N.N.R. and O.L.V. contributed in genome sequencing, assembling, annotation and analysis, in comparison of *M. primoryensis KMM* 3633^{T} porin with porins in different Data Bases, analysis of amino acides composition and theoretical porin properties. O.L.V. contributed in literature analysis, discussion of manuscript structure and in manuscript writing. All authors have read and agreed to the published version of the manuscript.

Funding: This work was supported by the Russian Foundation for Basic Research grant № 19-04-00318.

Acknowledgments: The authors of the article are grateful to Victor Emelyanenko (Institute for Biological Instrumentation, Institute of Theoretical and Experimental Biophysics, Pushchino, Russia) for a helpful discussion.

Conflicts of Interest: The authors declare that they have no conflict of interest.

Abbreviations

BLM—bilayer lipid membrane; CD—circular dichroism; LPS—lipopolysaccharide; POE—*n*-octylpolyoxyethylene; OM—outer membrane; SDS-PAGE—sodium dodecyl-sulfate polyacrylamide gel electrophoresis.

References

1. MacLeod, R.A. The question of the existence of Specific Marine Bacteria. *Bacteriol. Rev.* **1965**, *29*, 9–23. [CrossRef] [PubMed]
2. Welch, T.J.; Bartlett, D.H. Identification of a regulatory protein required for pressure-responsive gene expression in the deep-sea bacterium *Photobacterium profundum* species strain SS9. *Mol. Microbiol.* **1998**, *27*, 977–985. [CrossRef]

3. Wang, F.; Wang, J.; Jian, H.; Zhang, B.; Li, S.; Wang, F.; Zeng, X.; Gao, L.; Bartlett., D.H.; Yu, J.; et al. Environmental adaptation: Genomic analysis of the piezotolerant and psychrotolerant deep-sea iron reducing bacterium *Shewanella piezotolerans* WP3. *PLoS ONE* **2008**, *3*, e1937. [CrossRef]
4. Benner, R.; Benitez-Nelson, B.; Kaiser, K.; Amon, R.M.W. Export of Young Terrigenous Dissolved Oganic Carbon from Rvers to the Arctic Ocean. *Geophys. Res. Lett.* **2004**, *31*, L05305. [CrossRef]
5. Dong, C.; Bai, X.; Sheng, H.; Jiao, L.; Zhou, H.; Shao, Z. Distribution of PAHs and the PAH-degrading bacteria in the deep-sea sediments of the high-latitude Arctic Ocean. *Biogeosciences* **2015**, *12*, 2163–2177. [CrossRef]
6. Sable, R.; Parajuli, P.; Jois, S. Peptides, Peptidomimetics, and Polypeptides from Marine Sources: A Wealth of Natural Sources for Pharmaceutical Applications. *Mar. Drugs* **2017**, *15*, 124. [CrossRef] [PubMed]
7. Anwar, M.A.; Choi, S. Gram-negative marine bacteria: Structural features of lipopolysaccharides and their relevance for economically important diseases. *Mar. Drugs* **2014**, *12*, 2485–2514. [CrossRef]
8. Vorob'eva, E.V.; Krasikova, I.N.; Solov'eva, T.F. Influence of lipopolysaccharides and lipids A from some marine bacteria on spontaneous and *Escherichia coli* LPS-induced TNF-α release from peripheral human blood cells. *Biochemistry (Moscow)* **2006**, *71*, 759–766. [CrossRef]
9. Machado, H.; Sonnenschein, E.C.; Melchiorsen, J.; Gram, L. Genome mining reveals unlocked bioactive potential of marine Gram-negative bacteria. *BMC Genomics* **2015**, *16*, 158. [CrossRef]
10. Bartlett, D.H. Microbial adaptations to the psychrosphere/piezosphere. *J. Mol. Microbiol. Biotechnol.* **1999**, *1*, 93–100.
11. Lu, S.; Wang, J.; Chitsaz, F.; Derbyshire, M.K.; Geer, R.C.; Gonzales, N.R.; Gwadz, M.; Hurwitz, D.I.; Marchler, G.H.; Song, J.S.; et al. CDD/SPARCLE: The conserved domain database in 2020. *Nucleic Acids Res.* **2020**, *48*, 265–268. [CrossRef] [PubMed]
12. Nikaido, H. Molecular basis of bacterial outer membrane permeability revisited. *Microbiol. Mol. Biol. Rev.* **2003**, *676*, 593–656. [CrossRef]
13. Masi, M.; Pagès, J.M. Outer Membrane Porins. In *Bacterial Cell Walls and Membranes*; Kuhn, A., Ed.; Springer: Cham, Switzerland, 2019; Volume 92, pp. 79–123. [CrossRef]
14. Hancock, R.E. Resistance mechanisms in Pseudomonas aeruginosaand other nonfermentative Gram-negative bacteria. *Clin. Infect. Dis.* **1998**, *27*, 93–99. [CrossRef] [PubMed]
15. Xu, C.; Ren, H.; Wang, S.; Peng, X. Proteomic analysis of salt-sensitive outer membrane proteins of *Vibrio parahaemolyticus*. *Res. Microbiol.* **2004**, *155*, 835–842. [CrossRef] [PubMed]
16. Xu, C.; Wang, S.; Ren, H.; Lin, X.; Wu, L.; Peng, X. Proteomic analysis on the expression of outer membrane proteins of *Vibrio alginolyticus* at different sodium concentrations. *Proteomics* **2005**, *5*, 3142–3152. [CrossRef] [PubMed]
17. Novikova, O.D.; Khomenko, V.A.; Frolova, G.M.; Likhatskaya, G.N.; Romanenko, L.A.; Portnyagina, O.Y.; Kuznetsova, S.M.; Solovyeva, T.F. Porins from marine bacteria of the Genus *Pseudoalteromonas* (*Gammaproteobacteria*: *Pseudoalteromonadaceae*). *Russ. J. Mar. Biol.* **2013**, *39*, 58–64. [CrossRef]
18. Khomenko, V.A.; Likhatskaya, G.N.; Romanenko, L.A.; Portnyagina, O.Y.; Solovyeva, T.F.; Novikova, O.D. Protein composition of the cell envelope of the bacterium *Shewanella frigidimarina* Pi 2-35 (*Gammaproteobacteria: Shewanellaceae*). *Russ. J. Mar. Biol.* **2016**, *42*, 73–80. [CrossRef]
19. Sanchez-Amat, A.; Solano, F. Genus III. Marinomonas. In *Bergey's Manual of Systematic Bacteriology*; Brenner, D.J., Krieg, N.R., Staley, J.T., Garrity, G.M., Eds.; Springer: New York, NY, USA, 2005; pp. 284–289.
20. Espinosa, E.; Marco-Noales, E.; Gómez, D.; Lucas-Elío, P.; Ordax, M.; Garcías-Bonet, N.; Duarte, C.M.; Sanchez-Amat, A. Taxonomic study of *Marinomonas* strains isolated from the seagrass *Posidonia oceanica*, with descriptions of *Marinomonas balearica* sp. nov. and *Marinomonas pollencensis* sp. nov. *Int. J. Syst. Evol. Microbiol.* **2010**, *60*, 93–98. [CrossRef]
21. Gupta, P.; Chaturvedi, P.; Pradhan, S.; Delille, D.; Shivaji, S. *Marinomonas polaris* sp. nov., a psychrohalotolerant strain isolated from coastal sea water of the subantarctic Kerguelen islands. *Int. J. Syst. Evol. Microbiol.* **2006**, *56*, 361–364. [CrossRef]
22. Prabagaran, S.R.; Suresh, K.; Manorama, R.; Delille, D.; Shivaji, S. *Marinomonas ushuaiensis* sp. nov., isolated from coastal sea water in Ushuaia, Argentina, sub-Antarctica. *Int. J. Syst. Evol. Microbiol.* **2005**, *55*, 309–313. [CrossRef]
23. Romanenko, L.A.; Uchino, M.; Mikhailov, V.V.; Zhukova, N.V.; Uchimura, T. *Marinomonas primoryensis* sp. nov., a novel psychrophile isolated from coastal sea-ice in the Sea of Japan. *Int. J. Syst. Evol. Microbiol.* **2003**, *53*, 829–832. [CrossRef] [PubMed]

24. Hobb, R.I.; Fields, J.A.; Burns, C.M.; Thompson, S.A. Evaluation of procedures for outer membrane isolation from *Campylobacter jejuni*. *Microbiology* **2009**, *155*, 979–988. [CrossRef] [PubMed]
25. Garavito, R.M.; Rosenbusch, J.P. Isolation and crystallization of bacterial porin. *Methods Enzymol.* **1986**, *125*, 309–328. [CrossRef] [PubMed]
26. Schulte, A.; Ruamchan, S.; Khunkaewla, P.; Suginta, W. The outer membrane protein VhOmp of *Vibrio harveyi*: Pore-forming properties in black lipid membranes. *J. Membr. Biol.* **2009**, *230*, 101–111. [CrossRef] [PubMed]
27. Nitzan, A.; Gribun, A.; Pechatnikov, I. Characterization of Porins from the Outer Membrane of *Vibrio damsel*. *Curr. Microbiol.* **1999**, *38*, 1–8. [CrossRef] [PubMed]
28. Chen, R.; Krämer, C.; Schmidmayr, W.; Chen-Schmeisser, U.; Henning, U. Primary structure of major outer-membrane protein I (ompF protein, porin) of *Escherichia coli* B/r. *Biochem. J.* **1982**, *203*, 33–43. [CrossRef]
29. Capaldi, R.A.; Vandercooi, G. The low polarity of many membrane proteins. *Proc. Natl. Acad. Sci. USA* **1972**, *69*, 930–932. [CrossRef]
30. Hatch, F.T. Correlation of amino acid composition with certain characteristics of protein. *Nature* **1965**, *206*, 677–679. [CrossRef]
31. Manavalan, P.; Johnson, W.C. Sensitivity of circular dichroism to protein tertiary structure class. *Nature* **1973**, *305*, 831–832. [CrossRef]
32. Eisele, J.L.; Rosenbusch, J.P. *In vitro* folding and oligomerization of a membrane protein. Transition of bacterial porin from random coil to native conformation. *J. Biol. Chem.* **1990**, *265*, 10217–10220. [PubMed]
33. Sreerama, N.; Woody, R.W. Estimation of protein secondary structure from circular dichroism spectra: Comparison of CONTIN, SELCON, and CDSSTR methods with an expanded reference set. *Anal. Biochem.* **2000**, *287*, 252–260. [CrossRef]
34. Inouye, M. What is the outer membrane? In *Bacterial Outer Membranes: Biogenesis and Functions*; Inouye, M., Ed.; Wiley-Interscience: New York, NY, USA, 1979; pp. 5–12.
35. Vostrikova, O.P.; Kim, N.Y.; Likhatskaya, G.N.; Guzev, K.V.; Vakorina, T.I.; Khomenko, V.A.; Novikova, O.D.; Solov'eva, T.F. Structure and function of pore-forming proteins from bacteria of the genus *Yersinia*: I. Isolation and a comparison of physicochemical properties and functional activity of *Yersinia* porins. *Russ. J. Bioorg. Chem.* **2006**, *32*, 333–344. [CrossRef] [PubMed]
36. De Voe, L.W.; Oginsky, E.L. Cation interactions and biochemical composition of the cell envelope of a *Marine acterium*. *J. Bacteriol.* **1969**, *98*, 1368–1377. [CrossRef] [PubMed]
37. Forsberg, C.W.; Costerton, J.W.; MacLeod, R.A. Separation and localization of cell wall layers of a Gram-negative bacterium. *J. Bacteriol.* **1970**, *104*, 1338–1353. [CrossRef] [PubMed]
38. Nelson, J.D.; MacLeod, R.A. Distribution of lipopolysaccharide and the detection of a new subfraction in the cell envelope of a marine Pseudomonad. *J. Bacteriol.* **1977**, *94*, 1764–1777. [CrossRef]
39. Chistyulin, D.K.; Novikova, O.D.; Portnyagina, O.Y.; Khomenko, V.A.; Vakorina, T.I.; Kim, N.Y.; Isaeva, M.P.; Likhatskaya, G.N.; Solov'eva, T.F. Isolation and characterization of OmpF-like porin from *Yersinia ruckeri*. *Biochem. (Mosc.) Suppl. Ser. A Membr. Cell Biol.* **2012**, *6*, 235–242. [CrossRef]
40. Benz, R. Structure and function of porins from Gram negative bacteria. *Annu. Rev. Microbiol.* **1988**, *42*, 359–394. [CrossRef]
41. Simon, M.; Mathes, A.; Blanch, A.; Engelhardt, H. Characterization of a porin from the outer membrane of *Vibrio anguillarum*. *J. Bacteriol.* **1996**, *178*, 4182–4188. [CrossRef] [PubMed]
42. Gamage, D.G.; Gunaratne, A.; Periyannan, G.R.; Russell, T.G. Applicability of Instability Index for in vitro Protein Stability Prediction. *Protein Pept. Lett.* **2019**, *26*, 339–347. [CrossRef]
43. Angus, B.L.; Hancock, R.E.W. Outer membrane porinproteins F, P, and D1 of *Pseudomonas aeruginosa* and PhoE of *Escherichia coli*: Chemical cross-linking to reveal native oligomers. *J. Bacteriol.* **1983**, *155*, 1042–1051. [CrossRef]
44. Vachon, V.; Kristianson, D.N.; Coulton, J.W. Outer membrane porin protein of *Haemophilus influenza* type b: Pore size and subunit structure. *Can. J. Microbiol.* **1987**, *34*, 134–140. [CrossRef] [PubMed]
45. Sciacca, M.F.M.; Tempra, C.; Scollo, F.; Milardi, D.; La Rosa, C. Amyloid growth and membrane damage: Current themes and emerging perspectives from theory and experiments on Aβ and hIAPP. *Biochim. Biophys. Acta* **2018**, *1860*, 1625–1638. [CrossRef]
46. Cowan, S.W.; Schirmer, T.; Rummel, G.; Steiert, M.; Ghosh, R.; Pauptit, R.A.; Jansonius, J.N.; Rosenbusch, J.P. Crystal structures explain functional properties of two *E. coli* porins. *Nature* **1992**, *358*, 727–733. [CrossRef] [PubMed]

47. La Rosa, C.; Condorelli, M.; Compagnini, G.; Lolicato, F.; Milardi, D.; Do, T.N.; Karttunen, M.; Pannuzzo, M.; Ramamoorthy, A.; Fraternali, F.; et al. Symmetry-breaking transitions in the early steps of protein self-assembly. *Eur. Biophys. J.* **2020**, *49*, 175–191. [CrossRef] [PubMed]
48. Phale, P.S.; Philippsen, A.; Kiefhaber, T.; Koebnik, R.; Phale, V.P.; Schirmer, T.; Rosenbusch, J.P. Stability of trimeric OmpF porin: The contributions of the latching loop L2. *Biochemistry* **1998**, *37*, 15663–15670. [CrossRef]
49. Meng, G.; Fronzes, R.; Chandran, V.; Remaut, H.; Waksman, G. Protein oligomerization in the bacterial outer membrane. *Mol. Membr. Biol.* **2009**, *26*, 136–145. [CrossRef]
50. Naveed, H.; Jackups, R.; Liang, J. Predicting weakly stable regions, oligomerization state, and protein–protein interfaces in transmembrane domains of outer membrane proteins. *Proc. Natl. Acad. Sci. USA* **2009**, *106*, 12735–12740. [CrossRef] [PubMed]
51. Naveed, H.; Liang, J. Weakly stable regions and protein-protein interactions in beta-barrelmembrane proteins. *Curr. Pharm. Des.* **2014**, *20*, 1268–1273. [CrossRef] [PubMed]
52. Lugtenberg, B.; Meijers, J.; Peters, R.; van der Hoek, P.; van Alphen, L. Electrophoretic resolution of the major outer membrane protein of *Escherichia coli* K-12 into four bands. *FEBS Lett* **1975**, *58*, 254–258. [CrossRef]
53. Voulhoux, R.; Bos, M.P.; Geurtsen, J.; Mols, M.; Tommassen, J. Role of a highly conserved bacterial protein in outer membrane protein assembly. *Science* **2003**, *299*, 262–265. [CrossRef]
54. Gaal, O.; Medgyesi, G.A.; Vereczkey, L. *Electrophoresis in the Separation of Biological Macromolecules*; Akademia Kiado: Budapest, Hungary, 1980; p. 178.
55. Winder, A.F.; Gent, W.L.G. Correction of light-scattering errors in spectrophotometric protein determinations. *Biopolymers* **1971**, *10*, 1243–1251. [CrossRef] [PubMed]
56. Devenyi, T.; Gergely, J. *Amino Acids, Peptides and Proteins: Biochemical and Immunological Techniques in Protein Chemistry*; Elsevier Scientific Pub Co.: Amsterdam, The Netherlands, 1974; p. 632. [CrossRef]
57. Wilson, K. Unit 2.4 Preparation of genomic DNA from bacteria. In *Current Protocols in Molecular Biology*; Wiley Online Library: Hoboken, NJ, USA, 2001. [CrossRef]
58. Aziz, R.K.; Bartels, D.; Best, A.A.; DeJongh, M.; Disz, T.; Edwards, R.A.; Formsma, K.; Gerdes, S.; Glass, E.M.; Kubal, M.; et al. The RAST server: Rapid annotations using subsystems technology. *BMC Genomics* **2008**, *9*, 75. [CrossRef] [PubMed]
59. Overbeek, R.; Begley, T.; Butler, R.M.; Choudhuri, J.V.; Chuang, H.Y.; Cohoon, M.; de Crécy-Lagard, V.; Diaz, N.; Disz, T.; Edwards, R.; et al. The ubsystems approach to genome annotation and its use in the project to annotate 1000 genomes. *Nucleic Acids Res.* **2005**, *33*, 5691–5702. [CrossRef] [PubMed]
60. Käll, L.; Krogh, A.; Sonnhammer, E.L. Advantages of combined transmembrane topology and signal peptide prediction—The Phobius web server. *Nucleic Acids Res.* **2007**, *35* (Suppl. 2), W429–W432. [CrossRef]
61. Yu, N.Y.; Wagner, J.R.; Laird, M.R.; Melli, G.; Rey, S.; Lo, R.; Dao, P.; Sahinalp, S.C.; Ester, M.; Foster, L.J.; et al. PSORTb 3.0: Improved protein subcellular localization prediction with refined localization subcategories and predictive capabilities for all prokaryotes. *Bioinformatics* **2010**, *26*, 1608–1615. [CrossRef]
62. El-Gebali, S.; Mistry, J.; Bateman, A.; Eddy, S.R.; Luciani, A.; Potter, S.C.; Qureshi, M.; Richardson, L.J.; Salazar, G.A.; Smart, A.; et al. The Pfam protein families database in 2019. *Nucleic Acids Res.* **2019**, *47*, D427–D432. [CrossRef]
63. Saier, M.H., Jr.; Reddy, V.S.; Tsu, B.V.; Ahmed, M.S.; Li, C.; Moreno-Hagelsieb, G. The Transporter Classification Database (TCDB): Recent advances. *Nucleic Acids Res.* **2016**, *44*, D372–D379. [CrossRef]

Sample Availability: Samples of the compounds are available by agreement with the authors and management of the organization for conducting joint research from the authors.

molecules MDPI

Article

Studies on the Structure and Properties of Membrane Phospholipase A_1 Inclusion Bodies Formed at Low Growth Temperatures Using GFP Fusion Strategy

Svetlana I. Bakholdina [1,*], Anna M. Stenkova [2], Evgenia P. Bystritskaya [1], Evgeniy V. Sidorin [1], Natalya Yu. Kim [1], Ekaterina S. Menchinskaya [1], Tatiana Yu. Gorpenchenko [3], Dmitry L. Aminin [1,4], Nikita A. Shved [2,5] and Tamara F. Solov'eva [1,*]

[1] G.B. Elyakov Pacific Institute of Bioorganic Chemistry, Far-Eastern Branch of the Russian Academy of Sciences, Prospekt 100-let Vladivostoku 159, 690022 Vladivostok, Russia; belyjane@gmail.com (E.P.B.); sev1972@mail.ru (E.V.S.); natalya_kim@mail.ru (N.Y.K.); ekaterinamenchinskaya@gmail.com (E.S.M.); daminin@piboc.dvo.ru (D.L.A.)

[2] Department of Medical Biology and Biotechnology, FEFU Campus, School of Biomedicine, Far Eastern Federal University, Russky Island Ajax Bay 10, 690922 Vladivostok, Russia; stenkova@gmail.com (A.M.S.); nikitawayfarer@yandex.ru (N.A.S.)

[3] Federal Scientific Center of the East Asia Terrestrial Biodiversity, Far Eastern Branch of the Russian Academy of Sciences, Prospect 100-let Vladivostoku, 159, 690022 Vladivostok, Russia; gorpenchenko@biosoil.ru

[4] Department of Biomedical Science and Environmental Biology, Kaohsiung Medical University, 100, Shih-Chuan 1st Road, Kaohsiung 80708, Taiwan

[5] A.V. Zhirmunsky National Scientific Center of Marine Biology, Far Eastern Branch of the Russian Academy of Sciences, Palchevskogo str. 17, 690041 Vladivostok, Russia

* Correspondence: sibakh@mail.ru (S.I.B.); soltaf@mail.ru (T.F.S.); Tel.: +7-423-231-11-58 (S.I.B. & T.F.S.); Fax: +7-423-231-40-50 (S.I.B. & T.F.S.)

Citation: Bakholdina, S.I.; Stenkova, A.M.; Bystritskaya, E.P.; Sidorin, E.V.; Kim, N.Y.; Menchinskaya, E.S.; Gorpenchenko, T.Y.; Aminin, D.L.; Shved, N.A.; Solov'eva, T.F. Studies on the Structure and Properties of Membrane Phospholipase A_1 Inclusion Bodies Formed at Low Growth Temperatures Using GFP Fusion Strategy. *Molecules* **2021**, *26*, 3936. https://doi.org/10.3390/molecules26133936

Academic Editor: Carmelo Corsaro

Received: 2 May 2021
Accepted: 22 June 2021
Published: 28 June 2021

Abstract: The effect of cultivation temperatures (37, 26, and 18 °C) on the conformational quality of *Yersinia pseudotuberculosis* phospholipase A_1 (PldA) in inclusion bodies (IBs) was studied using green fluorescent protein (GFP) as a folding reporter. GFP was fused to the C-terminus of PldA to form the PldA-GFP chimeric protein. It was found that the maximum level of fluorescence and expression of the chimeric protein is observed in cells grown at 18 °C, while at 37 °C no formation of fluorescently active forms of PldA-GFP occurs. The size, stability in denaturant solutions, and enzymatic and biological activity of PldA-GFP IBs expressed at 18 °C, as well as the secondary structure and arrangement of protein molecules inside the IBs, were studied. Solubilization of the chimeric protein from IBs in urea and SDS is accompanied by its denaturation. The obtained data show the structural heterogeneity of PldA-GFP IBs. It can be assumed that compactly packed, properly folded, proteolytic resistant, and structurally less organized, susceptible to proteolysis polypeptides can coexist in PldA-GFP IBs. The use of GFP as a fusion partner improves the conformational quality of PldA, but negatively affects its enzymatic activity. The PldA-GFP IBs are not toxic to eukaryotic cells and have the property to penetrate neuroblastoma cells. Data presented in the work show that the GFP-marker can be useful not only as target protein folding indicator, but also as a tool for studying the molecular organization of IBs, their morphology, and localization in *E. coli*, as well as for visualization of IBs interactions with eukaryotic cells.

Keywords: recombinant phospholipase A_1; *Yersinia pseudotuberculosis*; inclusion bodies; fusion protein; green fluorescent protein

1. Introduction

The production of recombinant proteins in *Escherichia coli* is often accompanied by the formation of their insoluble aggregates, the so-called inclusion bodies (IBs), in the cell cytoplasm [1]. For a long time, it was believed that IBs consist of misfolded proteins, and the isolation of a recombinant protein from them in a functionally active form is a laborious and expensive procedure with a low yield [2]. However, it was later found that a protein in a native-like conformation can be present in IBs, and its amount can be controlled by manipulating the genetic and physiological parameters of expression [3]. A growing body of data confirms the fact that the protein within IBs adopts different conformational states ranging from unstructured, partly folded, native or native-like, to cross-β-sheet (amyloid-like) [4]. IBs containing a high percentage of properly folded protein can be functionally active and be used as nanomaterials in medicine and biotechnology [5]. The target protein from these IBs can be isolated under mild conditions without the use of renaturing steps and with high yield.

An important factor influencing the conformational quality of the protein in IBs is the expression temperature [6]. IBs synthesized at growth temperatures below the optimal 37 °C usually have a higher content of properly folded protein. The slower rate of protein production at low temperatures gives newly synthesized recombinant proteins more time to fold properly. Using low growth temperature, for a series of water-soluble globular proteins, including oligomeric ones, it was possible to obtain biologically active IBs with a high protein content in the native conformation, known as "nonclassical" IBs [7–9]. However, recently there was information that the globular protein L-asparaginase II forms "nonclassical" IBs, on the contrary, at elevated culture temperatures [10]. Moreover, it was reported that IBs formed at the optimal bacterial growth temperature contained a functionally active protein, and that the temperature regime did not affect the protein structure in IBs [11,12]. These facts demonstrate that the effect of the expression temperature on the conformational quality of IBs is ultimately determined by the nature of the protein, i.e., the intrinsic physicochemical properties of its amino acid sequence. At this time, the correlation between these factors affecting the structural organization of IBs is poorly understood. Most of the published research findings on this issue were performed on IBs of soluble globular proteins. IBs formed by membrane proteins are largely unexplored. In our work, the recombinant phospholipase A_1 of *Yersinia pseudotuberculosis* as a model integral membrane protein was expressed in *E. coli* at different growth temperatures and the resulting IBs were characterized. In addition, determination of the growth temperature, which promotes the formation of IBs with a high content of correctly folded protein, can also be useful for developing novel, improved methods for biotechnology production of active enzyme. Lipolytic enzymes, including phospholipase A_1, find wide application in food, cosmetic, and pharmaceutical industries [13].

The phospholipase A_1 from the outer membrane protein of *Y. pseudotuberculosis* (PldA—detergent-resistant phospholipase A_1, EC 3.1.1.32) is a key enzyme that renews intracellular lipids (phospholipids) and thereby affects the permeability of the outer membrane, the functioning of various membrane-bound proteins, and is also one of the pathogenic factors of Gram-negative bacteria [14]. It has a molecular weight of 31 kDa. The PldA molecule is a β-barrel formed by a β-sheet, consisting of 12 antiparallel amphiphilic β-strands. In an active form, PldA is a homodimer [14].

IBs are of interest not only as a potential source of a functionally active recombinant protein, but also as a model system for studying protein aggregation associated with so-called conformational diseases, including numerous neurodegenerative disorders and amyloidosis [15]. Information about the mechanisms of formation and disaggregation of IBs, as well as their molecular organization, is important for understanding the molecular basis of conformational diseases, and in the design of drugs for inhibiting or reversing protein aggregation. However, such data on IBs are limited. This fact is largely determined by the methodological difficulties of studying these conformationally heterogeneous aggregates.

Green fluorescent protein (GFP) and GFP-like proteins were shown to be unique genetically encoded fluorescent markers for basic and applied research in the field of molecular and cellular biology. Currently, GFP are widely used to monitor the expression and folding quality of a recombinant protein under different bacterial culture conditions [6]. For this purpose, GFP is combined with the target protein so that the formation of a chromophore is possible only with the correct folding of the target protein [16,17]. This approach significantly simplifies the process of searching for the expression conditions of a recombinant protein in the native conformation.

The aim of our work was to study the structure and properties of phospholipase A_1 IBs formed at the optimal expression temperature in terms of their conformational quality, using a GFP tag as an indicator of recombinant protein folding. The research was focused on the following specific objectives: studying the effect of expression temperature on the conformational quality of PldA-GFP IBs; evaluating the secondary structure and arrangement of protein molecules inside the IBs; and collecting information about the stability and behavior of IBs in denaturing agents, as well as their enzymatic activity and interaction with eukaryotic cells.

2. Results and Discussion

2.1. Effect of the Cultivation Temperature Downshift on the Expression of the PldA-GFP Fusion Protein

To synthesize the PldA-GFP fusion, we have constructed the recombinant plasmid containing sequences of *pldA* encoding *Y. pseudotuberculosis* membrane bound phospholipase A_1 and *gfp* encoding CopGFP based on pRSETa vector (Figure S1). The expression of the PldA-GFP at different cultivation temperatures (37, 26, and 18 °C) was induced by adding 0.1–1 mM IPTG followed by incubation for 3, 5 and 16 h. The amount of recombinant protein (based on intensity of the protein band on the SDS-PAGE) and the level of GFP fluorescence were determined for each experiment.

E. coli strain BL21(DE3) pLysS/PldA-GFP expressed the major protein in the region of 57 kDa corresponding to the calculated molecular weight of the expected PldA-GFP fusion protein. It was found that lowering the cultivation temperature from 37 to 26 °C significantly increased the recombinant protein production (Figure 1B). A further decrease in the growth temperature to 18 °C did not have a noticeable effect on the level of protein synthesis. The maximum GFP fluorescence was observed in *E. coli* grown at 18 °C, it decreased by almost four times during bacterial cultivation at 26 °C, and was close to the negative control—plasmidless cell level at 37 °C (Figure 1A). Such effects of cultivation temperature on the level of PldA-GFP expression and GFP fluorescence were observed at all studied time points (data not shown), while the IPTG concentration did not significantly change the level of recombinant protein synthesis and GFP fluorescence of the fusion protein (Figure 1).

Thus, the low cultivation temperature promoted an increase in GFP fluorescence in the chimeric protein and, consequently, an increase in the proportion of the correctly folded form of phospholipase A_1.

Fluorescence confocal microscopy was used to study the morphology of the recombinant protein PldA-GFP and its localization in the cell. Figure 2 shows that homogeneous fluorescence throughout the cell cytoplasm was observed for bacteria expressing GFP only (Figure 3A), which is characteristic of the GFP soluble form. At the same time, under the similar culture conditions, the cells containing the pRSETa/PldA-GFP plasmid accumulated the chimeric protein in the form of aggregates (inclusion bodies), as evidenced by inhomogeneous fluorescence (Figure 2B). A similar morphology of inclusion bodies formed by chimeric proteins with GFP and other fluorophores, or mutant GFPs prone to IB formation, has been described in several works [3,18,19].

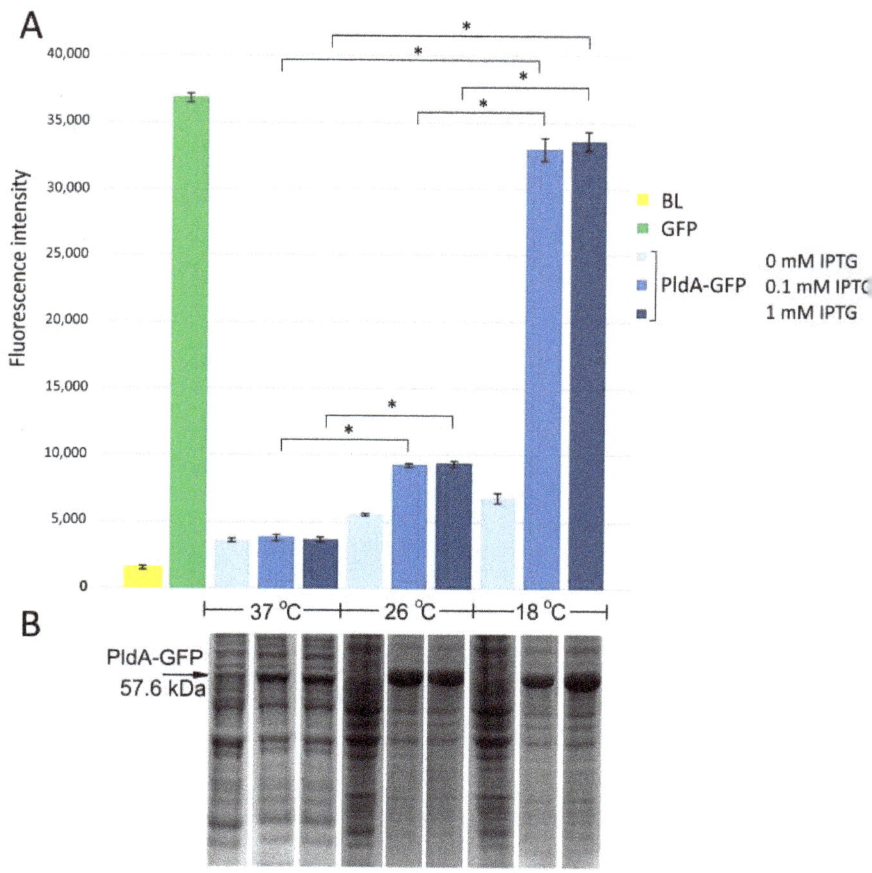

Figure 1. The level of GFP fluorescence in *E. coli* cells (**A**) and the expression of PldA-GFP fusion protein (**B**) in *E. coli* cell lysates after 16 h of incubation at different cultivation temperatures and IPTG concentrations. BL—plasmidless cells (negative control); GFP—cells expressing GFP protein (positive control). Fluorescence and bacterial mass for cell lysates were normalized according to OD_{600}. The experiments were performed in three biological replicates. Results show the mean ± standard deviation (SD). The asterisk (*) indicates a significant difference (p-value < 0.05) in the fluorescence level between indicated groups.

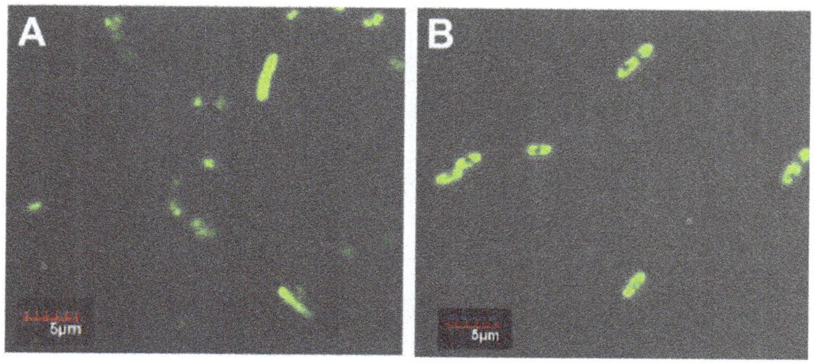

Figure 2. Fluorescence microscopy images of *E. coli* cells expressing GFP (**A**) and PldA-GFP (**B**).

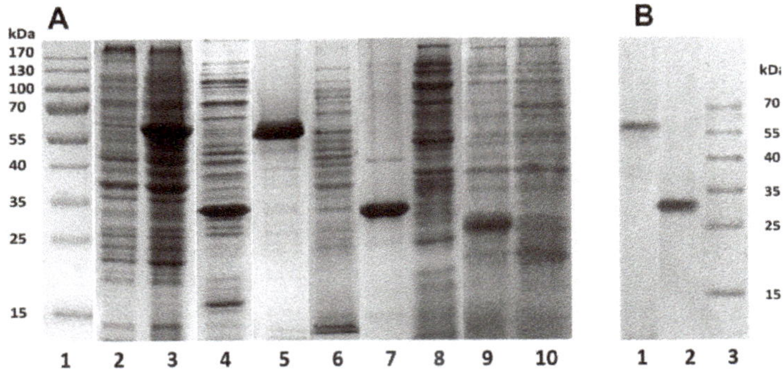

Figure 3. SDS-PAG electrophoresis (**A**) and Western blotting (**B**). (**A**): Whole cell lysates of *E. coli* BL plasmidless (2) and *E. coli* expressing PldA-GFP (3) and PldA (4); purified PldA-GFP IBs: pellet (5) and supernatant (6); purified PldA IBs: pellet (7) and supernatant (8); GFP supernatant (9) and pellet (10); molecular weight standard (1). (**B**): PldA-GFP IBs (1); PldA IBs (positive control) (2); molecular weight standard (3). Western blotting was performed using mouse polyclonal antiserum to recombinant PldA.

2.2. Isolation and Purification of PldA-GFP IBs, PldA IBs, and GFP

The proteins PldA-GFP, PldA, and GFP were expressed in *E. coli* at 18 °C, IPTG concentration of 0.1 mM, and cultivation time of 16 h (Figure 1). The bacterial cells were destroyed by lysozyme in combination with ultrasound, and the resulting suspensions were centrifuged. Using SDS-PAGE analysis, the recombinant proteins PldA-GFP and PldA were found only in insoluble fractions (pellets), while GFP was detected only in the supernatant (Figure 3A). These results confirm that in *E. coli* cells under mentioned expression conditions the chimeric protein and phospholipase A_1 form IBs, while GFP is in a soluble form.

PldA-GFP IBs and PldA IBs were purified from impurities by sequential treatment with detergent solutions: 0.1% Sarcosyl and 1% sodium deoxycholate to remove LPS [20]. Denaturing SDS-PAGE showed that PldA-GFP and PldA migrate in the gel as bands corresponding to proteins with molecular weights of 57 and 31 kDa, respectively (Figure 3A). Both IBs samples barely contain any concomitant proteins. It should be noted that PldA-GFP loses GFP fluorescence under denaturing (2% SDS, boiled) and semi-native (0.1% SDS without boiled) SDS-PAGE. Immunoblotting using antibodies to PldA found the only protein band with MW of 57 kDa, which belongs to PldA-GFP (Figure 3B).

Recombinant GFP (rGFP) was isolated from the supernatant obtained by centrifugation of the lysed *E. coli* cells under the conditions described above using gel chromatography on a Superdex 200HR column. The fluorescent fractions were pooled and considered as containing GFP in its native conformation, as the formation of a fluorophore occurs spontaneously after this protein folding [21].

According to gel filtration data, GFP in PBS (pH 7.4) has an apparent molecular weight of approximately 42 kDa and, therefore, exists under these conditions as a dimer. Under denaturing (2% SDS and boiling) and semi-native SDS-PAGE (0.1% SDS without boiling), GFP migrates as a monomer with an apparent molecular weight of about 26 kDa, and loses fluorescence (Figure 3). Therefore, purified GFP is a dimer that has a fluorescence spectrum with maxima ex/em at 482/502 nm, and may be converted to a non-fluorescent monomer in the presence of a denaturant (SDS).

In phosphate-buffered saline PldA-GFP IBs have the same fluorescence spectrum as rGFP, and therefore contain the recombinant protein in a native-like conformation. The fluorescence intensity of IBs, calculated per mg of GFP, was four times lower than that of GFP (Figure 4). Therefore, in IBs only 24, 5%, of the chimeric protein is in a fluorescently active form. However, this value is not entirely correct. It can be either overestimated, as according to CD spectroscopy data, recombinant GFP contains a certain amount of an inactive protein; and underestimated, as is known, the aggregation of GFP leads to the loss of a large part of its fluorescent activity [22].

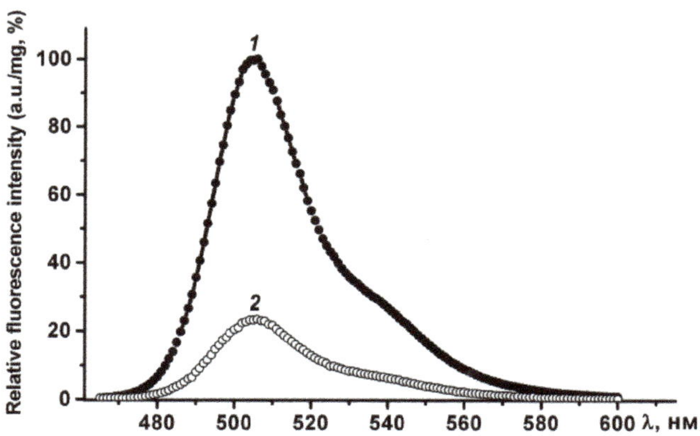

Figure 4. GFP fluorescence emission spectra of recombinant proteins in PBS: (1) GFP, (2) PldA-GFP IBs. The fluorescence intensity of GFP calculated per 1 mg of protein was taken as 100%.

2.3. *Stability of PldA-GFP IBs*

The IBs of the chimeric protein are rather stable in aqueous suspensions. According to the dynamic light scattering (DLS) data, IBs have a monomodal size distribution with an average hydrodynamic radius (R_H) of 412 nm in PBS at a concentration of 2 to 20 μg/mL (Figure 5A). However, repeated passing of the IBs suspension (20 μg/mL) through a 0.1 mm needle, followed by sonication to homogenize aggregates in PBS, leads to the appearance of a population of small particles (40–100 nm), which account for about 18% of the total volume, and a decrease in the average IBs radius (by volume) to 367 nm (Figure 5A). As was reported earlier, sonication can disturb the structure of IBs and leads to the loss of some protein from their surface and an increase in porosity [23].

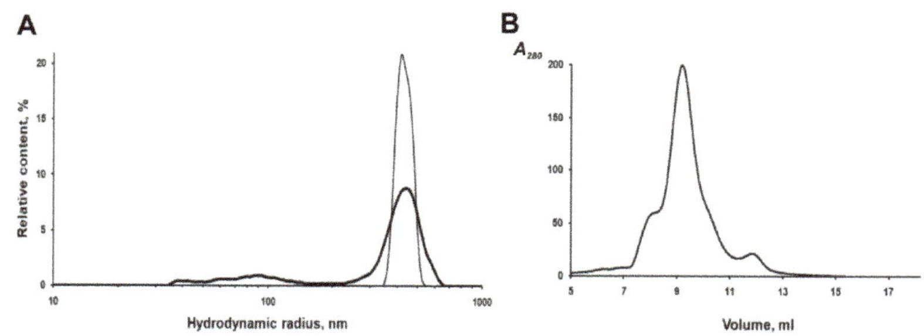

Figure 5. Characteristic of PldA-GFP IBs. (**A**) Particle size volume distribution of IBs in PBS (20 μg/mL): IBs intact (thin line), IBs after passing through a 0.1 mm needle and sonication (thick line); (**B**) Gel filtration chromatography of PldA-GFP solubilized in 0.1% SDS, 50 mM Tris-HCl buffer, pH 8.0, on a Superdex 200 HR column.

Stability in solutions of denaturants is an important characteristic of IBs, as it is directly related to their structural organization and conformation of the protein which forms them. As is known, IBs with the high protein content in native or native-like conformation are looser and more unstable than IBs, where the protein is mainly in the form of partially or improperly folded intermediates. They are usually dissolved in mild detergents and solutions with a low molar concentration of chaotropic agents [24]. The investigation of the stability of PldA-GFP IBs in SDS (0.01–0.1%) and urea (1–8 M) was carried out using the method of turbidimetry and GFP fluorescence.

As we can see from Figure 6A, the optical density at 350 nm of IBs solutions drops sharply with an increase in the concentration of the detergent from 0.02 to 0.03%, indicating a reduction in the number of large particles, and reaches zero at 0.05% SDS.

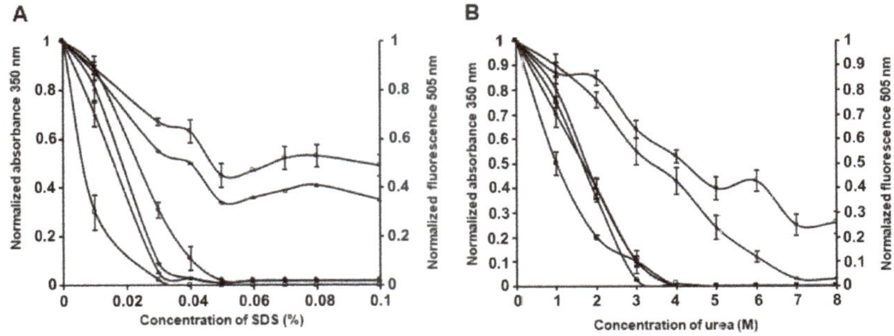

Figure 6. Solubility of PldA-GFP IBs in SDS (**A**) and urea (**B**). Dependence of turbidity signal (solid symbols) and GFP fluorescence (empty symbols) of IBs incubated in denaturant solutions for 5 min (circle), 1 h (triangle), 2 h (rhombus), and 24 h (square) on the denaturant concentration. The experiments were performed in three biological replicates. Results show the mean ± standard deviation (SD).

Simultaneously with the solubilization of IBs, their GFP fluorescence decreases. In 0.05% SDS, the fluorescence intensity of IBs is 47% of the initial value, and remains at this level for 60 min with an increase in the concentration of the detergent to 0.1%. Increasing the incubation time to 2 h leads to the GFP fluorescence of solubilized IBs falling to a low value in 0.03% and being completely absent in 0.05% SDS.

The results obtained show that the correctly folded chimeric protein, which is present in IBs, interacts with denaturant differently. One portion of the PldA-GFP fusion protein is solubilized from IBs with a simultaneous its denaturation and loss of activity, while the other portion goes into solution, retaining the native-like conformation and the fluorescent properties. This fact may be a consequence of the fluorescently active protein different localization within IBs [25] and, therefore, the availability for the detergent, as well as differences in its macromolecular and/or supramolecular structure. The latter assumption may be supported by gel chromatography data below about the structure of the fluorescent chimeric protein.

To characterize the fluorescently active protein, IBs were solubilized in 0.1% SDS and fractionated on a Superdex 200HR column. As shown in Figure 5B, fluorescent PldA-GFP was eluted from the column as three peaks, one major and two minor, corresponding to proteins with apparent molecular weights of ~600, 423 (major), and 120 kDa. As shown above, active GFP obtained by gel chromatography on Superdex 200HR is a dimer. Based on the data of gel chromatography, we can suppose that the fluorescent chimeric protein is a dimer (MW 114 kDa), which in 0.1% SDS exists mainly as an oligomer consisting of four dimers. It is possible that correctly folded PldA-GFP molecules are capable of forming more densely packed oligomeric structures in IBs and, therefore, more resistant to denaturants than the rest of the aggregate. As is known, the quaternary structure is an important factor of the conformational stability of fluorescent proteins [26].

The stability of PldA-GFP IBs in urea was also determined (Figure 6B). The turbidity of the IBs suspension decreases with an increase in the concentration of urea and the incubation time and drops to zero at a chaotropic agent concentration of 4 M. At the same time, in 4 M urea, the protein retains its fluorescence at the level of 53, 43, and 1.2% of the original value after incubation for 5 min, 2 h, and 24 h, respectively. These data are consistent with the findings of the study of IBs solubility in SDS. At the same time, urea is apparently a milder denaturing agent regarding the recombinant protein than SDS, as in 4 M urea, compared to 0.05% SDS (the minimum concentrations of denaturants providing, according to the data turbidimetry, complete dissolution of IBs), the recombinant protein retains its activity for a longer time. The retention of fluorescent activity by the recombinant protein solubilized from IBs with a short exposure in 0.05% SDS and 4 M urea suggests that the intermolecular interactions stabilizing IBs are less resistant to the denaturant than intramolecular contacts that provide the native conformation and activity of GFP. These results suggest that it is possible to select such solubilization conditions when the network of intermolecular contacts in IBs will be specifically disrupted, without denaturation of the correctly folded protein incorporated into the aggregates. As was demonstrated, maintaining a native-like secondary structure of the protein during its solubilization from IBs increases the yield of the recombinant protein during its refolding [27].

2.4. The Proteinase K Digestion of the IBs

PldA-GFP IBs were treated with proteinase K. Kinetics of proteolytic digestion of IBs was monitored by the measurement of turbidity at 350 nm and GFP fluorescence emission of the suspension over time. As it is shown in Figure 7, the turbidity of the IBs suspension decreases to its lowest level after 30 min of incubation, while the GFP fluorescence of the fusion protein remains practically unchanged.

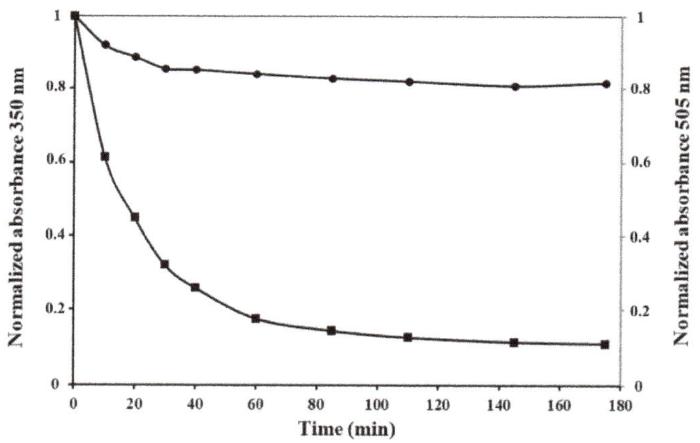

Figure 7. Proteinase K digestion of PldA-GFP IBs. Profiles of IBs fluorescence (circle) and turbidity (square).

The results presented here strongly suggest that more than one chimeric protein species with different protease resistance coexist within IBs. Non-fluorescent species of the recombinant protein, which are highly sensitive to proteolysis, appear to be misfolded or partially folded polypeptides. Proteinase K showed a strong preference for hydrolyzing unstructured protein regions [28].

The resistance of the active protein to proteinase K digestion suggests that it has a compact supramolecular structure, as the quaternary structure of GFP has been shown to be highly resistant to proteases [26]. However, it is also possible that proteinase K partially cleaves active protein, including GFP, without affecting the chromophore. As is known, the intensity of fluorescence of the chromophore can be changed insignificantly with major disturbances of the native structure of GFP [29]. The obtained data show the structural heterogeneity of PldA-GFP IBs. It can be assumed that compactly packed, properly folded, proteolytic resistant, and structurally less organized polypeptides susceptible to proteolysis can coexist in PldA-GFP IBs.

2.5. CD Spectroscopy

The secondary structure of the recombinant proteins was determined by CD spectroscopy in the far ultraviolet region. CD spectra of PldA-GFP, PldA and GFP in 0.03% SDS (exposure time from 10 min to 2 h and 24 h) and in 0.1% SDS (incubation time 2 and 24 h), as well as the spectrum of GFP in PBS were obtained (Figure 8A). According to turbidimetry data (Figure 6A), in these solutions of the detergent, almost complete solubilization of the recombinant protein from IBs is observed.

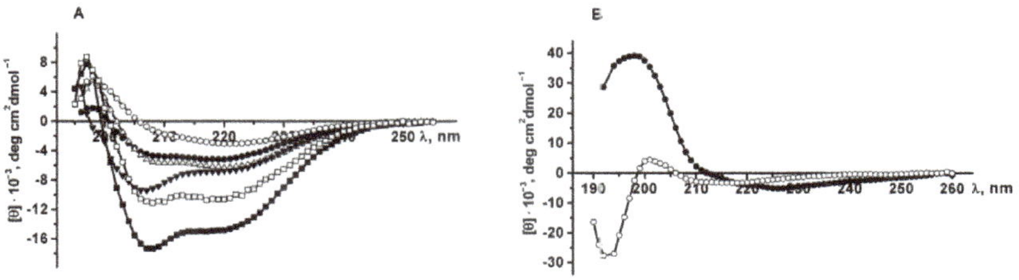

Figure 8. Far UV CD spectra of PldA-GFP, PldA and GFP. (**A**) PldA-GFP (•), GFP (Δ), and PldA (■): 0.03% SDS, 10 min; PldA-GFP (▼) and GFP (□): 0.1% SDS, 2 h; GFP: PBS (○); (**B**) PldA-GFP: 4 M urea, 2 h (•) and 24 h (○).

Recombinant GFP in PBS has a CD spectrum with a maximum at 198 nm and only one minimum at 220 nm, characteristic of proteins with β-pleated sheet structure (Figure 8A). As shown earlier [30], the *Y. pseudotuberculosis* phospholipase A_1, which, similar to GFP, has the cylindrical β-sheet structure, exhibits a CD spectrum with a positive band at 193 nm and a broad negative band centered at 215 nm. The CD spectra of the recombinant proteins PldA-GFP, PldA and GFP in SDS solutions are characterized by a maximum at 197 nm and two minima at 208–207 nm and 219–217 nm, and are typical for mixed α-β proteins. Increasing the SDS concentration and incubation time caused an increase in the spectra amplitude and a decrease in the peak ratio at 218 and 207 nm which indicates an increase in the content of the α-helix in the protein (Figure 8A).

Using the CDPro software [31] the content of secondary structure elements of the recombinant proteins was determined (Table 1).

Table 1. Content of secondary structure elements (%) in PldA-GFP, PldA and GFP, dissolved in SDS and PBS.

Sample	α-Helix, %	β-Sheet, %	β-Turns, %	Random coil, %
GFP, PBS	6.1	40.8	21.3	31.8
GFP, 0.03% SDS, 15 min	20.1	29.6	21.9	28.4
GFP, 0.1% SDS, 2 h	32.0	18.2	22.9	26.9
GFP, 0.1% SDS, 24 h	41.5	9.1	20.9	28.5
PldA-GFP, 0.1% SDS, 2 h	22.3	26.1	21.5	30.1
PldA-GFP, 0.1% SDS, 24 h	21.9	23.0	23.5	31.6
PldA-GFP, 0.03% SDS, 2 h	14.9	30.6	22.1	32.4
PldA, 0.03% SDS, 2 h	39.7	10.1	22.6	27.6
GFP, 0.01 M sodium phosphate, pH 7.5 [32]	20 ± 1	52 ± 2	16 ± 1	13 ± 1
PldA (OMPLA *E. coli*) *	7.7	61	11.9	19.2
PldA of *Y. pseudotuberculosis* (predicted secondary structure) [33]	6.5	56.7	29.0	7.8

* UniProtKB—POA921 (PA1-*E.coli*).

As can be seen from Table 1, in 0.03% SDS PldA-GFP, PldA, and GFP have a pronounced secondary structure including α-helices and β-pleated sheets. However, the content of α-helices and random coil structure in these recombinant proteins is higher than in the corresponding native proteins. These data suggest that in the studied proteins, molecules with a native-like structure, which are fluorescently active, coexist with partially and misfolded polypeptides or segments. An increase in SDS concentration and incubation time in detergent solutions leads to an increase in the content of α-helices and a decrease in the content of β-structure. At the same time, the content of regular (α-helix and β-sheet) and random coil structure remains practically unchanged, and is about 50% and 28–32%, respectively. The data obtained suggest that SDS induces the β-sheet to α-helix structural conversion in the recombinant proteins. A similar type of denaturation by SDS has been found for some globular proteins that have a β-sheet structure [34]. In addition, it has been shown earlier that SDS-denatured PldA *E. coli* has a non-native secondary structure with a high α-helix content [35].

As follows from the data presented in Table 1, PldA-GFP and PldA, dissolved from IBs with SDS at the same conditions, differ significantly in the content of β-structure, which is 3–4 times lower in PldA than in the fusion protein. Based on these data, it can be assumed that PldA-moiety in the chimeric protein, as compared to PldA, has a secondary structure that is closer to the native one, and is more resistant to denaturation by SDS. Thus, the use of GFP as a fusion partner appears to improve the folding of PldA when expressed in *E. coli*.

Comparative analysis of the spectral data of GFP and PldA-GFP solubilized in 0.1% SDS during incubation for 2 and 24 h showed that the structural changes of GFP induced by SDS, compared with those of PldA-GFP, occur more slowly and lead to a more significant decrease in the content of β-structure in the protein (Figure 8, Table 1). As a result, we can assume that GFP is more sensitive to SDS-denaturation than GFP-moiety in chimeric protein.

The secondary structure of PldA-GFP solubilized from IBs with 4 M urea was also determined. The CD spectrum of the recombinant protein in urea (incubation time 2 h) has a maximum at 197 nm, minima at 225 and 218 nm, and is specific for proteins with a β-structure (Figure 8B). The long-wavelength position of the negative band (225 nm) in the CD spectrum is probably due to the aggregated state of the solubilized protein. An increase in exposure time of the IBs in urea solution to 24 h leads to the widening of the PldA-GFP spectrum with a decrease in the ellipticity of the positive and negative (−5294 to −3220 deg cm^2 $dmol^{-1}$) bands, a shift in the positions of the minima to 217 and 211 nm, and the appearance of an intense negative band at 193 nm, which indicates an increase in the content of random coil structures in the protein. A quantitative analysis of CD spectra of PldA-GFP indicates a decrease in the content of both α-helix (8.8 to 0.6%) and β-sheet (34.7 to 31.7%) with an increase in random coil structure fraction (33.8 to 45.5%), when the incubation time of protein in urea increases. We can assume that urea not only dissolves the recombinant protein from IBs, but also denatures it. It should be noted that denaturation of *E. coli* PldA in the presence of urea is accompanied by the formation of a random coil structure [35]. Thus, urea induces an increase in the content of random coil conformation in PldA-GFP, while maintaining a high percentage of β-sheet structure. The protein species with a native-like β-sheet structure appear to be more resistance to urea, than misfolded or partially folded polypeptides.

2.6. Interaction of PldA-GFP IBs with Amyloid-Specific Dye Thioflavin T

Recent studies have shown that recombinant protein in IBs can have, along with other conformations, the structure of an intermolecular β-sheet (cross-β conformation), which is characteristic of amyloid proto-fibrils and fibrils [36]. As is known, the bacterial outer membrane proteins with the β-pleated sheet secondary structure have a high propensity to form amyloid-like structures [37]. We have previously shown the presence of amyloids in PldA IBs formed in *E. coli* at 37 °C [33].

A specific fluorescent dye, Thioflavin T (ThT), was used for the detection of amyloids in PldA-GFP IBs [38]. It should be noted that ThT allows detecting the amyloids in IBs in the presence of a protein with the native β-sheet structure, i.e., to differentiate the β-sheet conformation of the native protein from the structure of an intermolecular β-sheet in an aggregated form (amyloid fibrils) of the same protein [39]. This is important in the present study as the native-like structure of PldA is a β-sheet.

The ThT fluorescence intensity at 484 nm increases by more than 100 times in the presence of PldA-GFP IBs suspended in PBS, which indicates the existence of amyloid-like structures in IBs. An increase in the incubation time of IBs in PBS leads to a slight increase in the dye fluorescence intensity (Figure 9).

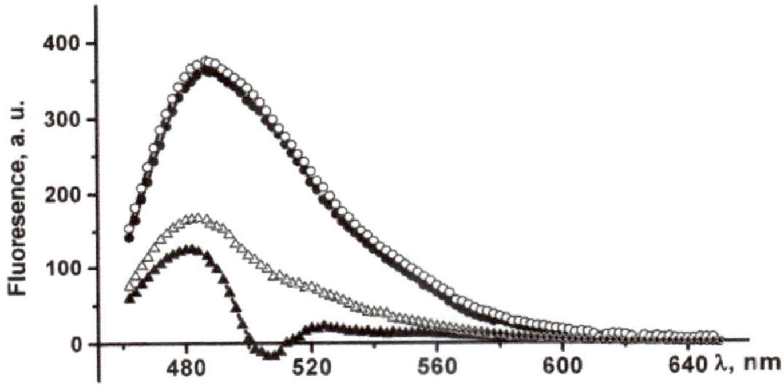

Figure 9. Difference fluorescence spectra of Thioflavin T bound to PldA-GFP IBs. IBs were incubated in PBS (triangle) and 0.04% SDS (circle) for 2 h (solid symbols) and 24 h (empty symbols).

IBs treated with 0.04% SDS also bind ThT, which is accompanied by a 300-fold increase in the dye fluorescence intensity (Figure 9). The incubation time of IBs in a detergent solution (10 min and 24 h) does not affect the fluorescence intensity of a bound ThT. A three-fold enhancement in the fluorescence of ThT in the presence of IBs incubated in SDS relative to untreated IBs may be due to several reasons: an increase in both the content of amyloids in IBs and their availability for dye binding, or changes in the structural and physical characteristics of the amyloids. The supposed changes in the amyloids in detergent-treated IBs seem to be quite realistic, as IBs solubilization occurs in the presence of SDS, and it has also been shown that SDS can induce the formation and modification of amyloid-like structures [40,41].

2.7. *Enzymatic Activity of Phospholipase A_1 Associated with and Released from IBs*

As is known, IBs with a high protein content in the native or native-like conformation are easily soluble in the presence of relatively low molar concentrations of chaotropic agents [24,42]. The chimeric protein was solubilized from PldA-GFP IBs in 4 M urea. As shown above, the efficiency of IBs solubilization under these conditions was rather high, and the isolated recombinant protein contained a high percentage of β-structure. Parallel to this, IBs were dissolved under extremely harsh conditions in a traditional manner using high concentration of denaturant—8 M urea. The specific enzymatic activity of the PldA-GFP IBs suspended in a Tris-HCl buffer solution and the isolated recombinant protein solubilized in urea was determined after their incubation in the presence of Triton X-100 micelles, which is necessary for the formation of a functionally active spatial structure of PldA [35].

No studied samples of isolated PldA-GFP fusion protein and IBs exhibited enzymatic activity. The lack of phospholipase A_1 activity is probably due to oligomerization of GFP, as it is known that the monomeric state of GFP is a necessary condition for maximum preservation of the native function of the target protein [43]. In addition, the properties of the linker sequence (flexibility, length, hydrophobicity, etc.) could affect the conformation and activity of the fusion protein [22].

To confirm the GFP inhibitory effect on the fusion protein enzymatic activity, PldA was expressed in *E. coli* at 18 °C and PldA IBs were isolated and purified. The PldA IBs suspended in a Tris-HCl buffer solution showed an enzymatic activity of 5 µmol/min per 1 mg. The detected IBs enzymatic activity despite its low level demonstrates the fundamental possibility of obtaining functionally active PldA IBs. The PldA, which was solubilized from IBs with 4 M urea, had a 5-fold higher activity (25–30 µmol/min per 1 mg). Attention is drawn to the fact that the recombinant protein in the functionally active form was obtained by passing the stage of complete unfolding in 8 M urea. It should be noted that PldA IBs expressed at 37 °C were inactive [33].

2.8. Interaction of PldA-GFP IBs with Neuroblastoma Cells

The ability of PldA-GFP IBs to penetrate eukaryotic cells was studied. It was found that IBs of the chimeric protein is not toxic to neuronal cells in the range of concentrations from 0.4 to 100 µg/mL (Figure 10).

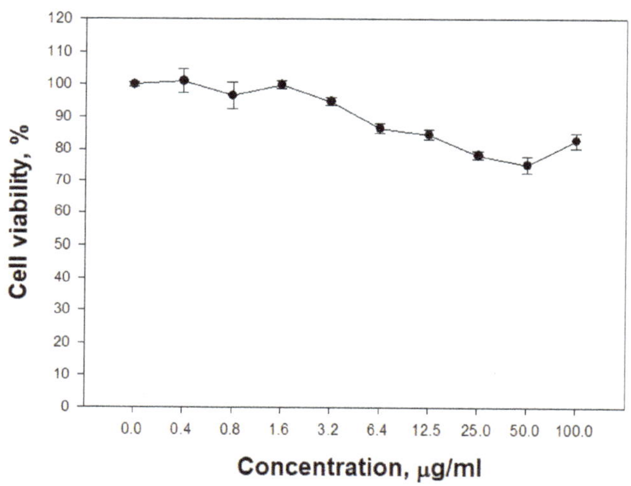

Figure 10. Cytotoxic activity of PldA-GFP IBs on mouse neuroblastoma Neuro-2a cells. The data are presented as the mean ± SEM values (n = 3).

However, it was found that IBs have the property to bind the neuroblastoma Neuro-2a cells and subsequently penetrate them (Figure 11A–C). According the analysis of electron microscopic images of Neuro-2a cells and PldA-GFP IBs, the size of the IBs of the chimeric protein is in the range of 800–1000 nm (Figure 11D,E).

According to fluorescence image analysis, the average amount of PldA-GFP IBs penetrating the neuroblastoma Neuro-2a cell in control (0 h) is not detectable (data not shown), in 1 h is 0.05 ± 0.03, in 4 h is 1.4 ± 0.5, while in 24 h this value increases almost 5 times to 5.5 ± 1.4 IBs particles per cell in comparison with 1 h (Figure 11F).

In summary, phospholipase A_1 fused with GFP was expressed at different growth temperatures (37, 26, and 18 °C). The conformational quality of PldA-GFP IBs, as reflected by its specific fluorescence emission, enhanced by producing them at low expression temperatures. In this regard, the membrane beta-barrel protein phospholipase A_1 is similar to most soluble globular proteins. The maximum content of the fusion protein in the native-like conformation was found in IBs produced at 18 °C. We studied the size, stability in denaturant solutions, and enzymatic and biological activity of these IBs, and the secondary structure of the recombinant protein forming them.

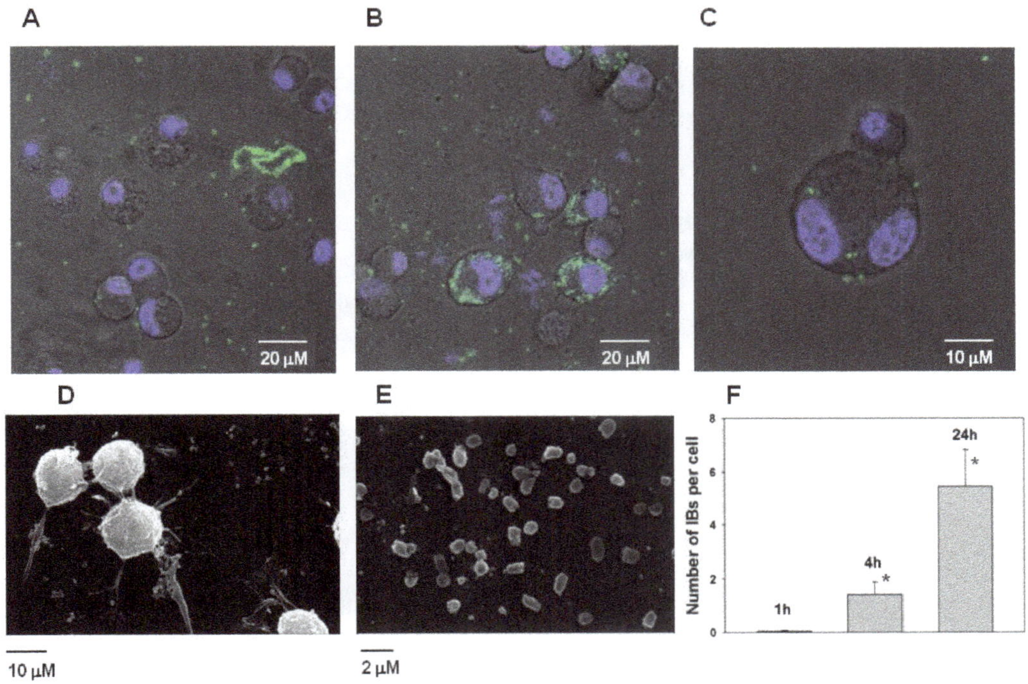

Figure 11. Fluorescence image of mouse neuroblastoma Neuro-2a cells after incubation with PldA-GFP IBs (20 μg/mL) for 4 (**A**) and 24 h (**B**). Images are obtained by superimposing the passing and fluorescent channels. Fluorescence of IBs is shown in green (GFP), cell nuclei are shown in blue (Hoechst 33342). An enlarged image (×5) of a neuroblastoma cell with included PldA-GFP IBs inside (**C**); Scanning electron microscopy of Neuro-2a mouse neuroblastoma cells with adsorbed PldA-GFP IBs (**D**); and magnified image (×5) of PldA-GFP IBs alone (**E**). The average amount of PldA-GFP IBs in a single neuroblastoma Neuro-2a cell after incubation for 1, 4, and 24 h (**F**). The fluorescent images of 150 randomly selected cells for each time point were inspected, and the amount of fluorescent IBs penetrating the cell surface was calculated and expressed as number of PldA-GFP IBs per cell. The data are presented as the mean ± SEM values (n = 150); * $p < 0.05$ compared with 1 h.

PldA-GFP IBs have fluorescent activity which indicates the presence of the recombinant protein in a native-like conformation. According to the CD spectroscopy data, isolated PldA-GFP has a high β-sheet content. Amyloid-like structures were also found in the studied IBs. Solubilization of the chimeric protein from IBs with urea and SDS is accompanied by a modification of its secondary structure. The nature and degree of structural changes in the protein induced by these denaturants depends on the nature of the denaturant, its concentration, and incubation time of IBs. It was found that intermolecular interactions holding properly folded PldA-GFP molecules in IBs may be less resistant to denaturant action than intramolecular ones, which provide the spatial structure of the protein. This suggests that the recombinant protein can be released from IBs while maintaining the native-like conformation when using mild conditions for its solubilization.

The presence of a fluorescent marker in a properly folded chimeric protein provided additional information on the molecular organization of IBs. According to the data obtained, PldA-GFP IBs are structurally heterogeneous: compactly packed, properly folded, proteolytic resistant and structurally less organized, more sensitive to the action of denaturants and proteinase K, protein molecules coexist in IBs. We also found differences in the structural organization of the fluorescently active polypeptides in IBs.

Comparative characteristics of PldA-GFP IBs and PldA IBs, produced in *E. coli* under the same conditions, showed that fusion with GFP enables more correct folding of PldA, but negatively affects its enzymatic activity. This fact is possibly due to the GFP type and the linker structure, used in these experiments. At the same time, IBs formed by PldA at 18 °C had enzymatic activity. It was firstly found that a membrane protein embedded in IBs showed functional activity. As can be assumed from the data obtained, PldA fused with a monomeric type of GFP will form enzymatically active IBs at low temperatures; this remains to be clarified in our further studies.

Chimeric protein IBs is not toxic to eukaryotic cells and has the property of penetrating into neuroblastoma cells.

In the course of the work, it was shown that the GFP marker can be useful not only as an indicator of the target protein folding, but also as a tool for studying the structural organization of IBs, their morphology and localization in *E. coli* cells, as well as for visualizing IBs interaction with eukaryotic cells.

The culture of neuroblastoma cells was selected for experiments as eukaryotic cells do not have the ability to phagocytosis and, therefore, do not absorb PldA IBs in this way. In addition, Neuro-2a neuroblastoma cells are brain tumor cells. Our data confirm the possibility of delivering functionally active proteins to tumor cells using IBs, including for antitumor therapy.

3. Materials and Methods

3.1. Construction of Chimeric Insertion

The coding sequence of phospholipase PldA *Y. pseudotuberculosis* strain 488 (GenBank: MW848438) without signal sequence was amplified with primers: PldA-NdeI 5′-CCG**CATATG**GAAGCAACGATTGAAAAGATTC-3′ (forward) and PldA-Linker-R 5′-gttaattaaaccagcaccgtcaccAAGGACATCGTTCAACATGATAC-3′ (reverse). The pTurboGFP plasmid (Evrogen, Russia; GenBank: ASW25889.1) was used as a template for *gfp* amplification with primers: GFP-linker-F 5′-ggtgacggtgctggtttaattaacGCAGAAATCTATAACAAA GATGG-3′ (forward) and GFP-R 5′-TGAT**CTCGAG**TTATTCTTCACCGGCATCTGCATC CG-3′ (reverse). Bases in uppercase letters indicate *pldA* and *gfp* specific regions. Two restriction sites, *NdeI* and *XhoI*, (in bold letters) were incorporated at the 5′ and 3′ ends of the resulting insertion. In order to connect and improve the assembly of *pldA* and *gfp* we added the sequence coding flexible linker (GDGAGLIN) to the *pldA* reverse primer and the *gfp* forward primer (in lowercase letters). The gene fusion was assembled by II-rounded asymmetric PCR. At the first round, the *pldA* and *gfp* fragments were amplified in total volume of 20 µLeach. Reaction mixes included 250 µM of each dNTP, 0.2 U of Q5 High-Fidelity DNA Polymerase (Promega, Madison, WI, USA), 5× Q5 Reaction Buffer, 0.05 µM forward primer, and 0.5 µM reverse primer (for *pldA* production), or 0.5 µM forward primer and 0.05 µM reverse primer (for *gfp* production). The PCR conditions were as follows: initial denaturation at 95 °C for 5 min, then 21 cycles of 94 °C for 30 s, 55 °C for 30 s, and 72 °C for 40 s, followed by a final extension step at 72 °C for 2 min. In the second round, the reaction products were mixed, then an additional 0.2 U of Q5 High-Fidelity DNA Polymerase was added, and the combined reaction mix was amplified for 15 cycles at 94 °C for 30 s, 55 °C for 30 s, and 72 °C for 1 min, followed by a final extension step at 72 °C for 5 min. After amplification, the PCR fragment corresponding to the expected length of the chimeric insertion was purified and used for cloning.

3.2. PCR and Sequencing

PCR amplification of the *pldA* gene was performed using the primers PldA_NdeI 5′-CCG**CATATG**GAAGCAACGATTGAAAAGATTC-3′ and PldA_XhoI-Rev 5′-A**CTCGAG**TT AAAGGACATCGTTCAAC-3′, which included restriction sites for NdeI and *XhoI* (in bold letters). The expected amplicon size was 833 bp. PCR amplification of the *gfp* gene was performed using the primers GFP-F 5′-CCG**CATATG**GAGAGCGACGAGAGCGGCCTGCC CG-3′ and GFP-R 5′-TGAT**CTCGAG**TTATTCTTCACCGGCATCTGCATCCG-3′. The expected amplicon size was 715 bp. The PCR conditions were as follows: initial denaturation at 95 °C for 5 min, then 25 cycles of 94 °C for 15 s, 55 °C for 10 s, and 72 °C for 30 s, followed by a final extension step at 72 °C for 5 min. PCR-colonies was performed with standard T7-promoter/T7-terminator primers. The PCR fragments were evaluated on a 1.5% agarose gel stained with ethidium bromide. For subsequent cloning or sequencing, unincorporated primers and dNTPs were removed from PCR products with a GeneJET PCR Purification Kit (Thermo Fisher Scientific, Inc., Waltham, MA, USA). Sequencing was performed on a 3130xl Genetic Analyzer (Applied Biosystems; Thermo Fisher Scientific, Inc., Waltham, MA, USA) according to the manufacturer's instructions.

3.3. Molecular Cloning

The plasmid vector pRSETa (Invitrogen; Thermo Fisher Scientific, Inc., Waltham, MA, USA), designed for gene expression in prokaryotic cells such as *E. coli* BL21 (DE3) pLysS strain, was used. Vector and inserts (pldA, gfp, and pldA-gfp) were digested by *NdeI* and *XhoI* restrictases (Thermo Scientific). Then, inserts were ligated into vectors. The ligated products were transformed into *E. coli* TOP10′F cells (Invitrogen) and the resultant transformants were selected on carbenicillin plates and subjected to PCR-colony with T7 promoter primers. Several of the PCR-positive clones were inoculated into 3 mL test-tube cultures and allowed to grow overnight at 37 °C in a shaker at 200 rpm. Then, plasmid DNA were extracted from these cultures with GeneJET Plasmid Miniprep Kit (Thermo Fisher Scientific Inc., Waltham, MA, USA). Appropriate insertions of genes into the pRSETa vector were verified by DNA sequencing.

3.4. Expression

For protein expression, pRSETa/pldA-GFP, pRSETa/mPldA, and pRSETa/GFP constructs were transformed into *E. coli* BL21(DE3)pLysS competent cells. Single colonies were picked and inoculated in 10 mL of LB broth containing carbenicillin (100 mg/mL) and chloramphenicol (35 mg/mL) and grown overnight at 37 °C with shaking at 200 rpm. After 12 h, 250 mL of LB broth containing the same antibiotics was inoculated with 2% (v/v) of overnight grown primary culture and the culture was kept in incubator set at 37 °C with constant shaking at 200 rpm. IPTG induction was conducted at different concentrations (from 0.1 to 1 mM) when culture OD_{600} reached 0.5–0.7. The culture was further maintained at different temperatures, 37, 26, and 18 °C, for 3, 5, and 16 h. Before and after induction, approximately equal numbers of cells from various cultures (normalized according to OD_{600} values) were lysed in sample buffer and analyzed by SDS–PAGE. The GFP fluorescence of the chimeric protein PldA-GFP in *E.coli* cells was measured on the FL-600 Fluorescence/Absorbance Plate Reader (Bio TEK Instruments, Winooski, VT, USA). Absorption measurements (OD_{600}) and fluorescence intensity measurements (λexc. 488 nm/λexp. 530 nm) were made on 100 μLcell suspension. Plasmidless *E. coli* BL21(DE3) strain was used as negative control, and *E.coli* BL21(DE3) with pRSETa/GFP as positive control under all expression conditions. The *E. coli* BL21(DE3)pLysS/mPldA strain was expressed at 18 °C for 16 h.

All measurements were performed in three biological replicates. Results were expressed as the mean ± standard deviation (SD). The Student's t-test was used to evaluate the data. Statistical significance was considered for p-values < 0.05.

3.5. Isolation of IBs and GFP

PldA-GFP IBs and PldA IBs were isolated as described in previous work [33]. Briefly, wet bacterial biomass was lysed by lysozyme and exposed to ultrasound. IBs were sedimented from the cell biomass lysate by centrifugation at 4500× g for 30 min. IBs pellets was sequentially treated with 10 mM Tris-HCl buffer (pH 8.0) containing 0.1% *N*-Lauroylsarcosine (Sarcosyl) and with the same buffer containing 1% sodium deoxycholate and 1 M urea for 15 min, with the aim to achieve a more thorough purification from contaminants (proteins, LPS, and nucleic acids). The protease inhibitor PMSF (1 mM) was used at all stages of IBs isolation. IBs obtained by this scheme were stored at −70 °C for a maximum of 30 days.

E. coli cells expressing GFP were lysed by lysozyme and exposed to ultrasound under the conditions described in previous work [33] and centrifugated at 4500× *g* for 30 min. GFP was isolated from supernatant by size-exclusion chromatography on a Superdex 200HR column in PBS with an elution rate of 0.5 mL/min using a FPLC system (Amersham Pharmacia Biotech).

3.6. Determination of PldA-GFP and GFP Oligomeric Structure

The molecular weight of GFP (in PBS) and PldA-GFP (in 0.1% SDS, 50 mM Tris-HCl buffer, pH 8.0) was estimated by size-exclusion chromatography on a Superdex 200 column 10/300 GL using a FPLC system (Amersham Pharmacia Biotech) with an elution rate of 0.5 mL/min. The column was calibrated using proteins with known molecular weights ("ICN Biomedicals", Costa Mesa, CA, USA): apoferritin, bovine serum albumin, egg albumin, chymotrypsinogen, myoglobin, and cytochrome *c* (480, 67, 45, 24, 18, and 13 kDa, respectively). The relative error of the determination of the molecular weight was 5%.

3.7. Fluorescence Confocal Microscopy

Bacterial cells were pre-diluted in PBS to a concentration of 1×10^6 cells/mL and 20 µL of this solution was placed between two circle cover slips (Thermo Scientific) with a thickness of 0.08 to 0.12 mm. GFP fluorescence was visualized using a confocal microscope assembled based on Olympus FV1200 system equipped with a 488 nm argon laser with EX DM 405/488 and gallium arsenide phosphide (GaAsP) detector with filter set (DM 570, BA 505–540) and a 100× objective lens (UAPON100XOTIRF, Olympus). Transmission channel was visualized using 635 nm diode laser and transmission light detector. All images were obtained using Olympus FLUOVIEW software v.4.1.a with 4 us/pix exposure, 1600 × 1600 resolution and 4 kalman for digital noise elimination.

The murine neuroblastoma cell line Neuro-2a (CCL-131^{TM}) was purchased from American Type Culture Collection ATCC (ATCC). A cell suspension in DMEM medium (BioloT, Russia) containing 10% fetal bovine serum and 1% penicillin/streptomycin solution (BioloT, Russia) was introduced into the wells of chamber for confocal microscopy at a concentration of 1×10^3 cells per well and left for adhesion for 24 h in a CO_2-incubator (5% CO_2, 37 °C). After that, a suspension of PldA-GFP IBs in PBS (20 µg/mL) was added to the cells and the chambers were additionally incubated for 0, 1, 4, and 24 h. The accumulation of IBs in Neuro-2a cells was studied using laser scanning confocal microscopy and scanning electron microscopy. The fluorescent images of 150 randomly selected cells for each time point were inspected and the amount of fluorescent IBs penetrating the cell surface was calculated and expressed as number of PldA-GFP IBs per cell.

Fluorescence images of PldA-GFP IBs and neuroblastoma Neuro-2a cells were obtained using an LSM 710 LIVE AxioObserver laser scanning confocal microscope (Carl Zeiss GmbH, Jena, Germany). Cell nuclei were stained using a Hoechst 33,342 fluorescent dye (Invitrogen, USA) at a concentration of 5 µM. Fluorescence was excited at 488 nm, and emission was recorded at 493–652 nm. Processing and subsequent analysis of cell images was performed using the ZEN 2011 software (Carl Zeiss GmbH, Jena, Germany).

3.8. Scanning Electron Microscopy

Neuroblastoma cells on cover slides were fixed in 4% (*v/v*) glutaraldehyde in 0.1 M phosphate buffer (pH 7.2) for 2 h at room temperature. After dehydrating the cells by passing them through an ethanol series, the ethanol was replaced by isoamyl acetate. Then, parts of the cover glass with cells were dried under Critical point dryers K850 equipment (Quorum Technologies, London, UK), mounted on stubs, and coated with carbon. The samples were examined at the Instrumental Centre of Biotechnology and Gene Engineering of FSCEATB FEB RAS using a Scanning electron microscope EVO (Carl Zeiss, Jena, Germany).

3.9. Cytotoxic Activity

The cytotoxic activity of PldA-GFP IBs was determined by MTT method. After incubation of mouse neuroblastoma Neuro-2a cells with IBs in 96-well microplate for 24 h at 37 °C and 5% CO_2, the supernatant was replaced with pure medium. Then, 5 mg/mL MTT reagent solution (Sigma-Aldrich, St. Louis, MO, USA) was added to each well and microplate was incubated for additional 4 h, after which SDS-HCl solution (1 g SDS/10 mL dH_2O/17 µL 6 N HCl) was added and incubated at 37 °C for 4–18 h. Absorption was measured at a wavelength of 570 nm using a Multiskan FC spectrophotometer (Thermo Scientific, Canada). The cytotoxic activity of IBs was expressed as the concentration of EC_{50} at which the metabolic activity of cells is inhibited by 50%.

3.10. Dynamic Light Scattering (DLS)

The size of PldA-GFP IBs was determined using the DLS technique with a ZetaSizer Nano ZS instrument (Malvern, UK) equipped with a He-Ne-laser (λ 633 nm, 4 mW) at 173° angle. The hydrodynamic radius (R_H) of the protein particles was calculated using the software supplied with the instrument. IBs samples (0.1–0.6 mg/mL) were suspended in PBS, pH 7.5, by passing 10 times through a syringe microneedle and incubation at room temperature for a specified time with mixing. Data was accumulated for 5–60 min. Measurements were conducted in a 10 mm × 10 mm cuvette. Time of data accumulation for the correlation function was selected automatically via the instrument software, and it was 5–30 min. All measurements were replicated 2–3 times.

3.11. Solubility of PldA-GFP Ibs in Urea and SDS

IBs were suspended in 50 mM Tris-HCl buffer, pH 8.0, containing SDS from 0.01 to 0.1% or urea from 1 to 8 M, and incubated for a specified time (from 5 min to 24 h). Turbidity of the samples was measured at 350 nm with a µQuant spectrophotometer (BioTEK Instruments, Inc., USA). Fluorescence intensity of the PldA-GFP IBs solutions in urea and SDS was recorded with a FL-600 Fluorescence/Absorbance Plate Reader (BioTEK Instruments, USA) with excitation wavelength at 495 nm and emission at 530 nm. The experiments were performed in three biological replicates. Results show the mean ± standard deviation (SD).

3.12. Proteolytic Digestion of Inclusion Bodies

Purified PldA-GFP IBs were diluted to 1 OD at 350 nm in 980 mL of 50 mM Tris-HCl, 150 mM NaCl buffer of pH 8.0. Proteolytic digestion of IBs was initiated by adding 40 mL proteinase K (stock, 1 mg/mL) to the inclusion body solution (at 40 µg/mL final concentration). Proteolytic digestion was monitored for 120 min by measuring the changes in OD at 350 nm and in GFP fluorescence intensity (λexc. 488 nm/λexp. 530 nm).

3.13. The Thioflavin T Assay

To assess the presence of amyloid protein structure in IBs, we used the amyloid-specific dye Thioflavin T [34]. IBs were suspended in PBS (10 mM phosphate buffer, pH 7.4, containing 137 mM NaCl), or PBS containing 0.04% SDS, and incubated for a specified time (from 10 min to 24 h). The final reaction mixture for ThT assay consisted of 20 µg/mL protein and 20 µM ThT. The fluorescence measurements of the IBs samples were carried out at low SDS concentrations (0.004%) in order to avoid the influence of the detergent on the fluorescence intensity of ThT [44]. The samples were incubated for 10 min at room temperature. The fluorescence spectra were recorded in the range from 460 to 650 nm at λex = 440 nm. The background intensity from Raman scatter and ThT free samples were subtracted from each measurement sample of fluorescence intensity. Buffer controls did not show any significant ThT fluorescence.

3.14. Enzymatic Activity Measurement

The IB samples suspended in 50 mM Tris-HCl buffer (pH 8.0) or dissolved in the same buffer containing 4 or 8 M urea, were diluted 10 times with buffer solution (20 mM Tris-HCl, pH 8.3, 10 mM Triton X-100, 0.87 M urea) to a protein concentration of 10–40 μg/mL and incubated for 16 h at room temperature. The enzymatic activity of PldA was determined by the quantitative analysis of free oleic acid C18:1 formed during the hydrolysis of the substrate 1,2-dioleoyl-snglycero-3-phosphatidylcholine (Sigma-Aldrich, St. Louis, MO, USA) by GLC as described earlier [33]. The specific activity of PldA was expressed in μmol of the acid formed in 1 min per 1 mg of protein (IU/mg).

3.15. SDS-PAGE and Western Blotting

Whole cell-lysate proteins and recombinant proteins were separated by electrophoresis on 12% PAG under the standard denaturing conditions according to the method of Laemmli [45] with prior heating of the sample (5 min at 100 °C) in a buffer containing 2% SDS, as well as under non-denaturing conditions, using the method of semi-native PAGE [46]. In the latter case, the samples were dissolved at 0 °C in a buffer containing 0.1% SDS, without β-mercaptoethanol, and separated on a gel without SDS at 10 mA for 2.5 h at 5 °C. A set of colored proteins (Fermentas, Lithuania) with molecular weights of 10, 15, 25, 35, 40, 55, 70, 100, 130, and 170 kDa were used as markers. The proteins separated in the gel were stained with Coomassie R-250 in 10% acetic acid and 30% methanol. GFP fluorescence after SDS-PAGE was recorded using the VersaDoc imaging system (Bio-Rad, USA). The determination of the recombinant proteins molecular weight was carried out from the graph of the linear relationship between log MW of marker proteins (from 15 to 70 kDa) and relative migration distance (Rf) of these proteins according to the formula y = −0.0147x + 2.0684 (R2 > 0.994).

The localization of PldA in PldA-GFP IBs was assessed by Western blotting. After SDS-PAGE, proteins were transferred from non-stained gel on nitrocellulose membrane (0.2 μm, Merk Millipore, Burlington, MA, USA) by equipment for semi-dry transfer at the current of 0.8 mA/cm^2 overnight at 4 °C according to the standard procedure [47]. Immunodetection was carried out by the protein detection system SNAP i.d. according to the instruction of the manufacturing company (Merk Millipore, Burlington, MA, USA). A murine polyclonal antiserum to recombinant PldA was prepared as described in [48]. HRP-Goat Anti-Mouse antibodies (Invitrogen, Waltham, MA, USA) were used according to the instruction of the manufacturing company. Antigen-antibody complexes were identified on nitrocellulose membrane by the hydrogen peroxide detection with 3,3′-diaminobenzidine for 20 min at the room temperature.

3.16. Spectroscopic Methods

3.16.1. Circular Dichroism (CD)

CD spectra were recorded on Chirascan Plus CD spectrometer (Applied Photophysics, UK) in quartz cuvettes with an optical path length of 0.1 and 1 cm for the far-UV, or peptide for the near-UV or aromatic regions. The measurements were performed at room temperature (22 °C). Protein solutions in urea and detergent were centrifuged at 15,000× *g* for 20 min before taking the spectra. The presented data are the average of two independent measurements, each averaged of three scans. Each averaged CD spectrum was corrected for the buffer baseline by subtracting an averaged buffer CD signal over the same wavelength region. The content of the secondary structure elements in the recombinant proteins was calculated according to Sreerama method from software package CD*Pro* [31].

3.16.2. UV-Spectra

Absorption spectra were recorded using a UV-Visible spectrophotometer UV-1601 PC (Shimadzu, Japan) in quartz cuvettes with an optical path length of 1 cm, 0.1 cm, 0.01 cm in the range from 190 to 400 nm. Protein concentration of GFP in the PBS and PldA and PldA-GFP in the urea and SDS solutions was determined (following centrifugation at 15,000× *g* for 20 min) from the UV spectra at absorption maximum of 280 nm assuming that the $A_{1cm}^{0.1\%}$ value was 2.74—for *Y.pseudotuberculosis* phospholipase A_1; 1. 36—for PldA-GFP and 1.34—for GFP.

3.16.3. Fluorescence Spectra

The ThT and GFP fluorescence spectra were recorded on a RF-5301 PC spectrofluorophotometer (Shimadzu, Japan) in quartz cuvettes with an optical path length of 1 cm in the range from 460 to 650 nm at $\lambda ex = 450$ nm. The excitation and emission slit widths were set at 5 nm. The measurements of GFP fluorescence activity of the recombinant proteins were performed in the range of low concentrations (0.1–1.1 mg/mL), where the intensity of GFP fluorescence is directly proportional to its concentration. For comparison, all the GFP fluorescence data were normalized by the maximum fluorescence signal.

4. Conclusions

Despite their fundamental and practical importance, IBs are not well understood. To date, there are many unresolved and controversial issues regarding the conformational states and arrangement of protein molecules inside the IBs, as well as the exact nature of intermolecular interactions. Various biophysical methods have been used to characterize proteins aggregated in IBs, including NMR one, which enables detailed structural information at the specific residue level up to the three-dimensional structure of the protein to be obtained. However, no significant progress has been made in these studies so far.

Here, we used GFP as a fusion tag to provide further opportunities for structural studies of IBs. Therefore, some details of the inner organization IBs were revealed. The data obtained indicate the presence of highly structured clusters of correctly folded protein molecules inside the IBs, which can differ from one another in the degree of order in the structure. Intramolecular interactions that maintain the native-like conformation of the protein in these clusters may be more resistant to denaturants than intermolecular contacts in IBs.

According to our data, PldA-GFP IBs formed at low temperatures contain a noticeable percentage of the protein in the native conformation, and the recombinant protein solubilized from IBs has a high content of β-structure. This fact suggests that IB formation occurs at a relatively late stage of protein folding, when the degree of protein structural organization is high enough and proteins retain most of their secondary structure during aggregation. The results of the structural analysis of PldA-GFP IBs contribute to understanding the mechanisms of IB formation and their structure.

The accumulation of knowledge about IBs formed by proteins of different nature under various expression conditions provides a deeper understanding of their molecular organization, opening possibilities for controlling their structure and the process of formation. Further progress in the study of IBs will require new methodological approaches, as well as powerful biophysical research methods.

Supplementary Materials: The following are available online: Figure S1. Recombinant plasmid for the expression of the PldA-GFP fusion protein.

Author Contributions: S.I.B. was in charge of overall direction and planning and performed experiments (isolation and characterization of recombinant proteins); N.Y.K. carried out experiments (spectroscopic methods); A.M.S. and E.P.B. performed molecular genetic experiments; E.V.S. performed experiments (DLS, FPLC); E.S.M., T.Y.G. and N.A.S. performed bioactivity testing supervised by D.L.A.; and T.F.S. conducted project administration, literature analysis, data analysis, and interpretation, conceived the study, and was in charge of overall direction and planning. All authors discussed the results and contributed to the final manuscript. All authors have read and agreed to the published version of the manuscript.

Funding: This research received no external funding.

Institutional Review Board Statement: All the animal experiments were carried out in accordance with the interstate standard "Guide for the maintenance and care of laboratory animals (GOST 33215-2014)", which takes into account the main provisions of the international document ETS N 123, Appendix A of the European Convention for the protection of Vertebrate animals used for experimental and other scientific purposes Guidelines for accommodation and care of animals (Article 5 of the Convention) and in accordance with the animal care protocols and rules, approved for pre-clinical studies in Russian Federation and permitted by PIBOC Animal Ethics Committee. All animals were housed in standard facility under controlled environmental conditions at temperature 22 ± 2 °C and 12-h light-dark cycle, balanced ad libitum diet for laboratory animals in accordance with the Russian GOST P 50258-92 protocol and water ad libitum. 6- to 8-weeks old female BALB/c mice weighing 20 ± 2 g for the experiments were use.

Informed Consent Statement: Informed consent was obtained from all subjects involved in the study.

Data Availability Statement: Data is contained within the article and supplementary material.

Acknowledgments: The study was carried out on the equipment of the Collective Facilities Center «The Far Eastern Center for Structural Molecular Research (NMR/MS) PIBOC FEB RAS».

Conflicts of Interest: The authors declare that they have no conflicts of interest.

Sample Availability: Samples of the compounds are available by agreement with the authors and management of the organization for conducting joint research from the authors.

Abbreviations

PldA—detergent-resistant phospholipase A_1; GFP—green fluorescent protein; PldA-GFP—chimeric protein; IB—inclusion body; CD—circular dichroism; DLS—dynamic light scattering; LPS—lipopolysaccharide; SDS-PAGE—sodium dodecyl-sulfate polyacrylamide gel electrophoresis; ThT—Thioflavin T; R_H—hydrodynamic radius; PBS—phosphate-buffered saline, pH 7.5; and IPTG—isopropyl β-D-1-thiogalactopyranoside.

References

1. Villaverde, A.; Mar Carrió, M. Protein aggregation in recombinant bacteria: Biological role of inclusion bodies. *Biotechnol. Lett.* **2003**, *25*, 1385–1395. [CrossRef]
2. Vallejo, L.F.; Rinas, U. Strategies for the recovery of active proteins through refolding of bacterial inclusion body proteins. *Microb. Cell Fact.* **2004**, *3*, 11. [CrossRef] [PubMed]
3. Garcia-Fruitos, E.; Gonzalez-Montalban, N.; Morell, M.; Vera, A.; Ferraz, R.M.; Aris, A.; Ventura, S.; Villaverde, A. Aggregation as bacterial inclusion bodies does not imply inactivation of enzymes and fluorescent proteins. *Microb. Cell Fact.* **2005**, *4*, 27. [CrossRef] [PubMed]
4. Wang, L. Towards revealing the structure of bacterial inclusion bodies. *Prion* **2009**, *3*, 139–145. [CrossRef]
5. Peternel, S.; Liovic, M. Production of recombinant proteins in bacteria: The inclusion bodies formation and their use in biomedicine. *Recent Patents Biomed. Eng.* **2010**, *3*, 153–161. [CrossRef]
6. Groot, N.S.; Ventura, S. Effect of temperature on protein quality in bacterial inclusion bodies. *FEBS Lett.* **2006**, *580*, 6471–6476. [CrossRef] [PubMed]
7. Schwaighofer, A.; Ablasser, S.; Lux, L.; Kopp, J.; Herwig, C.; Spadiut, O.; Lendl, B.; Slouka, C. Production of active recombinant hyaluronidase inclusion bodies from *Apis mellifera* in *E. coli* Bl21(DE3) and characterization by FT-IR spectroscopy. *Int. J. Mol. Sci.* **2020**, *21*, 3881. [CrossRef] [PubMed]
8. Jevsevar, S.; Gaberc-Porekar, V.; Fonda, I.; Podobnik, B.; Grdadolnik, J.; Menart, V. Production of nonclassical inclusion bodies from which correctly folded protein can be extracted. *Biotechnol. Prog.* **2005**, *21*, 632–639. [CrossRef]
9. Shirano, Y.; Shibata, D. Low temperature cultivation of *Escherichia coli* carrying a rice lipoxygenase L-2 cDNA produces a soluble and active enzyme at a high level. *FEBS Lett.* **1990**, *271*, 128–130. [CrossRef]
10. Singh, A.; Upadhyay, V.; Singh, A.; Panda, A.K. Structure-function relationship of inclusion bodies of a multimeric protein. *Front. Microbiol.* **2020**, *11*, 876. [CrossRef]
11. Carvajal, P.; Gibert, J.; Campos, N.; Lopera, O.; Barbera, E.; Torne, J.M.; Santos, M. Activity of maize transglutaminase over expressed in *Escherichia coli* inclusion bodies: An alternative to protein refolding. *Biotechnol. Prog.* **2011**, *27*, 232–240. [CrossRef]
12. Margreiter, G.; Schwanninger, M.; Bayer, K.; Obinger, C. Impact of different cultivation and induction regimes on the structure of cytosolic inclusion bodies of TEM1-beta-lactamase. *Biotechnol. J.* **2008**, *3*, 1245–1255. [CrossRef]
13. Ramrakhiani, L.; Chand, S. Recent progress on phospholipases: Different sources, assay methods, industrial potential and pathogenicity. *Appl. Biochem. Biotechnol.* **2011**, *164*, 991–1022. [CrossRef] [PubMed]
14. Istivan, T.S.; Coloe, P.J. Phospholipase A in Gram-negative bacteria and its role in pathogenesis. *Microbiology* **2006**, *152*, 1263–1274. [CrossRef] [PubMed]
15. Gregersen, N.; Bross, P.; Vang, S.; Christensen, J.H. Protein misfolding and human disease. *Annu. Rev. Genom. Hum. Genet.* **2006**, *7*, 103–124. [CrossRef] [PubMed]
16. Waldo, G.S.; Standish, B.M.; Berendzen, J.; Terwilliger, T.C. Rapid protein-folding assay using green fluorescent protein. *Nat Biotechnol.* **1999**, *17*, 691–695. [CrossRef]

17. Drew, D.E.; von Heijne, G.; Nordlund, P.; de Gier, J.W. Green fluorescent protein as an indicator to monitor membrane protein overexpression in *Escherichia coli*. *FEBS Lett.* **2001**, *507*, 220–224. [CrossRef]
18. Wu, W.; Xing, L.; Zhou, B.; Lin, Z. Active protein aggregates induced by terminally attached self-assembling peptide ELK16 in *Escherichia Coli*. *Microb. Cell Fact.* **2011**, *10*, 9. [CrossRef]
19. Raghunathan, G.; Munussami, G.; Moon, H.; Paik, H.-J.; An, S.S.A.; Kim, Y.-S.; Kang, S.; Lee, S.-G. A variant of green fluorescent protein exclusively deposited to active intracellular inclusion bodies. *Microb. Cell Fact.* **2014**, *13*, 68. [CrossRef]
20. Frankel, S.; Sohn, R.; Leinwand, L. The use of sarkosyl in generating soluble protein after bacterial expression. *Proc. Natl. Acad. Sci. USA* **1991**, *88*, 1192–1196. [CrossRef]
21. Heim, R.; Prasher, D.; Tsien, R. Wavelength mutations and posttranslational autoxidation of green fluorescent protein. *Proc. Natl. Acad. Sci. USA.* **1994**, *91*, 12501–12504. [CrossRef]
22. Huang, Z.; Zhang, C.; Chen, S.; Ye, F.; Xing, X.-H. Active inclusion bodies of acid phosphatase PhoC: Aggregation induced by GFP fusion and activities modulated by linker flexibility. *Microb. Cell Fact.* **2013**, *12*, 25. [CrossRef]
23. Peternel, Š.; Komel, R. Isolation of biologically active nanomaterial (inclusion bodies) from bacterial cells. *Microb. Cell Fact.* **2010**, *9*, 66. [CrossRef]
24. Peternel, S.; Jevsevar, S.; Bele, M.; Gaberc-Porekar, V.; Menart, V. New properties of inclusion bodies with implications for biotechnology. *Biotechnol. Appl. Biochem.* **2008**, *49*, 239–246. [CrossRef] [PubMed]
25. García-Fruitós, E.; Arís, A.; Villaverde, A. Localization of functional polypeptides in bacterial inclusion bodies. *Appl. Environ. Microbiol.* **2007**, *73*, 289–294. [CrossRef] [PubMed]
26. Stepanenko, O.V.; Verkhusha, V.V.; Shavlovskiĭ, M.M.; Aleĭnikova, T.D.; Uverskiĭ, V.N.; Kuznetsova, I.M.; Turoverov, K.K. The Role of quarternary structure in fluorescent protein stability. *Tsitologiia* **2005**, *47*, 1017–1027.
27. Khan, R.H.; Rao, K.B.; Eshwari, A.N.; Totey, S.M.; Panda, A.K. Solubilization of recombinant ovine growth hormone with retention of native-like secondary structure and its refolding from the inclusion bodies of *Escherichia coli*. *Biotechnol. Prog.* **1998**, *14*, 722–728. [CrossRef] [PubMed]
28. Bocharova, O.V.; Makarava, N.; Breydo, L.; Anderson, M.; Salnikov, V.V.; Baskakov, I.V. Annealing prion protein amyloid fibrils at high temperature results in extension of a proteinase K-resistant core. *J. Biol. Chem.* **2006**, *281*, 2373–2379. [CrossRef]
29. Melnik, S.; Povarnitsyna, T.V.; Melnik, T.N. Can the fluorescence of green fluorescent protein chromophore be related directly to the nativity of protein structure? *Biochem. Biophys. Res. Commun.* **2009**, *390*, 1167–1170. [CrossRef]
30. Bakholdina, S.I.; Tischenko, N.M.; Sidorin, E.V.; Isaeva, M.P.; Likhatskaya, G.N.; Dmitrenok, P.S.; Kim, N.Y.; Chernikov, O.V.; Solov'eva, T.F. Recombinant phospholipase A_1 of the outer membrane of psychrotropic *Yersinia pseudotuberculosis: Expression, purification, and characterization*. *Biochemistry* **2016**, *81*, 47–57.
31. Sreerama, N.; Woody, R.W. Estimation of protein secondary structure from circular dichroism spectra: Comparison of CONTIN, SELCON and CDSSTR methods with an expanded reference set. *Anal. Biochem.* **2000**, *287*, 252–260. [CrossRef] [PubMed]
32. Vissera, N.V.; Hinka, M.A.; Borsta, J.W.; van der Krogta, G.N.M.; Vissera, A.J.W.G. Circular dichroism spectroscopy of fuorescent proteins. *FEBS Lett.* **2002**, *521*, 31–35. [CrossRef]
33. Bakholdina, S.I.; Sidorin, E.V.; Khomenko, V.A.; Isaeva, M.P.; Kim, N.Y.; Bystritskaya, E.P.; Pimenova, E.A.; Solov'eva, T.F. The Effect of conditions of the expression of the recombinant outer membrane phospholipase A_1 from *Yersinia pseudotuberculosis* on the structure and properties of inclusion bodies. *Russ. J. Bioorganic Chem.* **2018**, *44*, 178–187. [CrossRef]
34. Rasmussen, H.; Enghild, J.J.; Otzen, D.E.; Pedersen, J.S. Unfolding and partial refolding of a cellulase from the SDS-denatured state: From β-sheet to α-helix and back. *BBA Gen. Subj.* **2020**, *1864*, 129434. [CrossRef] [PubMed]
35. Dekker, N.; Merck, K.; Tommassen, J.; Verheij, H.M. In vitro folding of *Escherichia coli* outer-membrane phospholipase A. *Eur. J. Biochem.* **1995**, *232*, 214–219. [CrossRef]
36. Shivu, B.; Seshadri, S.; Li, J.; Oberg, K.A.; Uversky, V.N.; Fink, A.L. Distinct β-sheet structure in protein aggregates determined by ATR-FTIR spectroscopy. *Biochemistry* **2013**, *52*, 5176–5183. [CrossRef] [PubMed]
37. Gupta, A.; Mahalakshmi, R. Single-residue physicochemical characteristics kinetically partition membrane protein self-assembly and aggregation. *J. Biol. Chem.* **2020**, *295*, 1181–1194. [CrossRef]
38. Radko, S.P.; Khmeleva, S.A.; Suprun, E.V.; Bodoev, N.V.; Archakov, A.I.; Shumyantseva, V.V.; Kozin, S.A.; Makarov, A.A. Physico-chemical methods for studing β-amyloid aggregation. *Biochem. Suppl. Ser. B Biomed. Chem.* **2015**, *9*, 258–274. [CrossRef] [PubMed]
39. Ajmal, M.R.; Nusrat, S.; Alam, P.; Zaidi, N.; Badr, G.; Mahmoud, M.H.; Rajpoot, R.K.; Khan, R.H. Differential mode of interaction of Thioflavin T with native β structural motif in human a 1-acid glycoprotein and cross beta sheet of its amyloid: Biophysical and molecular docking approach. *J. Mol. Struct.* **2016**, *1117*, 208–217. [CrossRef]
40. Khan, J.M.; Qadeer, A.; Chaturvedi, S.K.; Ahmad, E.; Abdul Rehman, S.A.; Gourinath, S.; Khan, R.H. SDS can be utilized as an amyloid inducer: A case study on diverse proteins. *PLoS ONE* **2012**, *7*, e29694. [CrossRef] [PubMed]
41. Yamamoto, S.; Hasegawa, K.; Yamaguchi, I.; Tsutsumi, S.; Kardos, J.; Goto, Y.; Gejyo, F.; Naiki, H. Low concentrations of sodium dodecyl sulfate induce the extension of beta 2-microglobulin-related amyloid fibrils at a neutral pH. *Biochemistry* **2004**, *43*, 11075–11082. [CrossRef]
42. Singh, A.; Upadhyay, V.; Singh, S.M.; Garg, L.C.; Panda, A.K. Protein recovery from inclusion bodies of *Escherichia coli* using mild solubilization process. *Microb. Cell Fact.* **2015**, *14*, 41. [CrossRef] [PubMed]
43. Olenych, S.G.; Claxton, N.S.; Ottenberg, G.K.; Davidson, M.W. The fluorescent protein color palette. *Curr. Protoc. Cell Biol.* **2007**, *21*, 5. [CrossRef]

44. Kumar, S.; Singh, A.K.; Krishnamoorthy, G.; Swaminathan, R. Thioflavin T displays enhanced fluorescence selectively inside anionic micelles and mammalian cells. *J. Fluoresc.* **2008**, *18*, 1199–1205. [CrossRef] [PubMed]
45. Laemmli, U.K. Cleavage of structural proteins during the assembly of the head of bacteriophage T 4. *Nature* **1970**, *227*, 680–685. [CrossRef] [PubMed]
46. Volhoux, R.; Bos, M.P.; Geurtsen, J.; Mols, M.; Tommassen, J. Role of a highly conserved bacterial protein in outer membrane protein assembly. *Science* **2003**, *299*, 262–265. [CrossRef]
47. Towbin, H.; Staehelin, T.; Gordon, J. Electrophoretic transfer of proteins from polyacrylamide gels to nitrocellulose sheets: Procedure and some applications. *Proc. Natl. Acad. Sci. USA* **1979**, *76*, 4350–4354. [CrossRef] [PubMed]
48. Davydova, L.; Bakholdina, S.; Barkina, M.; Velansky, P.; Bogdanov, M.; Sanina, N. Effect of elevated growth temperature and heat shock on lipid composition of inner and outer membranes of *Yersinia pseudotuberculosis*. *Biochimie* **2016**, *123*, 103–109. [CrossRef]

molecules

Brief Report

Are the Closely Related *Cobetia* Strains of Different Species?

Yulia Noskova [1], Aleksandra Seitkalieva [1], Olga Nedashkovskaya [1], Liudmila Shevchenko [1], Liudmila Tekutyeva [2], Oksana Son [2] and Larissa Balabanova [1,2,*]

[1] Laboratory of Marine Biochemistry, G.B. Elyakov Pacific Institute of Bioorganic Chemistry, Far Eastern Branch, The Russian Academy of Sciences, 690022 Vladivostok, Russia; noskovaiulia@yandex.ru (Y.N.); sasha0788@inbox.ru (A.S.); oned2004@mail.ru (O.N.); lshev@piboc.dvo.ru (L.S.)

[2] Basic Department of Bioeconomy and Food Security, School of Economics and Management, Far Eastern Federal University, 690090 Vladivostok, Russia; tekuteva.la@dvfu.ru (L.T.); oksana_son@bk.ru (O.S.)

* Correspondence: balaban@piboc.dvo.ru

Abstract: Marine bacteria of the genus *Cobetia*, which are promising sources of unique enzymes and secondary metabolites, were found to be complicatedly identified both by phenotypic indicators due to their ecophysiology diversity and 16S rRNA sequences because of their high homology. Therefore, searching for the additional methods for the species identification of *Cobetia* isolates is significant. The species-specific coding sequences for the enzymes of each functional category and different structural families were applied as additional molecular markers. The 13 closely related *Cobetia* isolates, collected in the Pacific Ocean from various habitats, were differentiated by the species-specific PCR patterns. An alkaline phosphatase PhoA seems to be a highly specific marker for *C. amphilecti*. However, the issue of *C. amphilecti* and *C. litoralis*, as well as *C. marina* and *C. pacifica*, belonging to the same or different species remains open.

Keywords: *Cobetia amphilecti*; *Cobetia litoralis*; *Cobetia pacifica*; *Cobetia marina*, *Cobetia crustatorum*; identification markers; alkaline phosphatase PhoA

Citation: Noskova, Y.; Seitkalieva, A.; Nedashkovskaya, O.; Shevchenko, L.; Tekutyeva, L.; Son, O.; Balabanova, L. Are the Closely Related *Cobetia* Strains of Different Species? *Molecules* **2021**, *26*, 690. https://doi.org/10.3390/molecules26030690

Academic Editors: Pavel S. Dmitrenok and Francesco Epifano

Received: 18 December 2020
Accepted: 26 January 2021
Published: 28 January 2021

1. Introduction

Bacteria of the genus *Cobetia* are Gram-negative, aerobic and halotolerant and belong to the *Halomonadaceae* family. For the first time, a bacterium of the genus *Cobetia* was described in 1970 by Cobet et al. [1] and was originally assigned to the species *Arthrobacter marinus* sp. nov. In further studies, it was assigned to various genera, such as *Pseudomonas* [2], *Deleya* [3], and *Halomonas* [4]. Then, based on an analysis of the 23S and 16S rRNA sequences and amending the description of the species *Halomonas marina*, including new features, the authors [5] proposed assigning it to a new genus, *Cobetia* gen. nov., within the *Halomonadaceae* family. The type species is *Cobetia marina*, with the type strain LMG 2217 (= CIP 104,765 = IAM 14,107 = CECT 4278 = DSM 4741 = NCIMB 1877 = CCUG 49,558 = NBRC 102,605 = JCM 21,022 = ATCC 25374).

At present, a few species are characterized for marine bacteria of the genus *Cobetia*, namely: *C. marina* [5], *C. crustatorum* [6], *C. amphilecti* [7], *C. pacifica* [7], and *C. litoralis* [7]. The whole genome sequence analysis is presented only for three strains of the genus *Cobetia*, including a type strain *C. marina* JCM 21022^{T} [8], and the non-type strains *C. amphilecti* KMM 296 (formerly *C. marina* KMM 296) [9] and *Cobetia* sp. cqz5-12 [10]. At the same time, more and more information appears about *Cobetia* isolates, which are promising sources of unique enzymes and secondary metabolites degrading oil [11], bacterial biofilms [12], alginate [10,13], and phenol [14]. However, their species identification are complicated due to the high level of identity for their 16S rRNA genes and the absence of whole genome sequences for the type strains of known species, excluding from *C. marina* JCM 21022^{T} [8]. The 16S rRNA gene usually is highly specific to each bacterial species that makes it a good target for identification of both environmental and clinical bacterial isolates [15–17]. However, the 16S rRNA sequences were found to be indistinguishable for a few species [18].

The usage of 16S rRNA gene as a marker is limited for the closely related species, which have a high percentage of the sequence similarity and lack enough variations. Thus, a conventional method of 16S rRNA phylogeny has often failed to correctly identify *Vibrio* species [19]. Similarly, we found that the *Cobetia* species have 99–100% identity of their compared 16S rRNA sequences. In this regard, searching for the additional molecular markers and/or methods for the species identification of the *Cobetia* isolates is especially relevant.

In this work, we have applied the genome-based found species-specific coding sequences for the essential enzymes from each functional category and different structural families, such as nucleases, proteases, phosphatases, and phospholipase, as the additional molecular markers in the polymerase chain reaction (PCR)-based method, to differentiate the species of the closely related isolates of the marine bacteria *Cobetia*, collected in the Pacific Ocean from various habitats. In general, we have raised the issue of the possible species reclassification, due to their highly homologous 16S rRNA sequences, PCR-patterns with the use of the additional molecular markers, which were suggested here, and a high similarity between two whole genome sequences of the closely related non-type strains.

2. Results and Discussion

Thirteen strains of the marine bacteria, isolated from coastal seawater and sediments from the Sea of Japan, marine invertebrates, the mussel *Crenomytilus grayanus* from the Sea of Japan and the deep-water sponge *Esperiopsis digitate* from the Sea of Okhotsk, and the red algae *Ahnfeltia tobuchiensis* from the Sea of Okhotsk, which are deposited in the Collection of Marine Microorganisms (KMM, G.B. Elyakov Pacific Institute of Bioorganic Chemistry, Far Eastern Branch, Russian Academy of Sciences, http://www.piboc.dvo.ru/), were assigned to the genus *Cobetia* by physiological, biochemical and molecular genetic parameters, using sequencing and phylogenetic analysis of their 16S rRNA genes (Table 1). However, the strain-specific metabolic versatility and ecophysiological diversity of these *Cobetia* isolates could not allow distinguishing between their species [9,20,21]. Thus, our study showed that 7 from 13 strains have 99.86–100% identity of the 16S rRNA genes simultaneously to two type strains *C. marina* LMG 2217^T (JCM 21022^T) and *C. pacifica* KMM 3879^T (NRIC 0813^T), one strain has 100% identity to the type strain *C. crustatorum* JCM 15644^T, and three strains have 100% identity to the type strain *C. amphilecti* KMM 1561^T (NRIC 0815^T) (Table 1). The strain *Cobetia* sp. 2AS (KMM 7514), isolated from a coastal seawater, showed 99.93 and 99.86% similarities with *C. amphilecti* KMM 1561^T and *C. litoralis* KMM 3880^T, respectively (Table 1). From four independent replicates, half of the 16S rRNA DNA samples from the clones of the strain *Cobetia* sp. 29-18-1 (KMM 7000), isolated from the sponge, were of 100% identity with the type strain *C. amphilecti* KMM 1561^T (NRIC 0815^T), and the other half had 100% identity with the 16S rRNA gene of the type strain *C. litoralis* KMM 3880^T (NRIC 0814^T) (Table 1). Therefore, the EzBioCloud identification, based on the use of 16S rRNA gene sequences of the type strains [22], showed only one of the identical reference records in its database (Table S1). Furthermore, the results of similarity calculation by EzBioCloud 16S database showed that the 16S rRNA genes are completely identical for *C. marina* JCM 21022^T and *C. pacifica* KMM 3879^T (100% identity), and the only single nucleotide polymorphism (99.93% identity) is between the type strains *C. amphilecti* KMM 1561^T and *C. litoralis* KMM 3880^T (Tables S2 and S3).

The comparative genomics of *Cobetia* isolates is also impossible due to the absence of the type strains' whole genome sequences to date except for *C. marina* JCM 21022^T [8]. Moreover, the next-generation sequencing (NGS) solutions cannot always allow resolving this problem quickly without detailed bioinformatics analysis. Thus, the whole genome shotgun sequencing for the strain *Cobetia* sp. 2AS1, KMM 7005 (GenBank: JADAZN000000000.1) has led to the loss of its complete 16S rRNA gene, and consequently, indicated the same similarity (100%) of its partial sequence to both *C. amphilecti* KMM 1561^T and *C. litoralis* KMM 3880^T with the use of EzBioCloud identification system "Genome-based ID" [22].

Table 1. The species identification of *Cobetia* isolates from the Pacific Ocean.

Isolate	Isolation Source	Collection Number, KMM *	Results of Identification		16S rRNA Genbank ID
			16S rRNA, % Identity	PCR-Based Method	
Cobetia sp. 1AS1	Coastal seawater, Vostok Bay, Sea of Japan	KMM 7516	*C. amphilecti, 100%*	*C. amphilecti*	MW332480
C. amphilecti KMM 296	Coelomic liquid, mussel *Crenomytilus grayanus*	KMM 296	*C. amphilecti, 100%*	*C. amphilecti*	NZ_JQJA01000078
Cobetia sp. 29-18-1	Sponge *Esperiopsis digitata*, Sea of Okhotsk, Is. Sakhalin, Piltun bay, 107 m.	KMM 7000	*C. amphilecti, 100%/ C. litoralis, 100%*	*C. litoralis*	MW332487
Cobetia sp. 2AS	Sediments, Vostok Bay, Sea of Japan	KMM 7514	*C. amphilecti, 99.93%/ C. litoralis 99.86%*	*C. litoralis*	MW332483
Cobetia sp. 2AS1	Sediments, Vostok Bay, Sea of Japan	KMM 7005	*C. amphilecti, 100%*	*C. litoralis*	MW332484
Cobetia sp. 41-10Alg46	The red algae *Ahnfeltia tobuchiensis*, collected near Is. Paramushir, Kuril Isles, Sea of Okhotsk	KMM 6284	*C. marina/C. pacifica, 100%*	*C. marina*	MK587632
Cobetia sp. 2S	Coastal seawater, Vostok Bay, Sea of Japan	KMM 7508	*C. marina/C. pacifica, 99.86%*	*C. pacifica*	MW332481
Cobetia sp. 3AS	Coastal seawater, Vostok Bay, Sea of Japan	KMM 7515	*C. marina/C. pacifica, 99.86%*	*C. pacifica*	MW332482
Cobetia sp. 11Alg1	The red algae *A. tobuchiensis* (long-time cultivated), collected near Island Paramushir, Kuril Isles, Sea of Okhotsk	KMM 6816	*C. marina/C. pacifica, 100%*	*C. pacifica*	MW332485
Cobetia sp. 11Alg14	The red algae *A. tobuchiensis* (long-time cultivated), collected near Island Paramushir, Kuril Isles, the Sea of Okhotsk	KMM 6818	*C. marina/C. pacifica, 100%*	*C. pacifica*	MW332486
Cobetia sp. 3AK	Coastal seawater, Vostok Bay, Sea of Japan	KMM 7505	*C. marina/C. pacifica, 100%*	*C. pacifica*	MW332488
Cobetia sp. 41-10Alg146	The red alga A. tobuchiensis, collected near Is. Paramushir, Kuril Isles, Sea of Okhotsk	KMM 6731	*C. marina/C. pacifica, 100%*	*C. pacifica*	KC247358
Cobetia sp. 11Alg4	The red algae *A. tobuchiensis* (long-time cultivated), collected near Is. Paramushir, Kuril Isles, Sea of Okhotsk	KMM 6817	*C. crustatorum, 100%*	*C. crustatorum*	MW332489

* KMM—Collection of Marine Microorganisms, G.B. Elyakov Pacific Institute of Bioorganic Chemistry, Far Eastern Branch, Russian Academy of Sciences, http://www.piboc.dvo.ru/.

To clarify the species identity of this coastal seawater isolate *Cobetia* sp. 2AS1 (KMM 7005), we have tried to obtain PCR-products, using the sequences of structural genes encoding for the key metabolic hydrolases of the non-type strain *C. amphilecti* KMM 296 associated with a mussel, which were selected for the functional genomic studies for this species (GenBank: JQJA00000000.1). Putative coding DNA sequences (CDSs) for the following enzymes were selected as identification markers: alkaline phosphatases of the structural families PhoA (DQ435608) and PhoD (WP_043333989); EEP-like (DNaseI-like) nuclease (WP_084589364) and DNA/RNA non-specific (S1-like) endonuclease (WP_043334786), ATP-dependent protease Clp (KGA03297), phospholipase A (WP_084589432), and periplasmic serine peptidase with thrypsin-like peptidase domain of the Do/DeqQ family (KGA03014). The Do/DeqQ family serine peptidase has a chaperone function at low temperatures and proteolytic activity at elevated temperatures, thus protecting bacteria from thermal and other stresses [23]. Clp proteases are involved in a number of cellular processes, such as degradation of misfolded proteins, regulation of short-lived proteins and housekeeping removal of dysfunctional proteins, the control of cell growth, and targeting DNA-binding protein from starved cells [23]. The large EEP (exonuclease/endonuclease/phosphatase) domain superfamily (structural family: cl00490) includes a diverse set of proteins, including the ExoIII family of apurinic/apyrimidinic (AP) endonucleases, inositol polyphosphate 5-phosphatases (INPP5), and deoxyribonuclease I (DNaseI), which share a common catalytic mechanism of cleaving phosphodiester bonds, with the substrates range from nucleic acids to phospholipids and, probably, proteins [23]. The DNA/RNA endonuclease belonging to the structural cl00089: NUC superfamily can non-specifically cleave both double- and single-stranded DNA and RNA, whose domain may be present in phosphodiesterases [23]. Thus, all these enzymes are fundamental for bacteria survival [24–29]. If the strains *Cobetia* sp. 2AS1 (KMM 7005) and *C. amphilecti* KMM 296 are of the same species, they should have a high level of their CDS similarity and the same pattern of the presence and distribution of the PCR products by electrophoresis. In addition, the same PCR primers were also applied to all type strains and new isolates belonging to the genus *Cobetia* (Figures 1 and 2).

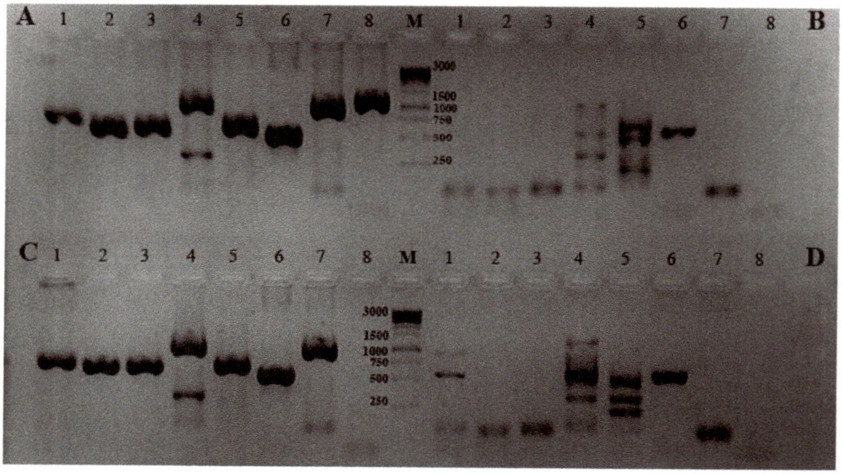

Figure 1. Gel-electrophoresis of PCR products of the new molecular markers: (**A**) ***C. amphilecti* KMM 1561^T**, *C. amphilecti* KMM 296 and KMM 7516; (**B**) ***C. marina* LMG 2217^T** and KMM 6284; (**C**) ***C. litoralis* KMM 3880^T**, KMM 7000, KMM 7005 and KMM 7514; (**D**) ***C. pacifica* KMM 3879^T**, KMM 6731, 7505, 7508, 7515, 6816, and 6818. **Lane numbers: 1**—DNA/RNA non-specific endonuclease precursor; **2**—DNA/RNA non-specific endonuclease; **3**—EEP-like (DNaseI-like) nuclease; **4**—alkaline phosphatase/phosphodiesterase PhoD; **5**—phospholipase A; **6**—ATP-dependent protease Clp; **7**—periplasmic serine peptidase Do/DeqQ; **8**—CmAP-like alkaline phosphatase PhoA; **M**—1 kb DNA ladder marker (Evrogen).

Figure 1 shows the results of gel-electrophoresis of the PCR products for the type strains of the *Cobetia* species: *C. amphilecti* NRIC 0815^T (KMM 1561^T) (A), *C. marina* LMG 2217^T (B), *C. litoralis* NRIC 0814^T (KMM 3880^T) (C), *C. pacifica* NRIC 0813^T (KMM 3879^T) (D), and *C. crustatorum* JCM 15644^T (Figures 1 and 2). The type strain of each species is characterized by an individual distribution of the PCR products that could allow classifying the new *Cobetia* isolates by these patterns (Figures 1 and 2). Because of the PCR-based method for the molecular differentiation, using the gene-specific primers and the genomic DNA of the strains under study, they can be divided into five groups belonging to the five species of the genus *Cobetia*. The type strain *C. amphilecti* NRIC 0815^T (KMM 1561^T) and the strains *C. amphilecti* KMM 296 and KMM 7516 had an identical pattern of distribution for the PCR products of all new eight molecular markers, which indicate their 100% homology accordingly to the results of 16S rRNA analysis (Figure 1A, Table 1 and Table S1).

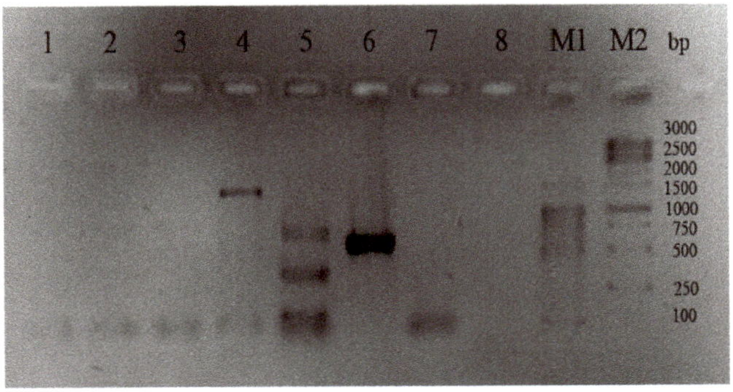

Figure 2. Gel electrophoresis of PCR products of the new molecular markers in the strains *C. crustatorum* JCM 15644^T and *C. crustatorum* KMM 6817. Lane numbers: **1**—DNA/RNA non-specific endonuclease precursor; **2**—DNA/RNA non-specific endonuclease; **3**—EEP-like (DNaseI-like) nuclease; **4**—alkaline phosphatase/phosphodiesterase PhoD. **5**—phospholipase A; **6**—ATP-dependent protease Clp; **7**—periplasmic serine peptidase Do/DeqQ; **8**—CmAP-like alkaline phosphatase PhoA; **M**—1 kb DNA ladder marker (Evrogen).

The next group of the strains, with an identical PCR pattern, includes two *C. marina* strains: LMG 2217^T and KMM 6284 (Figure 1B), which expectedly lost most PCR products due to the more distant relation to *C. amphilecti* by both their 16S rRNA genes and whole genome sequences [8,9].

Remarkably, the strain *Cobetia* sp. KMM 7000, and two strains KMM 7514 (2AS) and KMM 7005 (2AS1), which have higher homology by 16S rRNA sequences with the species *C. amphilecti* (Table 1 and Table S1), showed the complete identity of the PCR product distribution with the type strain *C. litoralis* KMM 3880^T (Figure 1C). However, both type strains of *C. amphilecti* and *C. litoralis* have been found to possess highly homologous CDSs for all the predicted hydrolases (Figure 1A,C), used here as the molecular markers, except for the well-studied highly active PhoA-like alkaline phosphatase CmAP (Figure 1A,C, lanes **8**) [30]. The question arose as to how close these strains can be to each other, taking into account the fact that their physiological parameters and the results of DNA:DNA hybridization allowed them to be attributed to different biological species [7]. A comparative analysis of the whole genome sequence of *Cobetia* sp. 2AS1 (KMM 7005) (GenBank: JADAZN000000000.1), with the use of the SEED Viewer at the Rapid Annotation using Subsytems Technology (RAST) server [31], confirmed that the most CDSs for the enzymes, used as the marker genes (Table S4; column **B**: 2373, 3075, 2144, 1612, 617, 1748, 322), are similar with those of *C. amphilecti* KMM 296 by 99.32–99.64%, except for phospholipase A1 (97.14%) and protease Cpl (100%). However, an alkaline phosphatase PhoA, structurally similar to the alkaline phosphatase CmAP from *C. amphilecti* KMM 296, is absent in *Cobetia* sp. 2AS1 (KMM 7005) (Table S4). A putative orthologue (alkaline phosphatase EC 3.1.3.1), which should carry a similar function, showed only 38.39% identity with CmAP (Table S4, column B: 1612). Generally, 94.84% from 3253 CDSs of *Cobetia* sp. KMM 7005 (GenBank: JADAZN000000000.1) showed an average similarity of 83.5% with CDSs of *C. amphilecti* KMM 296 (GenBank: JQJA00000000.1) that is approximately correspondent to 88% average nucleotide identity (ANI) of their genomes (Table S4).

The high percentage of 16S rRNA genes' (99.93–100%) and whole genomes' (88%) identities may mean that the strains of *C. amphilecti* and *C. litoralis* belong to the same species, but currently they are undergoing significant phenotypic and genotypic divergence because of adaptive evolution [32]. Possibly, the highly active alkaline phosphatase PhoA was acquired by the cosmopolite *Cobetia* strains during their trying colonization of an invertebrate digestive tract due to the putatively significant role of the enzyme in the relationship (symbiotic or pathogenic) between marine habitants, such as *C. amphilecti* KMM 296 and the mussel *C. grayanus* or *C. amphilecti* KMM 1561^T, and the eponymous sponge *Amphilectus digitatus* [7,9,20,24]. Meanwhile, their closely related strains of *C. litoralis*, including the type strain KMM 3880^T, were isolated predominantly from coastal sediments, therefore, they may not need such enzymatically active and specific alkaline phosphatase as CmAP [7,30]. The 16S rRNA heterogeneity of *C. litoralis* KMM 7000 may be an additional evidence of the species divergence due to the adaptation to colonization of marine invertebrates, which are the predominant habitats of the closely related strains of the species *C. amphilecti*, including KMM 296 and KMM 1561^T [7,20,21]. Thus, a squid-vibrio symbiosis is feasible by modulation of the bacterial symbiont lipid A signaling by the host alkaline phosphatases facilitating its colonization of the juvenal squid's light organ [33]. The urgent need for mineralization and repair of the invertebrate's exoskeleton can also be a key factor in symbiosis with a carrier of a highly efficient nonspecific phosphatase like CmAP [9,30,34]. However, the conclusion should be drawn only after sequencing the whole genomes of the type strains of *Cobetia* and elucidation of biological functions of their species- and strain-specific genes and proteins. In addition, such a high adaptability and metabolic versatility in various environmental conditions requires investigating the possible contribution of the bacterium to the toxicity or pathogenicity of shellfish, particularly, for the humans consuming them raw.

A similar situation may be with other closely related species *C. marina* and *C. pacifica*. The largest group of our isolates, assigned to the species *C. marina/C. pacifica* (KMM 7508, 7515, 6816, 6818, 7505 and 6731), have 100% identity by 16S rRNA with both species, but their results from the suggested PCR-based method correspond to the species-specific pattern inherent for the type strain *C. pacifica* KMM 3879^T (Figure 1D, Table 1 and Table S1). The dominant differences between these species were in the lanes 1, 4, and 5, indicating the differences in their PCR-targeted sequences (Figure 1B,D).

Finally, *C. crustatorum* KMM 6817 and the type strain *C. crustatorum* JCM 15644^T significantly differ from other groups of the *Cobetia* isolates in the number and location of the bands of PCR products that correspond to the 16S rRNA analysis results (Figure 2, Table 1 and Tables S1–S3).

Thus, the suggested molecular markers used in the PCR-based method allowed distinguishing the isolates of *C. marina* and *C. amphilecti* from the isolates of *C. pacifica* and *C. litoralis*, respectively. From seven isolates of indistinguishable 16S rRNA sequences, only one from the red algae seeds was identified as *C. marina* (KMM 6284) and the others were of *C. pacifica* (KMM 7508, 7515, 6816, 6818, 7505, and 6731), isolated from algae and coastal seawater (Table 1). Five isolates of *C. amphilecti* and *C. litoralis*, indistinguishable by 16S rRNA analysis, were assigned as two *C. amphilecti* (KMM 296 from the mussel and KMM 7516 from coastal seawater) and three *C. litoralis* (KMM 7005, KMM 7514 from sediments and KMM 7000 from the sponge). The strain *Cobetia* sp. KMM 6817 was easily assigned to the *C. crustatorum* species according to the results of both methods of analysis, which proves the validity of the suggested molecular markers (Table 1).

According to the results of the *Cobetia* species identification at this stage of investigation, there is a tendency for the predominant association of the Pacific Ocean populations of *C. pacifica*, *C. amphilecti* and *C. litoralis* with algae, invertebrates and sediments or coastal water, respectively (Table 1). However, to confirm these observations, a more extensive search for isolates from different habitats should be carried out.

3. Materials and Methods

3.1. Isolation of the Strains Belonging to the Genus Cobetia

Bacterial strains were isolated from the different marine environments, including coastal waters, sediments, seaweeds and animals, using the standard dilution-plating method. Strains KMM 6284 and KMM 6731 were directly isolated from the red alga *Ahnfeltia tobuchiensis* collected near Island Paramushir, Kuril Isles, the Sea of Okhotsk, and the Pacific Ocean. Strains KMM 6816, 6817 and 6818 were isolated from the same alga after its long-term continuous cultivation in the natural seawater for 6 months. Strain KMM 7000 was recovered from the sponge *Esperiopsis digitata*, collected near Island Sakhalin, Piltun

Bay, the Sea of Okhotsk, from depth of 107 m by plating on medium A containing (L^{-1}): 2 g Bacto peptone (Difco), 2 g Bacto yeast extract (Difco). 1 g casein hydrolysate, 0.2 g KH_2PO_4 and 0.1 g ferric ammonium citrate prepared with 50% (v/v) natural sea water and 50% (v/v) distilled water. For strains isolation, 0.1 mL of tissue homogenate was transferred onto plates of marine agar 2216 (MA; Difco) (strains KMM 6284, 6731, 6816, 6817, and 6818) or medium A or medium B. After primary isolation and purification, the strains were cultivated at 28 °C on the same medium and stored at –80 °C in marine broth (Difco) supplemented with 20% (v/v) glycerol. Strains KMM 7516, 7505, 7515 and 7508, and 7514 and 7005 were isolated from coastal seawater and sediment samples, respectively, collected from Vostok Bay, the Sea of Japan, by plating on medium B, containing: 1 g $(NH_4)_2SO_4$, 0.2 g $MgSO_4$ $7H_2O$, 1 g K_2HPO_4, 1 g KH_2PO_4, 0.02 g $CaCl_2$, 0.05 g $FeCl_3$, and 20 g NaCl in 1000 mL distilled water, with the addition of 1 mL per liter of sterile crude oil to the prepared medium. Strain *Cobetia* sp. KMM 296, previously identified as *Deleya marina* by Ivanova et al. [35], was received from the Collection of Marine Microorganisms (KMM, Russia). The type strains of the recognized species *Cobetia amphilecti* KMM 1561^T (NRIC 0815^T), *Cobetia litoralis* KMM 3880^T (NRIC 0814^T), *Cobetia pacifica* KMM 3379^T (NRIC 0813^T), and *Cobetia marina* LMG 2217^T were used as reference strains and kindly provided to us by the NODAI Culture Collection Center, (University of Agriculture, Tokyo, Japan) and from the BCCM/LMG Bacteria Collection (Universiteit Gent, Gent, Belgium), respectively [7]. The reference type strain *C. crustatorum* JCM 15644^T was kindly provided by the Japan Collection of Microorganisms (RIKEN BioResource Centre (BRC), Tsukuba, Japan).

3.2. Isolation and Analysis of DNA from the Strains

All strains of marine bacteria were cultivated in Petri dishes on sterile agar LB medium of the following composition (g/L): bacto-tryptone—10; yeast extract—5; NaCl—5; agar-agar—15; distilled water—0.98 l; the pH of the medium is 7.7. Five passages were made for each type of bacterium. Then, DNA was isolated from the grown colonies using the PureLink Genomic DNA Mini Kit protocol (Invitrogen). The isolated bacterial DNA was used to carry out a polymerase chain reaction (PCR), with universal primers to 16S rRNA sequences (BF/20: 5′-AGAGTTTGATCMTGGCTCA -3′; BR2/22: 5′- TACGGTTACCTTGT-TACGACTT -3′) or the primers corresponding to the coding DNA sequences (CDSs) for housekeeping enzymes used in this work as molecular markers for the *Cobetia* species identification (Table 2).

Table 2. Oligonucleotides for molecular differentiation of *Cobetia* species.

Name	Sequence	Molecular Marker *	Reference **
1CmNucF	5′–TATACCATGGACGATATTCGCTCGGCCGGCCGCAA-3′	DNA/RNA non-specific endonuclease precursor (1) and without leader peptide (2)	WP_043334786
CmNucR	5′–TATAGAGCTCTCAGTAACGTGATGGCGTACGACTG-3′		
2CmNucF	5′-TATACCATGGTATGGCAGGAGCGCGACTACCAGCA-3′		
CmNucR	5′–TATAGAGCTCTCAGTAACGTGATGGCGTACGACTG -3′		
CmEEPf	5′–TATACCATGGGACTCGACGAGACGGCACCTCCCCT -3′	exonuclease/ endonuclease/phosphatase (EEP)	WP_084589364
CmEEPr	5′–TATAGAGCTCTTATGCTAGCCCGATCGCCTTGCGGCA-3′		
CmPhoDf	5′–TATACCATGGAAGGACGGCGCCCGCGCATGCCCTC-3′	alkaline phosphatase/phosphodiesterase PhoD	WP_043333989
CmPhoDr	5′–TATAGAGCTCTTAGACACTGGCGGCGGCGGGGGTC-3′		
CmPLA_f	5′–TATACCATGGTACTCGATGAAAGCCTGGCCCAGCA-3′	phospholipase A	WP_084589432
CmPLA_r	5′–TATAGAGCTCTTAGGTCTCTGGCGAGCCGGCGAAG-3′		
Tryp_F	5′–TATACCATGGTACGTGAATTGCCCGACTTCACCCA-3′	periplasmic serine peptidase Do/DeqQ	KGA03014
Tryp_R	5′–TATACTCGAGTCACTTGTCGCTGTCGGCACGCATG-3′		
CmClp_F	5′-TATCCATGGTAAACGACTTCGACATCAAGAATGCT-3′	ATP-dependent caseinolytic protease Clp	KGA03297
CmClp_R	5′-TATAGAGCTCTCACTCCACGTCGGGACGGCGTTCC-3′		
X-PhoN_F	5′-TTAACCATGGCAGAGATCAAGAATGTCATTCTGAT-3′	alkaline phosphatase PhoA	DQ435608
CmAP_R	5′-TTAAGAATTCCTTCGCTACCACTGTCTTCAGATACTGTCC-3′		

* The molecular markers are the enzymes, predicted according to the structural classification of the National Center for Biotechnology Information (NCBI) Conserved Domain Database [23]; ** the reference genes' IDs are from the whole genome sequence annotation of the *C. amphilecti* KMM 296 (GenBank ID: JQJA00000000.1).

Reaction conditions for PCR of 16S rRNA sequence in DNA amplifier C1000TM Thermal Cycler (Bio-Rad Laboratories, Inc., California, USA) or Mastercycler gradient (Eppendorf, Hamburg, Germany): 10× buffer for polymerase, 50× mixture of Encyclo polymerases ("Encyclo PCR kit", Evrogen, Moscow), 50× mixture of dNTP (10 mM each), mixture of forward and reverse primers (5 μM each), and 20 ng DNA of a bacterial clone. The amplification process consists of the following stages: 30 PCR cycles × (15 s—95 °C; 30 s—55 °C; 1 min 30 s—72 °C), then incubation at 72 °C for 5 min. After amplification, PCR products were used for sequencing. The PCR products from each clone of each strain (four replicates) were sequenced and verified with the use of an ABI Prism 3130xl sequencer and Chromas program (version 2.5.1), respectively. Homology searches were performed against EzBioCloud 16S database using the Blast program to find sequences that provide significant alignment [22].

Reaction conditions for marker genes (Table 2) in DNA amplifier C1000TM Thermal Cycler (Bio-Rad Laboratories, Inc., California, USA) or Mastercycler gradient (Eppendorf, Hamburg, Germany): 10× Encyclo buffer, 50× Encyclo polymerase mixture (Encyclo PCR kit, Evrogen, Moscow), 50× dNTP mixture (10 mM each), a mixture of forward and reverse primers (5 μM each), and 20 ng DNA of a bacterial clone. The amplification process consisted of the following stages: 38 PCR cycles x (2 min—95 °C; 15 s—95 °C; 1 min 40 s—72 °C). After amplification, the PCR products from each clone of each strain (four replicates) were separated by gel electrophoresis in a 1% agarose gel stained with ethidium bromide. The PCR product visualization and documentation were performed with Herolab imaging system (Herolab GmbH Lab., Wiesloch, Germany). The PCR products for the species-specific coding sequencing regions of *C. amphilecti* and *C. litoralis* were confirmed by sequencing as described above.

Comparative analysis of the marker genes and whole genome sequences of *Cobetia* sp. 2AS1 (GenBank: JADAZN000000000.1) and *C. amphilecti* KMM 296 (GenBank: JQJA-00000000.1) was carried out by the SEED Viewer at the RAST server [31].

4. Conclusions

The gene-specific oligonucleotides corresponding to the coding DNA sequences for the enzymes responsible for the vital bacterial cell functions, such as EEP-like and DNA/RNA nonspecific nucleases, alkaline phosphatases PhoA and PhoD, proteases Cpl and Do/DeqQ family, and phospholipase A1, may be used for rapid molecular differentiation of the closely related species of the marine bacteria *Cobetia* in addition to the traditional 16S rRNA assay. Furthermore, the highly active alkaline phosphatase CmAP of the structural family PhoA is a highly specific marker for the species *C. amphilecti*, probably indicating the adaptability to the host–microbe relationships. However, the genus needs further study, including verification of the suggested molecular markers with the use of a higher number of isolates to develop the MLST-based method of identification. Primarily, comparative genomics of the type strains of *Cobetia* should be carried out to a final decision about the valid interspecies phylogeny of the genus, particularly towards the closely related species (*C. amphilecti* and *C. litoralis; C. marina* and *C. pacifica*), as well as about their genetic drivers and causes of such divergent evolution within the genus.

Supplementary Materials: The following are available online, Table S1: Similarity calculation of 16S rRNA gene sequences of *Cobetia* isolates by EzBioCloud 16S database, Table S2: Similarity calculation of 16S RNA gene sequence of the type strain *Cobetia marina* LMG 2217T, Table S3: Similarity calculation of 16S RNA gene sequence of the strain *Cobetia amphilecti* KMM 296, Table S4: Comparative analysis of coding DNA sequences of *Cobetia* sp. 2AS1 (KMM 7005) and *C. amphilecti* KMM 296.

Author Contributions: Conceptualization, L.B.; methodology, L.B. and O.N.; software, L.B.; validation, L.B. and O.N.; investigation, Y.N., A.S. and L.S.; resources, L.T., O.N., L.S.; data curation, L.B.; writing—original draft preparation, Y.N., A.S. and O.N.; writing—review and editing, L.B.; visualization, Y.N., A.S.; supervision, O.S.; project administration, O.S.; funding acquisition, L.T. All authors have read and agreed to the published version of the manuscript.

Funding: This research received no external funding.

Institutional Review Board Statement: Not applicable.

Informed Consent Statement: Not applicable.

Data Availability Statement: The data presented in this study are available on request from the corresponding author.

Conflicts of Interest: The authors declare no conflict of interest.

Sample Availability: Samples of the isolates and DNA are available from the authors.

References

1. Cobet, A.B.; Wirsen, C.J.; Jones, G.E. The effect of nickel on a marine bacterium, *Arthrobacter marinus* sp.nov. *J. Gen. Microbiol.* **1970**, *62*, 159–169. [CrossRef] [PubMed]
2. Baumann, L.; Baumann, P.; Mandel, M.; Allen, R.D. Taxonomy of aerobic marine eubacteria. *J. Bacteriol.* **1972**, *110*, 402–429. [CrossRef] [PubMed]
3. Baumann, L.; Bowditch, R.D.; Baumann, P. Description of *Deleya* gen. nov. created to accommodate the marine species *Alcaligenes aestus*, *A. pacificus*, *A. cupidus*, *A. venustus*, and *Pseudomonas marina*. *Int. J. Syst. Bacteriol.* **1983**, *33*, 793–802. [CrossRef]
4. Dobson, S.J.; Franzmann, P.D. Unification of the genera *Deleya* (Baumann et al. 1983), *Halomonas* (Vreeland et al. 1980), and *Halovibrio* (Fendrich 1988) and the species *Paracoccus halodenitrificans* (Robinson and Gibbons 1952) into a single genus, *Halomonas*, and placement of the genus *Zymobacter* in the family *Halomonadaceae*. *Int. J. Syst. Bacteriol.* **1996**, *46*, 550–558. [CrossRef]
5. Arahal, D.R.; Castillo, A.M.; Ludwig, W.; Schleifer, K.H.; Ventosa, A. Proposal of *Cobetia marina* gen. nov., comb. nov., within the family *Halomonadaceae*, to include the species *Halomonas marina*. *Syst. Appl. Microbiol.* **2002**, *25*, 207–211. [CrossRef]
6. Kim, M.S.; Roh, S.W.; Bae, J.W. *Cobetia crustatorum* sp. nov., a novel slightly halophilic bacterium isolated from traditional fermented seafood in Korea. *Int. J. Syst. Evol. Microbiol.* **2010**, *60*, 620–626. [CrossRef]
7. Romanenko, L.A.; Tanaka, N.; Svetashev, V.I.; Falsen, E. Description of *Cobetia amphilecti* sp. nov., *Cobetia litoralis* sp. nov. and *Cobetia pacifica* sp. nov., classification of *Halomonas halodurans* as a later heterotypic synonym of *Cobetia marina* and emended descriptions of the genus *Cobetia* and *Cobetia marina*. *Int. J. Syst. Evol. Microbiol.* **2013**, *63*, 288–297. [CrossRef] [PubMed]
8. Tang, X.; Xu, K.; Han, X.; Mo, Z.; Mao, Y. Complete genome of *Cobetia marina* JCM 21022T and phylogenomic analysis of the family *Halomonadaceae*. *J. Oceanol. Limnol.* **2018**, *36*, 528–536. [CrossRef]
9. Balabanova, L.A.; Golotin, V.A.; Kovalchuk, S.N.; Babii, A.V.; Shevchenko, L.S.; Son, O.M.; Kosovsky, G.Y.; Rasskazov, V.A. The Genome of the marine bacterium *Cobetia marina* KMM 296 isolated from the mussel *Crenomytilus grayanus* (Dunker, 1853). *Russ. J. Mar. Biol.* **2016**, *42*, 106–109. [CrossRef]
10. Cheng, W.; Yan, X.; Xiao, J.; Chen, Y.; Chen, M.; Jin, J.; Bai, Y.; Wang, Q.; Liao, Z.; Chen, Q. Isolation, identification, and whole genome sequence analysis of the alginate-degrading bacterium *Cobetia* sp. cqz5-12. *Sci. Rep.* **2020**, *2*, 10920. [CrossRef] [PubMed]
11. Guo, P.; Cao, B.; Qiu, X.; Lin, J. Draft genome sequence of the crude oil-degrading and biosurfactant-producing strain *Cobetia* sp. QF-1. *Genome Announc.* **2018**, *6*, e01456-17. [CrossRef] [PubMed]
12. Balabanova, L.; Podvolotskaya, A.; Slepchenko, L.; Eliseikina, M.; Noskova, Y.; Nedashkovskaya, O.; Son, O.; Tekutyeva, L.; Rasskazov, V. Nucleolytic enzymes from the marine bacterium *Cobetia amphilecti* KMM 296 with antibiofilm activity and biopreservative effect on meat products. *Food Control.* **2017**, *78*, 270–278. [CrossRef]
13. Moriya, H.; Takita, Y.; Matsumoto, A.; Yamahata, Y.; Nishimukai, M.; Miyazaki, M.; Shimoi, H.; Kawai, S.J.; Yamada, M. *Cobetia* sp. bacteria, which are capable of utilizing alginate or waste *Laminaria* sp. for poly(3-hydroxybutyrate) synthesis, isolated from a marine environment. *Front. Bioeng. Biotechnol.* **2020**, *25*, 974. [CrossRef] [PubMed]
14. Mei, R.; Zhou, M.; Xu, L.; Zhang, Y.; Su, X. Characterization of a pH-tolerant strain *Cobetia* sp. SASS1 and its phenol degradation performance under salinity condition. *Front. Microbiol.* **2019**, *4*, 2034. [CrossRef]
15. Coenye, T.; Vandamme, P. Intragenomic heterogeneity between multiple 16S ribosomal RNA operons in sequenced bacterial genomes. *FEMS Microbiol. Lett.* **2003**, *228*, 45–49. [CrossRef]
16. Weisburg, W.G.; Barns, S.M.; Pelletier, D.A.; Lane, D.J. 16S ribosomal DNA amplification for phylogenetic study. *J. Bacteriol.* **1991**, *173*, 697–703. [CrossRef]
17. Franco-Duarte, R.; Černáková, L.; Kadam, S.; Kaushik, K.S.; Salehi, B.; Bevilacqua, A.; Corbo, M.R.; Antolak, H.; Dybka-Stępień, K.; Leszczewicz, M.; et al. Advances in chemical and biological methods to identify microorganisms-from past to present. *Microorganisms* **2019**, *7*, 130. [CrossRef]
18. Liu, W.; Li, L.; Khan, M.A.; Zhu, F. Popular molecular markers in bacteria. *Mol. Gen. Mikrobiol. Virusol.* **2012**, 3, 14–17. [CrossRef]
19. Ashok Kumar, J.; Vinaya Kumar, K.; Avunje, S.; Akhil, V.; Ashok, S.; Kumar, S.; Sivamani, B.; Grover, M.; Rai, A.; Alavandi, S.V.; et al. Phylogenetic relationship among brackishwater *Vibrio* Species. *Evol. Bioinform. Online.* **2020**, *16*, 1176934320903288. [CrossRef]
20. Ivanova, E.; Christen, R.; Sawabe, T.; Alexeeva, Y.; Lysenko, A.; Chelomin, V.; Mikhailov, V. Presence of ecophysiologically diverse populations within *Cobetia marina* strains isolated from marine invertebrate, algae and the environments. *Microbes Environ.* **2005**, *20*, 200–207. [CrossRef]
21. Balabanova, L.; Nedashkovskaya, O.; Podvolotskaya, A.; Slepchenko, L.; Golotin, V.; Belik, A.; Shevchenko, L.; Son, O.; Rasskazov, V. Data supporting functional diversity of the marine bacterium *Cobetia amphilecti* KMM 296. *Data Brief* **2016**, *8*, 726–732. [CrossRef] [PubMed]
22. Yoon, S.H.; Ha, S.M.; Kwon, S.; Lim, J.; Kim, Y.; Seo, H.; Chun, J. Introducing EzBioCloud: A taxonomically united database of 16S rRNA gene sequences and whole-genome assemblies. *Int. J. Syst. Evol. Microbiol.* **2017**, *67*, 1613–1617. [CrossRef] [PubMed]

23. Lu, S.; Wang, J.; Chitsaz, F.; Derbyshire, M.K.; Geer, R.C.; Gonzales, N.R.; Gwadz, M.; Hurwitz, D.I.; Marchler, G.H.; Song, J.S.; et al. CDD/SPARCLE: The conserved domain database in 2020. *Nucleic Acids Res.* **2020**, *8*, D265–D268. [CrossRef] [PubMed]
24. Plisova, E.Y.; Balabanova, L.A.; Ivanova, E.P.; Kozhemyako, V.B.; Mikhailov, V.V.; Agafonova, E.V.; Rasskazov, V.A. A highly active alkaline phosphatase from the marine bacterium *Cobetia*. *Mar. Biotechnol.* **2005**, *7*, 173–178. [CrossRef]
25. Noskova, Y.; Likhatskaya, G.; Terentieva, N.; Son, O.; Tekutyeva, L.; Balabanova, L. A novel alkaline phosphatase/phosphodiesterase, CamPhoD, from marine bacterium *Cobetia amphilecti* KMM 296. *Mar. Drugs.* **2019**, *17*, 657. [CrossRef]
26. Jiang, N.; Tu, Z.; Zhang, Y.; Li, J.; Feng, Y.; Yang, N.; Sang, X.; Chen, Q. Identification and characterization of DNA endonucleases in Plasmodium falciparum 3D7 clone. *Malar. J.* **2018**, *17*, 232. [CrossRef]
27. Zhang, Y.; Chen, T.; Zheng, W.; Li, Z.H.; Ying, R.F.; Tang, Z.X.; Shi, L.E. Active sites and thermostability of a non-specific nuclease from *Yersinia enterocolitica* subsp. palearctica by site-directed mutagenesis. *Biotechnol. Biotechnol. Equip.* **2018**, *32*, 1306–1316. [CrossRef]
28. Liu, Q.; Wang, X.; Qin, J.; Cheng, S.; Yeo, W.-S.; He, L.; Ma, X.; Liu, X.; Li, M.; Bae, T. 2017 The ATP-dependent protease ClpP inhibits biofilm formation by regulating Agr and Cell wall hydrolase Sle1 in *Staphylococcus aureus*. *Front. Cell. Infect. Microbiol.* **2017**, *7*, 181. [CrossRef]
29. Sutto-Ortiz, P.; Camacho-Ruiz, M.; Kirchmayr, M.R.; Camacho-Ruiz, R.M.; Mateos-Díaz, J.C.; Noiriel, A.; Carrière, F.; Abousalham, A.; Rodríguez, J.A. Screening of phospholipase A activity and its production by new actinomycete strains cultivated by solid-state fermentation. *PeerJ* **2017**, *5*, e3524. [CrossRef]
30. Golotin, V.; Balabanova, L.; Likhatskaya, G.; Rasskazov, V. Recombinant production and characterization of a highly active alkaline phosphatase from marine bacterium *Cobetia marina*. *Mar. Biotechnol. (NY)* **2015**, *17*, 130–143. [CrossRef]
31. Overbeek, R.; Olson, R.; Pusch, G.D.; Olsen, G.J.; Davis, J.J.; Disz, T.; Edwards, R.A.; Gerdes, S.; Parrello, B.; Shukla, M.; et al. The SEED and the Rapid Annotation of microbial genomes using Subsystems Technology (RAST). *Nucleic Acids Res.* **2014**, *42*, D206–D214. [CrossRef] [PubMed]
32. Vos, M. A species concept for bacteria based on adaptive divergence. *Trends Microbiol.* **2011**, *19*, 1–7. [CrossRef] [PubMed]
33. Rader, B.A.; Kremer, N.; Apicella, M.A.; Goldman, W.E.; McFall-Ngai, M.J. Modulation of symbiont lipid A signaling by host alkaline phosphatases in the squid-vibrio symbiosis. *mBio.* **2012**, *1*, e00093-e12. [CrossRef] [PubMed]
34. Xie, M.Ø.; Olderøy, M.; Zhang, Z.; Andreassen, J.-P.; Strandd, B.L.; Sikorski, P. Biocomposites prepared by alkaline phosphatase mediated mineralization of alginate microbeads. *RSC Adv.* **2012**, *2*, 1457–1465. [CrossRef]
35. Ivanova, E.P.; Mikhailo, V.V.; Plisova, E.J.; Balabanova, L.A.; Svetashev, V.V.; Vysockyi, M.V.; Stepanenko, V.I.; Rasskazov, V.A. Characterization of the marine bacterium *Deleya marina* producing highly active alkaline phosphatase and associated with the mussel *Crenomytilus grayanus*. *Russ. J. Mar. Biol.* **1994**, *20*, 340–545.

molecules

Article

Differential Expression of *Yersinia pseudotuberculosis* General Porin Genes during Short- and Long-Term Antibiotic Stresses

Evgeniya Bystritskaya [1], Nadezhda Chernysheva [1], Anna Stenkova [2], Konstantin Guzev [1], Alexander Rakin [3] and Marina Isaeva [1,2,*]

1 G.B. Elyakov Pacific Institute of Bioorganic Chemistry, Far Eastern Branch, Russian Academy of Sciences, 159, Pr. 100 Let Vladivostoku, 690022 Vladivostok, Russia; belyjane@gmail.com (E.B.); chernysheva.nadezhda@gmail.com (N.C.); k.guzev@gmail.com (K.G.)
2 School of Biomedicine, Far Eastern Federal University, 8 Sukhanova St., 690090 Vladivostok, Russia; stenkova@gmail.com
3 Friedrich-Loeffler-Institut, Federal Research Institute for Animal Health, Institute for Bacterial Infections and Zoonoses, Naumburger Str. 96a, D-07743 Jena, Germany; rakin@gmx.de
* Correspondence: issaeva@gmail.com; Tel.: +7-423-231-11-68

Abstract: Here, we investigated general porin regulation in *Yersinia pseudotuberculosis* 488, the causative agent of Far Eastern scarlet-like fever, in response to sublethal concentrations of antibiotics. We chose four antibiotics of different classes and measured gene expression using qRT-PCR and GFP reporter systems. Our data showed temporal regulation of the general porin genes *ompF* and *ompC* caused by antibiotic stress. The porin transcription initially decreased, providing early defensive response of the bacterium, while it returned to that of the untreated cells on prolonged antibiotic exposure. Unlike the major porin genes, the transcription of the alternative porin genes *ompX* and *lamB* was increased. Moreover, a short-term *ompR*- and *marA*-mediated porin regulation was observed. The main finding was a phenotypic heterogeneity of *Y. pseudotuberculosis* population manifested in variable porin gene expression under carbenicillin exposure. This may offer adaptive fitness advantages for a particular bacterial subpopulation.

Keywords: porin gene expression; *Yersinia pseudotuberculosis*; antibiotic stress; phenotypic heterogeneity

Citation: Bystritskaya, E.; Chernysheva, N.; Stenkova, A.; Guzev, K.; Rakin, A.; Isaeva, M. Differential Expression of *Yersinia pseudotuberculosis* General Porin Genes during Short- and Long-Term Antibiotic Stresses. *Molecules* **2021**, *26*, 3956. https://doi.org/10.3390/molecules26133956

Academic Editor: Mark Brönstrup

Received: 2 June 2021
Accepted: 23 June 2021
Published: 28 June 2021

1. Introduction

The global rise of antibiotic resistance in pathogens represents a serious threat to medicine, as it is the predominant cause of treatment failure and increased mortality [1]. Recent evidence suggests that sublethal concentrations of antimicrobials frequently found in the human body and nature play an important role in the development of antibiotic resistance [2]. Initially, susceptible bacterial populations survive and propagate under such conditions due to adaptive gene expression responses that have been observed in a variety of bacterial species [1,3–5]. In Gram-negative bacteria, altering the gene expression of outer membrane proteins or porins, resulting in decreased membrane permeability, is considered one of the first defense mechanisms of adaptation to antibiotic stress [1].

Yersinia pseudotuberculosis, a Gram-negative bacterium belonging to the *Enterobacteriaceae* family, is a human pathogen causing pseudotuberculosis infection. In Russia and Japan, it causes outbreaks of the disease known as Far Eastern scarlet-like fever (FESLF), with serious systemic inflammatory symptoms [6].

Y. pseudotuberculosis produces two major nonspecific porins, OmpF and OmpC, which we study intensively [7–11]. These porins consist of 16-stranded β-barrel trimers, each of which forms a central channel [12]. The channels control the permeability of the cell envelope for low molecular compounds, including β-lactam, tetracycline, chloramphenicol, and fluoroquinolone antibiotics [13–16]. Thus, the porins provide cell defense against certain antibiotics and subsequently mediate antimicrobial resistance by downregulating their gene expression or inducing beneficial mutations [17]. Moreover, the reversible phenotypical response offers advantages over conventional irreversible mutations [18]. Furthermore, transcriptional porin regulation is considered a rapid adaptive mechanism to environmental and antibiotic stresses occurring within the first 60 min [19]. The porin-mediated stress response has been widely studied in various enterobacteria, including *Escherichia coli*, *Klebsiella pneumonia*, *Salmonella enterica*, *Serratia marcescens*, and others [20–22]. Still, little is known about the effect of antibiotic stress on the general porin regulation in *Y. pseudotuberculosis*. Most research focuses on the effect of lethal antibiotic concentrations on bacterial physiology and resistance, while subinhibitory antimicrobial doses have been mostly disregarded [23]. *Y. pseudotuberculosis* genetic response to low concentrations of antibiotics is of particular interest because successful treatment of pseudotuberculosis infection largely depends on antibiotic therapy to maintain lethal drug concentrations in different human tissues.

Therefore, we aim to investigate the role of general porin regulation in *Y. pseudotuberculosis* adaptation to sublethal concentrations of four antibiotics utilizing porin channels for cell entrance.

2. Results and Discussion

2.1. Antibiotics and *Y. pseudotuberculosis* 488 Susceptibility

Y. pseudotuberculosis 488 (= strain 117), serotype O:1b, was isolated from a patient with FESLF in the Russian Far East region. To elucidate *Y. pseudotuberculosis* porin regulation under antibiotic stress, we chose carbenicillin, tetracycline, kanamycin, and chloramphenicol. These antibiotics have been shown to use porin channels to enter into a bacterial cell [13,14]. Several studies have demonstrated that porin regulation, resulting in decreased penetration of antimicrobials into the cell, contributes to bacterial adaptive resistance [24]. We determined the MIC ranges of the selected antibiotics for *Y. pseudotuberculosis* 488 at 27 °C (optimal for bacterial cultivation) and 37 °C (human body temperature) since it is known that porin expression is tightly regulated by environmental factors [25,26], predominantly by temperature [27,28]. Such regulation is shown for *Y. pseudotuberculosis* strains in the interactive RNA atlas www.pathogenex.org (*ompF* YPK_2649; *ompC* YPK_2839).

Y. pseudotuberculosis 488 showed high susceptibility to the tested antibiotics (Table S1). However, cells cultivated at 27 °C exhibited a 2-fold increased resistance to kanamycin and tetracycline compared with 37 °C, whereas the MICs of carbenicillin and chloramphenicol were not affected by incubation temperature. To induce antibiotic stress, we used subinhibitory concentrations (sub-MICs) of antibiotics corresponding to $\frac{1}{2}$ MICs. Such sublethal antibiotic concentrations reflect the conditions that bacteria may encounter in the natural environments and the human body [29]. Moreover, exposure to sublethal concentration has been reported to play a protective role in the bacterial cell against a wide range of antimicrobials and alter the expression of various bacterial genes that can lead to nonspecific resistance [22].

2.2. An Early Transcriptional Response of General Porin Genes to Antibiotic Stress

Rapid modulation of porin gene expression is the first line of bacterial defense against harmful compounds, including antibiotics, which increases the chances of survival and adaptation of microorganisms to stressful conditions [19].

To study the early porin response in *Y. pseudotuberculosis* 488, we treated bacterial cells by sub-MICs of antibiotics for an hour, followed by measuring the level of *ompF* and *ompC* transcripts using qRT-PCR. The results obtained (Figure 1) indicate that kanamycin and tetracycline cause a transcription decrease in both general porins, regardless of temperature. *ompF* and *ompC* expression were downregulated 2.3- to 4-fold and 2.2- to 4.4-fold, respectively, in the presence of these antibiotics. It is not surprising because the reduced porin expression is considered to be a common strategy for protection against antibiotics. However, tetracycline and kanamycin downregulated only OmpF synthesis and upregulated OmpC production in *Escherichia coli* [30–33]. Nevertheless, Agafitei et al. have shown that reduced expression of the OmpC porin contributes to kanamycin resistance [34]. Viveiros et al. also found a coupled downregulation of the *E. coli* OmpF with the OmpC during exposure to increasing concentrations of tetracycline [24].

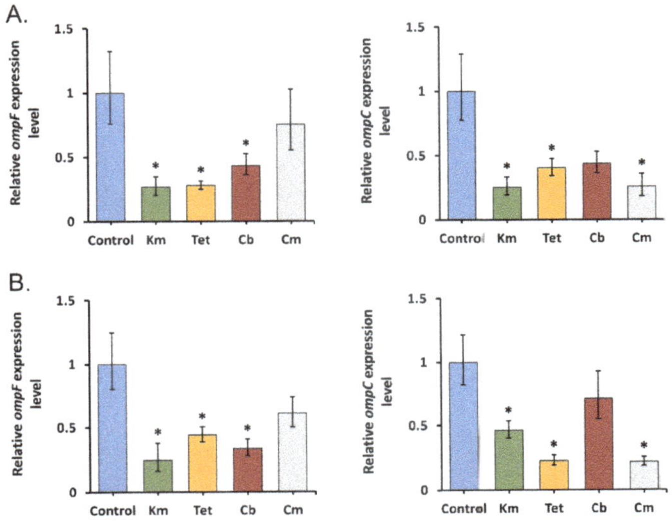

Figure 1. Relative expression levels of *Y. pseudotuberculosis ompF* and *ompC* after short-term antibiotic exposure detected by qRT-PCR. (**A**) 27 °C incubation temperature; (**B**) 37 °C incubation temperature. All results are expressed as mean ± standard deviation from three independent experiments. An asterisk indicates *p*-value < 0.05 vs. respective untreated control. Significance was calculated using one-way ANOVA. Km—kanamycin, Tet—tetracycline, Cb—carbenicillin, and Cm—chloramphenicol.

Interestingly, two other antibiotics caused different transcriptional responses of the porins in *Y. pseudotuberculosis* 488. After one-hour exposure of cells to carbenicillin, the expression level of *ompF* showed a 2.3- to 3-fold decrease, while the expression of *ompC* was not changed significantly (Figure 1). This result supports the findings that OmpF is the preferred route of entry for β-lactams [35]. We suppose that a decrease of *ompF* mRNA level observed at both temperatures could be a protective response of *Y. pseudotuberculosis* cells to the carbenicillin stress. This opinion agrees with the previous finding that lack or inhibition of OmpF contributes to increased resistance to β-lactams [36,37]. It was also unexpected that under short-term chloramphenicol stress, the *ompF* expression was not significantly altered, whereas the *ompC* expression was downregulated 3.9 to 4.6 times (Figure 1). Chloramphenicol has been reported, unlike β-lactams, to utilize both porins equally well [16]. However, there is no observation to indicate any effect of a reduced level of OmpC alone on the penetration of this antibiotic [38].

Considering the qRT-PCR results, we assumed that the general porins OmpF and OmpC of *Y. pseudotuberculosis* are involved in the early response to various antibiotic stresses caused by sublethal concentrations and provide the cells with the first defense mechanism by reducing the transcription level of one or both of their genes depending on the antibiotic type.

2.3. Transcriptional Response of General Porin Genes to Prolonged Antibiotic Stress

Several studies consider the modulation of porin gene expression as a rapid but temporary response of the cell to stress, which develops within the first 60 min [19,23]. However, in the case of antibiotic treatment, this mechanism can be replaced by more specific ones over time [1]. The data on the porin gene expression in time series are important for a comprehensive understanding of *Y. pseudotuberculosis* defense against different antimicrobials.

To evaluate the effect of prolonged antibiotic exposure on the porin gene expression, *Y. pseudotuberculosis* 488 was cultivated in the presence of sub-MICs of these antibiotics for 16 h. In contrast to the early porin response, here we did not see significant changes in the expression of *ompF* and *ompC* genes in most samples treated with antibiotics (Figure 2). The data obtained indicate that transcription of general porins in *Y. pseudotuberculosis* cells transiently decreased within the first hour and subsequently returned to levels comparable to those of the untreated cells.

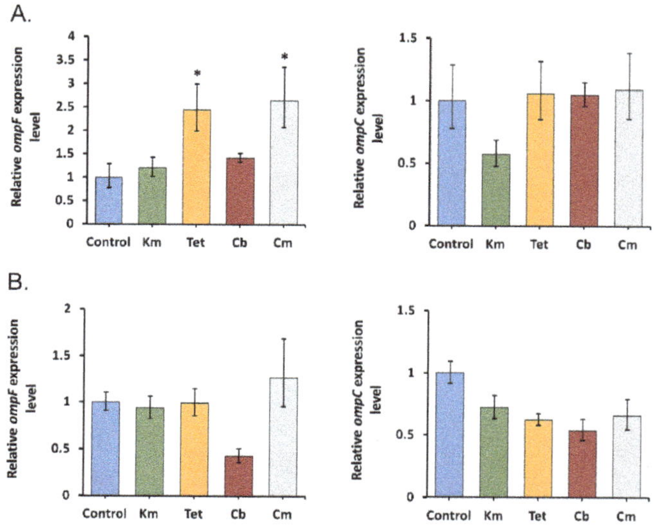

Figure 2. Relative expression levels of *Y. pseudotuberculosis ompF* and *ompC* after prolonged antibiotic exposure detected by qRT-PCR. (**A**) 27 °C incubation temperature; (**B**) 37 °C incubation temperature. All results are expressed as mean ± standard deviation from three independent experiments. An asterisk indicates *p*-value < 0.05 vs. respective untreated control. Significance was calculated using one-way ANOVA. Km—kanamycin, Tet—tetracycline, Cb—carbenicillin, and Cm—chloramphenicol.

Previously, Viveiros et al. observed that *ompF* and *ompC* transcript levels in *E. coli* remained comparable to the untreated controls after prolonged exposure to tetracycline [24]. However, post-translational *marA*-mediated porin regulation to antibiotic stress has been shown to inhibit their synthesis.

Surprisingly, we found a 2.4- and 2.6-fold increase in the expression of *ompF* under prolonged tetracycline and chloramphenicol stresses at 27 °C, which no one had reported earlier. At the same time, incubating cells with these two antibiotics at 37 °C did not affect porin transcription. One of the possible reasons for the *ompF* upregulation in the presence of tetracycline and chloramphenicol could be the decreased expression of alternative porins.

We hypothesized that observed changes in the porin regulation under antibiotic stress over time could be caused by several factors. For example, stationary-phase bacteria must respond and adapt to a variety of environmental stresses (nutrient exhaustion, pH changes, oxygen limitation, and others). In such a case, the regulation of porins provides the balance of outer membrane permeability between self-defense and competitiveness [39,40]. The significantly reduced expression of general porins, protecting cells from antibiotics within the first hour, might entail high metabolic costs at the stationary phase, since vital nutrients are simultaneously excluded.

It should be noted that in addition to the transcriptional regulation of porin genes, post-transcriptional regulation might play a major role in the physiological adaptation to prolonged antibiotic exposure, as was mentioned previously [24].

Moreover, the acquisition of mutations in porin genes, leading to loss of proteins, size, and conductance modification of their channels or decreasing their expression level, is known as another common mechanism for reducing the sensitivity of bacteria to antibiotics. It was observed that porin-related mutations substantially influence resistance to β-lactams, fluoroquinolones, tetracycline, and chloramphenicol [1]. However, we performed *ompF* and *ompC* sequence analysis of *Y. pseudotuberculosis* samples after prolonged antibiotic treatment and did not reveal any changes within the porin-encoding parts and their regulatory regions.

2.4. *ompR* and *marA* Effect on Porin Response under Antibiotic Stress

Next, we investigated the contribution of transcriptional factors in the regulation of porin genes.

The best-described input into controlling porin levels involves OmpR, which regulates *ompC* and *ompF* directly through OmpR–promoter DNA association [41]. The OmpR/EnvZ two-component system may play a major role in antibiotic resistance by modulating porin expression [42]. It was shown that OmpR deficiency in *Y. pseudotuberculosis* and *Y. enterocolitica* reduces bacterial sensitivity to the β-lactams, tetracycline, and polymyxin B, caused by the lack of OmpF and OmpC porins [43,44].

qRT-PCR analysis revealed differential expression of *ompR* in *Y. pseudotuberculosis* depending on the duration of antibiotic stress (Figure 3). Short exposure to antibiotics excluding carbenicillin and tetracycline at 27 °C slightly reduced (1.7- to 2.3-fold) its transcription associated with the downregulation of porin genes. On the contrary, prolonged antibiotic stress increased the level of *ompR* transcripts up to 5.3-fold; however, this did not affect the expression of *ompF* and *ompC*. A similar observation was described in the work of Viveiros et al., where the response of the *ompC* and *ompF* genes, despite the increased expression of *ompR*, remained comparable to their levels in *E. coli* cells not exposed to tetracycline [24]. Moreover, it was shown that OmpR positively controls the expression of the AcrAB-TolC efflux pump involved in the adaptive response of the *Y. enterocolitica* strain to different chemical stressors [45]. It is tempting to speculate that a high level of *ompR* transcription in *Y. pseudotuberculosis* may be required to induce this efflux pump system as another general defense mechanism against antibiotics.

Multiple studies have shown that general porin expression is post-transcriptionally regulated by global regulator MarA, involved in multidrug resistance response during antibiotic stress [46,47]. *marA* locus can directly or indirectly increase the level of *micF*, a non-coding RNA stress response gene, causing a post-transcriptional downregulation of *ompF* mRNA and reducing the OmpF synthesis [47]. Two MarA-like proteins have been reported in *Yersinia pestis*: MarA47 and MarA48 [48,49].

To test their involvement in the regulation of *Y. pseudotuberculosis* porin expression under short- and long-term exposure to antibiotics, we measured the levels of *marA47* and *marA48* mRNA using qRT-PCR (Figure 3). We found different activities of *marA* transcripts depending on the time of antibiotic exposure. A short treatment of cells with the antibiotics caused upregulation of at least one *marA* gene, except for chloramphenicol, which did not affect the *marA* expression. In the case of long-term exposure to the tested antibiotics, the level of *marA* expression remained comparable to those of the control samples.

The obtained results led us to conclude that regulation of *Y. pseudotuberculosis* general porins mediated by MarA was observed only for early antibiotic stress caused by sublethal concentrations of tetracycline, kanamycin, and carbenicillin. During long-term antibiotic presence, *marA* transcription remained unaltered and therefore MarA did not play a significant role in porin regulation at the post-transcriptional level.

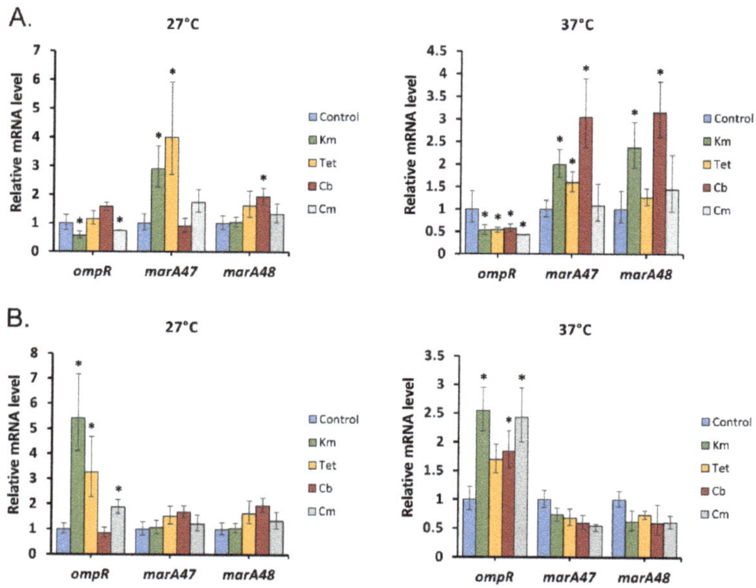

Figure 3. Relative expression levels of *Y. pseudotuberculosis ompR*, *marA47*, and *marA48* under antibiotic stress detected by qRT-PCR. (**A**) 1 h antibiotic exposure, (**B**) 16 h antibiotic exposure. All results are expressed as mean ± standard deviation from three independent experiments. An asterisk indicates *p*-value < 0.05 vs. respective untreated control. Significance was calculated using one-way ANOVA. Km—kanamycin, Tet—tetracycline, Cb—carbenicillin, and Cm—chloramphenicol.

2.5. Transcriptional Response of Y. pseudotuberculosis Alternative Porins to Prolonged Antibiotic Stress

In addition to major porins, *Enterobacteriaceae* can express alternative outer membrane proteins that may contribute to the adaptation of bacteria to antibiotic stress by modulating OmpF and OmpC levels or directly changing the cell permeability [35,50].

Therefore, to better understand the mechanism of *Y. pseudotuberculosis* adaptation, we investigated the transcriptional response of the genes coding for OmpX, OmpA, LamB, and OmpY porins to prolonged antibiotic stress.

Using qRT-PCR, we observed the increased activity of *ompX* transcripts (1.7–3.2-fold) in the presence of most antibiotics (Figure 4), which is consistent with previously reported results [50]. It is believed that overexpression of OmpX may impair the normal assembly of major porins, which leads to their subsequent degradation by Deg proteases [51]. From this, we hypothesized that OmpF and OmpC expression of *Y. pseudotuberculosis* could be regulated by this mechanism at the post-translational level.

We also observed the induction of *lamB* transcription in the presence of kanamycin, carbenicillin, and chloramphenicol at 37 °C (Figure 4), when its initial mRNA level was significantly lower compared to 27 °C (data not shown). It is interesting to note that overexpression of LamB previously found for *Enterobacter aerogenes*-resistant isolates was associated with a low amount of its major porins [52].

The transcription of *ompA*, playing a more important role in the maintenance of membrane integrity, rather than antibiotic transport [35], remained unaltered in *Y. pseudotuberculosis*, except for chloramphenicol exposure, where the *ompA* expression increased up to 2.5-fold (Figure 4).

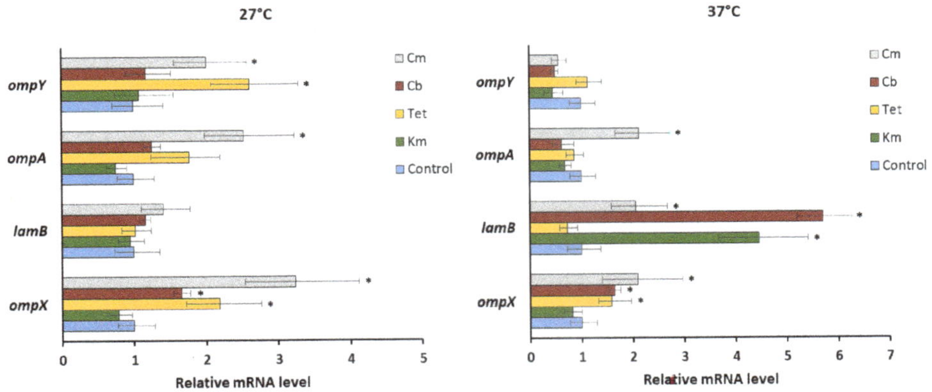

Figure 4. Relative expression levels of *Y. pseudotuberculosis ompX*, *lamB*, *ompA*, and *ompY* under prolonged antibiotic stress detected by qRT-PCR. All results are expressed as mean ± standard deviation from three independent experiments. An asterisk indicates p-value < 0.05 vs. respective untreated control. Significance was calculated using one-way ANOVA. Km—kanamycin, Tet—tetracycline, Cb—carbenicillin, and Cm—chloramphenicol.

Previously, we described a new quiescent porin OmpY in *Y. pseudotuberculosis* initially predicted from the genomic data [53]. To date, our knowledge of such porins has been very limited. Both the functional role of OmpY and the conditions significantly affecting its expression in *Yersinia* are still unknown. qRT-PCR analysis revealed *ompY* upregulation under chloramphenicol and tetracycline exposure at 27 °C, while at 37 °C its level did not change (Figure 4). Several studies have shown that the expression of quiescent porins was induced by antibiotics; however, the impact of such regulation on antibiotic susceptibility remains questionable [20,54].

2.6. *ompF and ompC Expression within Y. pseudotuberculosis Population*

Current experimental evidence suggests that bacterial cultures are constituted of heterogeneous subpopulations [55] with differential responses to environmental stresses [56,57]. For instance, Sánchez-Romero and Casadesús have observed the population heterogeneity of *S. enterica* cells in the expression level of *ompC* under kanamycin exposure [57]. Moreover, a low level of *ompC* correlates with high kanamycin resistance of the cell.

Our qRT-PCR results based on average C_t values did not reveal any significant transcriptional response of porins to long-term antibiotic exposure for most samples. They said nothing about the variability of *ompF* and *ompC* expression in the *Y. pseudotuberculosis* population or the amount of their gene products in the cell. To answer these questions, we first developed fluorescent reporter strains of *Y. pseudotuberculosis* 488. The GFP reporter systems were constructed by fusion of the *ompF*/*ompC* regulatory regions, including binding sites for transcription and translation factors (5′UTRs and signal sequences for sRNAs, C-terminal YQF amino acids for proteases), to the GFP reporter gene in a low copy number plasmid pACYC184 (Figure S1). Thus, the fluorescence of *Y. pseudotuberculosis* transformants potentially not only provided a high-resolution method for assaying *ompF*/*ompC* promoter activity but also indirectly indicated OmpF and OmpC amounts. We excluded chloramphenicol from the experiments due to the plasmid containing the gene for its resistance. However, we additionally verified that the presence of this antibiotic in the medium did not affect the level of sample fluorescence (Figure S2).

There was little evidence of any change in the mean fluorescence intensity, and hence OmpF and OmpC expression, after prolonged exposure to tetracycline and kanamycin (Figure 5), while qRT-PCR revealed the *ompF* upregulation in the presence of tetracycline at 27 °C (Figure 2). This difference indicates an additional post-transcriptional regulation of OmpF.

A.

B.

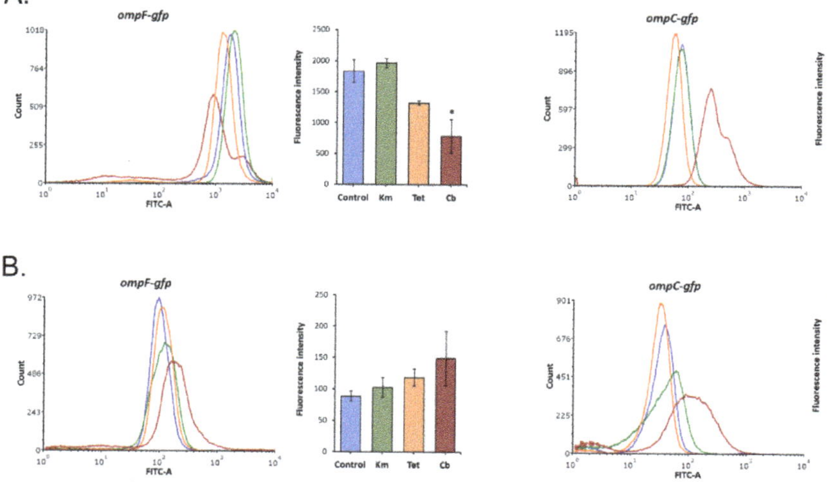

Figure 5. GFP fluorescence intensity in *Y. pseudotuberculosis* 488 transformed with *ompF*/*ompC* promoter-fused GFP reporter constructs following 16 h antibiotic treatment monitored by flow cytometry. (**A**) 27 °C incubation temperature; (**B**) 37 °C incubation temperature. All results are expressed as mean ± standard deviation between three experimental trials. An asterisk indicates p-value < 0.05 vs. respective control. Significance was calculated using one-way ANOVA. The histograms from a single representative experiment are shown. Control—without antibiotic treatment, Km—kanamycin, Tet—tetracycline, and Cb—carbenicillin.

After treatment with sub-MIC of carbenicillin at 27 °C, there was a 2.3-fold decrease in the *ompF* level and, on the contrary, a 2.5–3-fold increase in the *ompC* level. This is consistent with the previous observation that adaptation to β-lactams can involve increased expression of OmpC relative to OmpF since an increase in OmpC proportion hinders their penetration [37]. It is worth noting that carbenicillin inhibited the expression of *ompF* only at 27 °C. We observed that this transcript was prevailing compared to *ompC* in the untreated samples at both incubation temperatures; however, at 37 °C, its level decreased by more than 18 times (Figure S3). This initially low amount of *ompF* transcripts and additional nutrient deficiencies due to the stationary growth phase could prevent a further decrease in its expression at 37 °C under carbenicillin stress.

The distribution of GFP fluorescence within the sample cultures under kanamycin and tetracycline exposure was comparable to the untreated controls. Only carbenicillin presence led to the appearance of two *Y. pseudotuberculosis* subpopulations with different fluorescence intensities, most notably at 27 °C (Figure 6). F1 subpopulation responded to carbenicillin stress by 1.9-fold inhibition of *ompF* expression, while the cells from F2 increased it by 1.8 times. *ompC* was upregulated under carbenicillin exposure in both subpopulations; however, in C2, its level was significantly higher. Moreover, most of the cells formed subpopulations with a lower level of both *ompF* and *ompC* expression (71% and 78% of the total cell amount, respectively; Figure 6C).

We thus propose that modification of the bacterial envelope by reduced *ompF*/increased *ompC* expression confers adaptive fitness advantages to carbenicillin in a fraction of the bacterial population.

Previously it was shown that heterogenous responses improve the survival of bacterial cultures under the effect of stress factors [58]. Moreover, gene expression variability, including porin genes, is considered to underly adaptive resistance in phenotypically heterogeneous microbial populations [57,59].

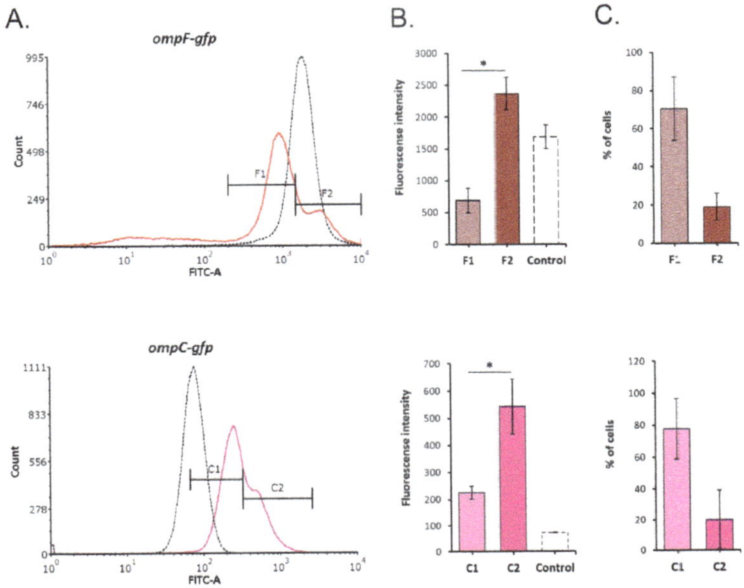

Figure 6. Fluorescence heterogeneity of *Y. pseudotuberculosis* 488 transformed with *ompF/ompC* promoter-fused GFP reporter constructs following treatment with a sublethal dose of carbenicillin monitored by flow cytometry. (**A**) Fluorescence histograms; (**B**) fluorescent intensity of subpopulations; (**C**) the percentage of cells in each subpopulation. All results are expressed as mean ± standard deviation between three experimental trials. An asterisk indicates *p*-value < 0.05 between groups. Significance was calculated using one-way ANOVA. The histograms from a single representative experiment are shown. F1, F2, and C1, C2—subpopulations of *Y. pseudotuberculosis* 488 with low/high GFP fluorescence.

3. Materials and Methods

3.1. Bacterial Strain and Growth Conditions

Y. pseudotuberculosis 488 strain (= 117 strain, O:1b serotype) isolated from a patient with FESLF in the Russian Far East region was used in the present study.

To study short-term porin response to antibiotic stress, bacterial cells were grown in LB broth overnight at 27 °C with shaking. The next day, overnight cultures were diluted at a ratio of 1:50 into fresh LB medium and incubated with agitation at 27 °C (optimal growth temperature) and 37 °C (host temperature) to mid-exponential growth phase (OD_{600} = 0.4–0.6). Then, the antibiotics kanamycin (Sigma-Aldrich, St. Louis, MA, USA), tetracycline (Sigma-Aldrich, St. Louis, MA, USA), chloramphenicol (Sigma-Aldrich, St. Louis, MA, USA), and carbenicillin (Sigma-Aldrich, St. Louis, MA, USA) were added to the media in sublethal concentrations (0.5 MIC). After a 1 h antibiotic treatment (180 rpm, 27 °C and 37 °C), the cells were mixed with RNAprotect Bacteria Reagent (Qiagen, Hilden, Germany) and centrifuged.

To study porin response to long-term antibiotic exposure, overnight cultures were diluted and inoculated into LB broth media to obtain a final concentration of 5×10^5 CFU/mL. Bacteria were incubated with sublethal concentrations of antibiotics (kanamycin, tetracycline, chloramphenicol, and carbenicillin) for 16 h at 27 °C and 37 °C. RNAprotect Bacteria Reagent (Qiagen, Hilden, Germany) was added to the culture samples.

Three independent bacterial cultures for each test or control condition were prepared as biological replicates for RNA isolation.

3.2. MIC Determination

The MIC for antibiotics against *Y. pseudotuberculosis* strain was determined by the broth microdilution method in a 96-well microplate. Tetracycline (Sigma-Aldrich, St. Louis, MA, USA), carbenicillin (Sigma-Aldrich, St. Louis, MA, USA), kanamycin (Sigma-Aldrich, St. Louis, MA, USA), and chloramphenicol (Sigma-Aldrich, St. Louis, MA, USA) were tested in the concentration range of 0.12–256 µg/mL obtained after a series of 2-fold dilutions in the LB broth. Bacterial colonies were cultured in LB broth at 27 °C to reach McFarland standard 0.5. The suspensions were further diluted and inoculated into the wells with antibiotics to obtain a final concentration of 5×10^5 CFU/mL. The plates were incubated for 16 h at 27 °C and 37 °C. The MIC value was determined as the lowest concentration of an antibiotic at which visible growth was inhibited.

3.3. Gene Expression Measurement by qRT-PCR

The expression of *ompF*, *ompC*, *ompA*, *ompX*, *ompY*, *ompR*, *marA47*, and *marA48* genes in *Y. pseudotuberculosis* 488 was studied using qRT-PCR. Total RNA was isolated from the prepared bacterial pellets using the Aurum total RNA mini kit (Bio-Rad, Hercules, CA, USA) according to the manufacturer's protocol. The RNA concentration and purity were assessed by electrophoresis and a microvolume spectrophotometer (Implen GmbH, Munich, Germany). Of the purified total RNA, 2 µg was reverse-transcribed into cDNA with MMLV RT Kit (Evrogen, Moscow, Russia) and random hexamer primers (Evrogen, Moscow, Russia). The cDNAs were subsequently used to quantify the relative level of *ompF*, *ompC*, *ompA*, *ompX*, *ompY*, *ompR*, *marA47*, and *marA48* by qRT-PCR in Light Cycler 96 (Roche, Basel, Switzerland). The nucleotide sequences of the studied genes are listed in Table S2. The gene-specific primers used in this experiment are listed in Table S3.

RT-PCR was carried out with HS GoTaq Polymerase (Promega, Madison, WI, USA) and the dye Eva Green (Biotium, Fremont, CA, USA). The following thermal cycling parameters were used for the reaction: initial denaturation at 95 °C for 8 min; and 40 cycles of amplification: 95 °C for 15 s, 55 °C for 10 s, and 72 °C for 20 s followed by fluorescence reading. A melting curve was drawn at the end to evaluate the specificity of the PCR. Quantification for each target gene expression was determined by the $2^{-\Delta\Delta CT}$ method [60] using the 16S rDNA gene as a reference.

qRT-PCR was carried out on three independent biological replicates, each containing two technical replicates. The results are presented in Table S4 as mean ± SD. One-way ANOVA was performed to assess the differences between the means of the test and control groups, with a p-value of ≤ 0.05 considered significant.

3.4. Porin Gene Sequencing

The coding and regulatory parts of *ompF* and *ompC* genes were amplified using OmpF/C-Bam_for and OmpF/C-Hind_rev primers and sequenced with internal primers (Table S3) to cover all regions. DNA sequencing was done with an ABI 3130 xl automated sequencer (Applied Biosystems, Waltham, MA, USA) and the ABI Prism dye terminator sequencing kit (Applied Biosystems, Waltham, MA, USA) according to the manufacturer's directions.

3.5. GFP Reporter System Construction

The recombinant plasmids pACYC184-F-GFP and pACYC184-C-GFP containing the *Y. pseudotuberculosis ompF* and *ompC* regulatory regions fused to GFP were constructed as previously described [7]. To obtain the coding region of GFP, the plasmid DNA pTurboGFP (Evrogen, Moscow, Russia) was digested with *Bam*HI and *Hind*III (Fermentas, Waltham, MA, USA), and the resulting *gfp* fragment was dephosphorylated with CAP alkaline phosphatase (Fermentas, Waltham, MA, USA). The *ompF*/*ompC* promoter, and terminator regions were amplified using specific primers, which contained *Bam*HI and *Hind*III restriction sites, respectively (Table S3). The resulting PCR products were digested with the same endonucleases and ligated together (separately for *ompF* and *ompC*) with *gfp* in a three-way ligation. Next, ligation mixtures were used as templates for amplifying *ompF-gfp* and *ompC-gfp* reporter constructions with OmpF-Bam_for/OmpF-Hind_rev or OmpC-Bam_for/OmpC-Hind_rev primers (Table S3). The PCRs were carried out using the following reagents: 2 U of GoTaq DNA polymerase (Promega, Madison, WI, USA), 1× buffer for GoTaq DNA polymerase, 0.2 mM dNTPs, 0.5 μM primers, and 25–50 ng of the template. The following reaction conditions were used: 1 cycle of 5 min at 95 °C for denaturation, 30 cycles of 20 s at 94 °C, 30 s at 55 °C, 1 min 30 s at 72 °C, and 1 final cycle of 5 min at 72 °C. The PCR fragments of expected lengths (1300–1400 bp) were cut from the agarose gel and phosphorylated with T4 polynucleotide kinase (Fermentas, Waltham, MA, USA). The resulting reporter cassettes were inserted in the dephosphorylated vector pACYC184 via *Eco*RV (Figure S1). Recombinant plasmids were transformed into *E. coli* DH5alpha strain by electroporation. The colonies were selected on LB agar plates supplemented with chloramphenicol. All constructs were verified by PCR and DNA sequencing with pACYCSal_seq and pACYCEcV_seq primers (Table S2). The final recombinant plasmids were introduced into competent cells of the *Y. pseudotuberculosis* 488 using electroporation.

3.6. Flow Cytometry

To obtain comparable measurements with flow cytometry, the *Y. pseudotuberculosis* 488 containing recombinant plasmid pACYC184-F-GFP or pACYC184-C-GFP were grown in the same conditions as for qRT-PCR analysis. All cultures were incubated in 96-well plates without/with a sublethal concentration of antibiotics (Tet, Km, and Cb) and the addition of chloramphenicol (10 μg/mL) at 37 °C and 27 °C with shaking. After 16 h of cultivation, the cells were washed with PBS, pH 7, and diluted to an OD_{600} of 0.01. The flow cytometry measurements were made with a FACSCanto flow cytometer (BD Biosciences, San Jose, CA, USA). The excitation beam for the GFP was set at 488 nm, and the emission signal was captured with a 530/30 nm bandpass filter. The signals were amplified with the logarithmic mode for SSC, FSC, and fluorescence. For each sample, 100,000 events were recorded. The data were captured with the BD FACSdiva software (version 6.1.2) (BD Biosciences, San Jose, CA, USA), and for data analysis FCS Express 6 (De Novo Software Inc., Los Angeles, CA, USA) was used. The mean fluorescence value of the whole population was chosen as a global indicator of porin expression level in each bacterial culture. *Y. pseudotuberculosis* 488 without plasmids was used as a negative control to exclude background fluorescence.

4. Conclusions

In summary, we revealed temporal regulation of *Y. pseudotuberculosis* generalporin genes *ompF* and *ompC* caused by sublethal concentrations of different antibiotics. The porin transcription initially decreased, providing early defensive response of the bacterium, while it returned to that of the untreated cells on prolonged antibiotic exposure. Unlike the major porin genes, the transcription of the alternativeporin genes *ompX* and *lamB* was increased. Such modulation of porin transcription reflects the model of *Y. pseudotuberculosis* physiological adaptation to antibiotic exposure.

The main finding was a phenotypic heterogeneity of *Y. pseudotuberculosis* population manifested in variable porin gene expression under carbenicillin exposure. To verify whether such phenotypic heterogeneity offers a competitive advantage and contributes to low carbenicillin susceptibility for a particular *Y. pseudotuberculosis* subpopulation, future research is warranted.

Supplementary Materials: The following are available online, Figure S1: GFP reporter constructions; Figure S2: GFP fluorescence intensity in *Y. pseudotuberculosis* 488 transformed with *ompC/ompF* promoter-fused GFP reporter constructs under increasing Cm exposure; Figure S3: GFP fluorescence intensity in *Y. pseudotuberculosis* 488 transformed with *ompF/ompC* promoter-fused GFP reporter constructs at 27 °C and 37 °C incubation temperature; Table S1: MIC of antibiotics against *Y. pseudotuberculosis* 488; Table S2: Nucleotide sequences of *Y. pseudotuberculosis* 488 genes used in the study; Table S3: Oligonucleotide primers used in the study; Table S4: Average Ct values for *ompF*, *ompC*, and 16S rRNA of *Y. pseudotuberculosis* 488.

Author Contributions: Conceptualization, M.I. and A.R.; methodology, M.I., A.S., N.C., and E.B.; software, E.B.; investigation, E.B., N.C., K.G., and A.S.; resources, M.I. and A.R.; data curation, M.I.; writing—original draft preparation, E.B. and M.I.; writing—review and editing, M.I. and A.R.; supervision, M.I.; funding acquisition, M.I. and E.B. All authors have read and agreed to the published version of the manuscript.

Funding: This research received no external funding.

Institutional Review Board Statement: Not applicable.

Informed Consent Statement: Not applicable.

Data Availability Statement: Unpublished data are available from the authors.

Acknowledgments: We are grateful to the anonymous reviewers and editors that have improved our manuscript.

Conflicts of Interest: The authors declare no conflict of interest.

References

1. Fernandez, L.; Hancock, R.E. Adaptive and mutational resistance: Role of porins and efflux pumps in drug resistance. *Clin. Microbiol. Rev.* **2012**, *25*, 661–681. [CrossRef]
2. Andersson, D.I.; Hughes, D. Microbiological effects of sublethal levels of antibiotics. *Nat. Rev. Microbiol.* **2014**, *12*, 465–478. [CrossRef]
3. Braoudaki, M.; Hilton, A.C. Adaptive resistance to biocides in *Salmonella enterica* and *Escherichia coli* o157 and cross-resistance to antimicrobial agents. *J. Clin. Microbiol.* **2004**, *42*, 73–78. [CrossRef] [PubMed]
4. Daikos, G.L.; Lolans, V.T.; Jackson, G.G. First-exposure adaptive resistance to aminoglycoside antibiotics in vivo with meaning for optimal clinical use. *Antimicrob. Agents Chemother.* **1991**, *35*, 117–123. [CrossRef]
5. Viveiros, M.; Jesus, A.; Brito, M.; Leandro, C.; Martins, M.; Ordway, D.; Molnar, A.M.; Molnar, J.; Amaral, L. Inducement and reversal of tetracycline resistance in *Escherichia coli* K-12 and the expression of proton gradient dependent multidrug efflux pump genes. *Antimicrob. Agents Chemother.* **2005**, *49*, 3578–3582. [CrossRef]
6. Somova, L.M.; Antonenko, F.F.; Timchenko, N.F.; Lyapun, I.N. Far Eastern Scarlet-Like Fever is a Special Clinical and Epidemic Manifestation of *Yersinia pseudotuberculosis* Infection in Russia. *Pathogens* **2020**, *9*, 436. [CrossRef]
7. Bystritskaya, E.P.; Stenkova, A.M.; Portnyagina, O.Y.; Rakin, A.V.; Rasskazov, V.A.; Isaeva, M.P. Regulation of *Yersinia pseudotuberculosis* major porin expression in response to antibiotic stress. *Mol. Gen. Microbiol. Virol.* **2014**, *29*, 63–68. [CrossRef]
8. Khomenko, V.A.; Portnyagina, O.Y.; Novikova, O.D.; Isaeva, M.P.; Kim, N.Y.; Likhatskaya, G.N.; Vostrikova, O.P.; Solov'eva, T.F. Isolation and characterization of recombinant OmpF-like porin from the *Yersinia pseudotuberculosis* outer membrane. *Russ. J. Bioorg. Chem.* **2008**, *34*, 162–168. [CrossRef]
9. Novikova, O.D.; Khomenko, V.A.; Emelyanenko, V.I.; Likhatskaya, G.N.; Zelepuga, E.A.; Kim, N.Y.; Isaeva, M.P.; Portnyagina, O.Y.; Vostrikova, O.P.; Sidorova, O.V.; et al. OmpC-like porin from *Yersinia pseudotuberculosis*: Molecular characteristics, physico-chemical and functional properties. *Biochem. Suppl. Ser. A Membr. Cell Biol.* **2011**, *5*, 263–277.
10. Stenkova, A.M.; Isaeva, M.P.; Shubin, F.N.; Rasskazov, V.A.; Rakin, A.V. Trends of the major porin gene (OmpF) evolution: Insight from the genus *Yersinia*. *PLoS ONE* **2011**, *6*, e20546. [CrossRef]
11. Stenkova, A.M.; Bystritskaya, E.P.; Guzev, K.V.; Rakin, A.V.; Isaeva, M.P. Molecular Evolution of the *Yersinia* Major Outer Membrane Protein C (OmpC). *Evol. Bioinform.* **2016**, *12*, 185–191. [CrossRef]
12. Cowan, S.W.; Schirmer, T.; Rummel, G.; Steiert, M.; Ghosh, R.; Pauptit, R.A.; Jansonius, J.N.; Rosenbusch, J.P. Crystal structures explain functional properties of two *E. coli* porins. *Nature* **1992**, *358*, 727–733. [CrossRef]
13. James, C.E.; Mahendran, K.R.; Molitor, A.; Bolla, J.-M.; Bessonov, A.N.; Winterhalter, M.; Pagès, J.-M. How beta-lactam antibiotics enter bacteria: A dialogue with the porins. *PLoS ONE* **2009**, *4*, e5453. [CrossRef]
14. Bafna, J.A.; Sans-Serramitjana, E.; Acosta-Gutiérrez, S.; Bodrenko, I.V.; Hörömpöli, D.; Berscheid, A.; Brötz-Oesterhelt, H.; Winterhalter, M.; Ceccarelli, M. Kanamycin uptake into *Escherichia coli* is facilitated by OmpF and OmpC porin channels located in the outer membrane. *ACS Infect. Dis.* **2020**, *6*, 1855–1865. [CrossRef]
15. Neves, P.; Berkane, E.; Gameiro, P.; Winterhalter, M.; De Castro, B. Interaction between quinolones antibiotics and bacterial outer membrane porin OmpF. *Biophys. Chem.* **2005**, *113*, 123–128. [CrossRef]
16. Mortimer, P.G.; Piddock, L.J. The accumulation of five antibacterial agents in porin-deficient mutants of *Escherichia coli*. *J. Antimicrob. Chemother.* **1993**, *32*, 195–213. [CrossRef]
17. Masi, M.; Réfrigiers, M.; Pos, K.M.; Pagès, J.M. Mechanisms of envelope permeability and antibiotic influx and efflux in Gram-negative bacteria. *Nat. Microbiol.* **2017**, *2*, 17001. [CrossRef] [PubMed]
18. Sousa, A.M.; Machado, I.; Pereira, M.O. Phenotypic switching: An opportunity to bacteria thrive. In *Science Against Microbial Pathogens: Communicating Current Research and Technological Advances*; Formatex Research Center: Badajoz, Spain, 2012; pp. 252–262.

19. Dupont, M.; James, C.E.; Chevalier, J.; Pagès, J.-M. An Early Response to Environmental Stress Involves Regulation of OmpX and OmpF, Two Enterobacterial Outer Membrane Pore-Forming Proteins. *Antimicrob. Agents Chemother.* **2007**, *51*, 3190–3198. [CrossRef]
20. Doménech-Sánchez, A.; Hernández-Allés, S.; Martínez-Martínez, L.; Benedí, V.J.; Albertí, S. Identification and characterization of a new porin gene of *Klebsiella pneumoniae*: Its role in beta-lactam antibiotic resistance. *J. Bacteriol.* **1999**, *181*, 2726–2732. [CrossRef]
21. Knopp, M.; Andersson, D.I. Amelioration of the fitness costs of antibiotic resistance due to reduced outer membrane permeability by upregulation of alternative porins. *Mol. Biol. Evol.* **2015**, *32*, 3252–3263. [CrossRef]
22. Chetri, S.; Singha, M.; Bhowmik, D.; Nath, K.; Chanda, D.D.; Chakravarty, A.; Bhattacharjee, A. Transcriptional response of OmpC and OmpF in *Escherichia coli* against differential gradient of carbapenem stress. *BMC Res. Notes* **2019**, *12*, 138. [CrossRef]
23. Dam, S.; Pagès, J.-M.; Masi, M. Stress responses, outer membrane permeability control and antimicrobial resistance in *Enterobacteriaceae*. *Microbiology* **2018**, *164*, 260–267. [CrossRef] [PubMed]
24. Viveiros, M.; Dupont, M.; Rodrigues, L.; Couto, I.; Davin-Regli, A.; Martins, M.; Pagès, J.-M.; Amaral, L. Antibiotic stress, genetic response and altered permeability of *E. coli*. *PLoS ONE* **2007**, *2*, e365. [CrossRef]
25. Deighan, P.; Free, A.; Dorman, C.J. A role for the *Escherichia coli* H-NS-like protein StpA in OmpF porin expression through modulation of *micF* RNA stability. *Mol. Microbiol.* **2000**, *38*, 126–139. [CrossRef]
26. Andersen, J.; Forst, S.A.; Zhao, K.; Inouye, M.; Delihas, N. The function of *micF* RNA. *micF* RNA is a major factor in the thermal regulation of OmpF protein in *Escherichia coli*. *J. Biol. Chem.* **1989**, *264*, 17961–17970. [CrossRef]
27. Begic, S.; Worobec, E.A. Regulation of *Serratia marcescens ompF* and *ompC* porin genes in response to osmotic stress, salicylate, temperature and pH. *Microbiology* **2006**, *152*, 485–491. [CrossRef]
28. Bystritskaya, E.; Stenkova, A.; Chistuylin, D.; Chernysheva, N.; Khomenko, V.; Anastyuk, S.; Novikova, O.; Rakin, A.; Isaeva, M. Adaptive responses of outer membrane porin balance of *Yersinia ruckeri* under different incubation temperature, osmolarity, and oxygen availability. *Microbiologyopen* **2016**, *5*, 597–603. [CrossRef] [PubMed]
29. Narimisa, N.; Amraei, F.; Kalani, B.S.; Mohammadzadeh, R.; Jazi, F.M. Effects of sub-inhibitory concentrations of antibiotics and oxidative stress on the expression of type II toxin-antitoxin system genes in *Klebsiella pneumoniae*. *J. Glob. Antimicrob. Resist.* **2020**, *21*, 51–56. [CrossRef] [PubMed]
30. Zhang, D.-F.; Jiang, B.; Xiang, Z.-M.; Wang, S.-Y. Functional characterisation of altered outer membrane proteins for tetracycline resistance in *Escherichia coli*. *Int. J. Antimicrob. Agents* **2008**, *32*, 315–319. [CrossRef]
31. Chang, C.C.; Yan, C.W.; Onal, R.; Zhang, Y. Genetic characterization and investigation of kanamycin susceptibility of *ompC* and *ompF* single gene deletion mutants of *Escherichia coli* K-12. *J. Exp. Microbiol. Immunol.* **2018**, *22*, 1–9.
32. Hu, W.-C.; MacDonald, R.; Oosthuizen, J.L.; Soeren, M.V. Sub-inhibitory kanamycin changes outer membrane porin ratios in *Escherichia coli* B23 by increasing the level of OmpC. *J. Exp. Microbiol. Immunol.* **2011**, *15*, 96–102.
33. Fei, Y.; Ma, V.; Maerkl, N.; You, P. The down regulation of E. coli OmpF in response to sub-inhibitory concentrations of kanamycin is not mediated by MarA. *J. Exp. Microbiol. Immunol.* **2012**, *16*, 101–107.
34. Agafitei, O.; Kim, E.J.; Maguire, T.; Sheridan, J. The role of *Escherichia coli* porins OmpC and OmpF in antibiotic cross resistance induced by sub-inhibitory concentrations of kanamycin. *J. Exp. Microbiol. Immunol.* **2010**, *14*, 34–39.
35. Choi, U.; Lee, C.-R. Distinct roles of outer membrane porins in antibiotic resistance and membrane integrity in *Escherichia coli*. *Front. Microbiol.* **2019**, *10*, 953. [CrossRef]
36. Moya-Torres, A.; Mulvey, M.R.; Kumar, A.; Oresnik, I.J.; Brassinga, A.K. The lack of OmpF, but not OmpC, contributes to increased antibiotic resistance in *Serratia marcescens*. *Microbiology* **2014**, *160 Pt 9*, 1882–1892. [CrossRef]
37. Igwe, J.C.; Olayinka, B.O.; Ehnimidu, J.O.; Onaolapo, J.A. Impact of outer membrane protein Ompc and Ompf on antibiotics resistance of *E. coli* isolated from UTI and diarrhoeic patients in Zaria, Nigeria. *Clin. Microbiol.* **2016**, *5*, 1000267.
38. Toro, C.S.; Lobos, S.R.; Calderón, I.; Rodríguez, M.; Mora, G.C. Clinical isolate of a porinless *Salmonella* typhi resistant to high levels of chloramphenicol. *Antimicrob. Agents. Chemother.* **1990**, *34*, 1715–1719. [CrossRef] [PubMed]
39. Fajardo-Lubián, A.; Zakour, N.L.B.; Agyekum, A.; Qi, Q.; Iredell, J.R. Host adaptation and convergent evolution increases antibiotic resistance without loss of virulence in a major human pathogen. *PLoS Pathog.* **2019**, *15*, e1007218. [CrossRef]
40. Ferenci, T.; Phan, K. How porin heterogeneity and trade-offs affect the antibiotic susceptibility of Gram-Negative bacteria. *Genes* **2015**, *6*, 1113–1124. [CrossRef]
41. Liu, X.; Ferenci, T. Regulation of porin-mediated outer membrane permeability by nutrient limitation in *Escherichia coli*. *J. Bacteriol.* **1998**, *180*, 3917–3922. [CrossRef]
42. Tierney, A.R.; Rather, P.N. Roles of two-component regulatory systems in antibiotic resistance. *Future Microbiol.* **2019**, *14*, 533–552. [CrossRef] [PubMed]
43. Flamez, C.; Ricard, I.; Arafah, S.; Simonet, M.; Marceau, M. Phenotypic analysis of *Yersinia pseudotuberculosis* 32777 response regulator mutants: New insights into two-component system regulon plasticity in bacteria. *Int. J. Med. Microbiol.* **2008**, *298*, 193–207. [CrossRef]
44. Nieckarz, M.; Raczkowska, A.; Jaworska, K.; Stefańska, E.; Skorek, K.; Stosio, D.; Brzostek, K. The role of OmpR in the expression of genes of the KdgR regulon involved in the uptake and depolymerization of oligogalacturonides in *Yersinia enterocolitica*. *Front. Cell Infect. Microbiol.* **2017**, *7*, 366. [CrossRef]
45. Raczkowska, A.; Trzos, J.; Lewandowska, O.; Nieckarz, M.; Brzostek, K. Expression of the AcrAB components of the AcrAB-TolC multidrug efflux pump of *Yersinia enterocolitica* is subject to dual regulation by OmpR *PLoS ONE* **2015**, *10*, e0124248. [CrossRef]

46. Masi, M.; Pagès, J.-M. Structure, function and regulation of outer membrane proteins involved in drug transport in *Enterobactericeae*: The OmpF/C—TolC case. *Open Microbiol. J.* **2013**, *7*, 22–33. [CrossRef]
47. Cohen, S.P.; McMurry, L.M.; Levy, S.B. MarA locus causes decreased expression of OmpF porin in multiple-antibiotic-resistant (Mar) mutants of *Escherichia coli*. *J. Bacteriol.* **1988**, *170*, 5416–5422. [CrossRef]
48. Lister, I.M.; Mecsas, J.; Levy, S.B. Effect of MarA-like proteins on antibiotic resistance and virulence in *Yersinia pestis*. *Infect. Immun.* **2010**, *78*, 364–371. [CrossRef]
49. Udani, R.A.; Levy, S.B. MarA-like regulator of multidrug resistance in *Yersinia pestis*. *Antimicrob. Agents Chemother.* **2006**, *50*, 2971–2975. [CrossRef] [PubMed]
50. Dupont, M.; Dé, E.; Chollet, R.; Chevalier, J.; Pagès, J.-M. *Enterobacter aerogenes* OmpX, a cation-selective channel mar- and osmo-regulated. *FEBS Lett.* **2004**, *569*, 27–30. [CrossRef] [PubMed]
51. Castillo-Keller, M.; Vuong, P.; Misra, R. Novel mechanism of *Escherichia coli* porin regulation. *J. Bacteriol.* **2006**, *188*, 576–586. [CrossRef]
52. Gayet, S.; Chollet, R.; Molle, G.; Pagès, J.-M.; Chevalier, J. Modification of outer membrane protein profile and evidence suggesting an active drug pump in *Enterobacter aerogenes* clinical strains. *Antimicrob. Agents Chemother.* **2003**, *47*, 1555–1559. [CrossRef]
53. Solov'eva, T.F.; Likhatskaya, G.N.; Khomenko, V.A.; Stenkova, A.M.; Kim, N.Y.; Portnyagina, O.Y.; Novikova, O.D.; Trifonov, E.V.; Nurminski, E.A.; Isaeva, M.P. A novel OmpY porin from *Yersinia pseudotuberculosis*: Structure, channel-forming activity and trimer thermal stability. *J. Biomol. Struct. Dyn.* **2011**, *28*, 517–533. [CrossRef]
54. Dam, S.; Pagès, J.-M.; Masi, M. Dual Regulation of the small RNA MicC and the quiescent porin OmpN in response to antibiotic stress in *Escherichia coli*. *Antibiotics* **2017**, *6*, 33. [CrossRef] [PubMed]
55. Ambriz-Aviña, V.; Contreras-Garduño, J.A.; Pedraza-Reyes, M. Applications of flow cytometry to characterize bacterial physiological responses. *BioMed Res. Int.* **2014**, *2014*, 461941. [CrossRef]
56. de Jong, I.G.; Veening, J.-W.; Kuipers, O.P. Single cell analysis of gene expression patterns during carbon starvation in *Bacillus subtilis* reveals large phenotypic variation. *Environ. Microbiol.* **2012**, *14*, 3110–3121. [CrossRef]
57. Sánchez-Romero, M.A.; Casadesús, J. Contribution of phenotypic heterogeneity to adaptive antibiotic resistance. *Proc. Natl. Acad. Sci. USA* **2014**, *111*, 355–360. [CrossRef]
58. Magdanova, L.A.; Golyasnaya, N.V. Heterogeneity as an adaptive trait of microbial populations. *Microbiology* **2013**, *82*, 1–10. [CrossRef]
59. Erickson, K.E.; Otoupal, P.B.; Chatterjee, A. Gene expression variability underlies adaptive resistance in phenotypically heterogeneous bacterial populations. *ACS Infect. Dis.* **2015**, *1*, 555–673. [CrossRef]
60. Livak, K.J.; Schmittgen, T.D. Analysis of relative gene expression data using real-time quantitative PCR and the $2^{-\Delta\Delta CT}$ method. *Methods* **2001**, *25*, 402–408. [CrossRef] [PubMed]

molecules

MDPI

Article

Genome-Wide Analysis of PL7 Alginate Lyases in the Genus *Zobellia*

Nadezhda Chernysheva, Evgeniya Bystritskaya, Galina Likhatskaya , Olga Nedashkovskaya and Marina Isaeva *

G.B. Elyakov Pacific Institute of Bioorganic Chemistry, Far Eastern Branch, Russian Academy of Sciences, 159, Pr. 100 let Vladivostoku, 690022 Vladivostok, Russia; chernysheva.nadezhda@gmail.com (N.C.); belyjane@gmail.com (E.B.); galin56@mail.ru (G.L.); oned2004@mail.ru (O.N.)
* Correspondence: issaeva@gmail.com; Tel.: +7-423-231-1168

Abstract: We carried out a detailed investigation of PL7 alginate lyases across the *Zobellia* genus. The main findings were obtained using the methods of comparative genomics and spatial structure modeling, as well as a phylogenomic approach. Initially, in order to elucidate the alginolytic potential of *Zobellia*, we calculated the content of polysaccharide lyase (PL) genes in each genome. The genus-specific PLs were PL1, PL6, PL7 (the most abundant), PL14, PL17, and PL40. We revealed that PL7 belongs to subfamilies 3, 5, and 6. They may be involved in local and horizontal gene transfer and gene duplication processes. Most likely, an individual evolution of PL7 genes promotes the genetic variability of the Alginate Utilization System across *Zobellia*. Apparently, the PL7 alginate lyases may acquire a sub-functionalization due to diversification between in-paralogs.

Keywords: *Zobellia*; genomes; polysaccharide lyase family 7; alginate utilization system; paralogs; orthologs

check for updates

Citation: Chernysheva, N.; Bystritskaya, E.; Likhatskaya, G.; Nedashkovskaya, O.; Isaeva, M. Genome-Wide Analysis of PL7 Alginate Lyases in the Genus *Zobellia*. *Molecules* **2021**, *26*, 2387. https://doi.org/10.3390/molecules26082387

Academic Editor: Changsheng Zhang

Received: 11 March 2021
Accepted: 15 April 2021
Published: 20 April 2021

1. Introduction

Marine algal polysaccharides are an important nutrient source for marine bacteria. To utilize these polysaccharides, which significantly differ from terrestrial plant ones, marine bacteria have developed unusual degradation mechanisms. Key players in the degradation of complex polysaccharides are the marine *Flavobacteriia* of the phylum *Bac-teroidetes* [1,2]. The microorganisms feature different polysaccharide utilization loci (PULs), which encode a set of enzymes and other proteins involved in the breakdown of specific polysaccharides. The first recognized PULs were alginolytic operons associated with alginate utilization in marine *Bacteroidetes* [3]. The current studies of PULs are based on the sequencing of the genomes of cultured bacteria and metagenomes as a valuable resource of CAZymes and CAZyme machineries [4–10], allowing us to expand our knowledge of the bacterial degradation of algal polysaccharides.

The major bacterial players in marine polysaccharide degradation have become the subject of extensive research studies using genomic and transcriptomic approaches. Representatives of the genus *Zobellia* are marine *Flavobacteriia* and they specialize in algal polysaccharide degradation. The genus *Zobellia* was proposed by Barbeyron et al. (2001) [11]. First, the genus included two species: *Zobellia galactanivorans* and *Zobellia uliginosa*. Later, three new species were added [12]: *Zobellia russellii*, *Zobellia laminariae*, and *Zobellia amurskyensis*. To date, the five species of the genus *Zobellia* have validly published names (as listed at https://lpsn.dsmz.de/genus/zobellia, February 2021), and all originate from marine environments.

Members of the *Zobellia* genus are frequently found associated with red and brown algae [13]. *Z. galactanivorans* Dsij^T, a marine species isolated from a red alga, was chosen as the type species [11]. Its genome exhibits a number of adaptive features such as consumption of algal polysaccharides, resistance to algal defense, and large amounts of CAZymes and sulfatases [14]. Numerous biochemical and structural studies have begun to unveil the complex enzyme systems for the degradation of various polysaccharides, such as agars, carrageenans, and laminarin [3,15–21].

In this work, we have applied genome-based approaches to investigate PL7 alginate lyases across the *Zobellia* genus.

2. Results and Discussion

2.1. Genomic Comparison of Zobellia Species

The characteristics of the ten publicly available *Zobellia* genomes and two unpublished ones (*Z. russellii*, KMM 3677^T, and *Z. barbeyronii*, KMM 6746^T, our data) were compared. Their genome sizes and GC content ranged from 4.92 to 5.52 Mb and from 36.7% to 42.8%, respectively. To clarify, the phylogenomic relationships of the *Zobellia* species, a phylogenetic tree of these genomes together with two other genomes of type strains from related genera, were built using PhyloPhlAn [22] based on 400 concatenated proteins. According to the genome tree (Figure 1a), all of the *Zobellia* strains clustered together, and the five subclades could be distinguished. The first subclade included a type strain of *Z. uliginosa* together with *Z. galactanivorans* strains, while other subclades consisted of type strains of *Z. laminariae* and '*Z. barbeyronii*', as well as *Zobellia* sp. strains (Asnod2-F07-B and Asnod3-E08-A). The separated subclades were formed by strains of *Z. russellii*, *Z. amurskyensis* strains, and *Zobellia* sp. (Asnod1-F08 and Asnod2-F02-B). This clustering indicates a closer genome sequence similarity of the strains within subclades. Interestingly, the subclade *Z. uliginosa*/*Z. galactanivorans* was at the base of the *Zobellia* clade, implying an evolutionary divergence from other species of the genus.

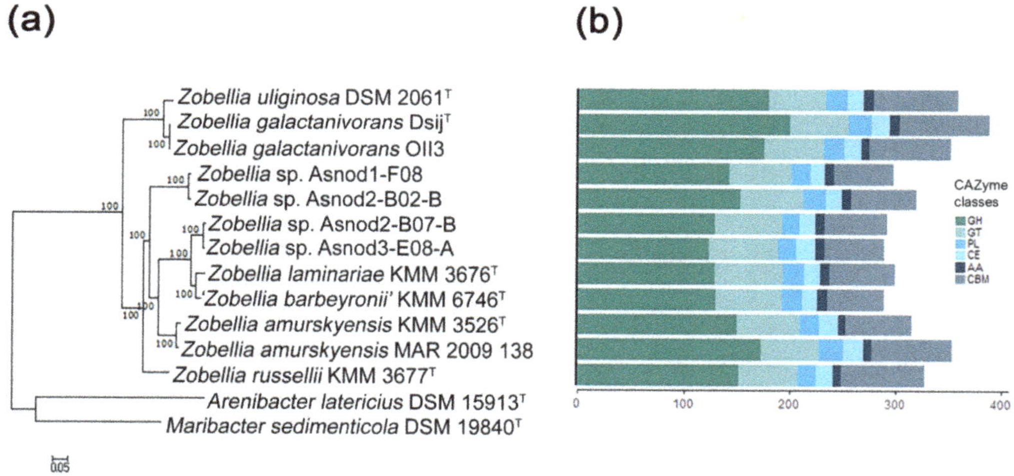

Figure 1. The phylogeny and CAZome of the genus *Zobellia*: (**a**) maximum-likelihood phylogeny based on 400 universal markers selected by PhyloPhlAn3.0 and reconstructed by RAxML with non-parametric bootstrapping using 100 replicates; (**b**) the bar plot showing the number of carbohydrate-active enzyme genes according to the CAZy classification for each strain. GH, glycoside hydrolase; GT, glycosyltransferase; PL, polysaccharide lyase; CE, carbohydrate esterase; AA, auxiliary activity; CBM, carbohydrate-binding module.

Previous studies have revealed that the *Zobellia* genomes are abundant with CAZymes genes, which encode the ability to efficiently degrade complex polysaccharide substrates [14,23]. It has been known that carbohydrate degraders are characterized by a high proportion of CAZymes found in their genomes (more than 5% of the predicted protein-coding genes). To identify CAZymes in the *Zobellia* genomes, we used the dbCAN2 meta server (http://cys.bios.niu.edu/dbCAN2, December 2020) [24].

We found that in all *Zobellia* genomes CAZymes account for more than 6% (from 6.4% in *Zobellia* sp. Asnod3-E08-A to 7.6% in *Z. galactanivorans* Dsij^T), which reflects the outstanding specialization of representatives of this genus in the degradation of polysaccharides. The total statistics of CAZyme classes predicted across the genomes are shown in Figure 1b. We found that *Z. galactanivorans* Dsij^T possesses the highest number of CAZymes (336), followed by *Z. amurskyensis* MAR 2009 138 (320), *Z. uliginosa* DSM 2061^T (315), and *Z. galactanivorans* OII3 (311). The smallest numbers of CAZymes were in *Zobellia* sp. Asnod3-E08-A (263) and Asnod2 B07BT (266). Notably, the genome of *Z. galactanivorans* Dsij^T encodes the largest and most diverse CAZYme repertoire, with approximately 60.7 CAZyme genes per Mb, in contrast to *Zobellia* sp. Asnod3-E08-A (52.5 CAZyme genes per Mb). However, these values indicate again a broad degradation potential conserved at the genus level [23].

Z. galactanivorans is well known to degrade alginates of brown algae [25] due to its alginolytic system including alginate lyases of distinct polysaccharide lyase families. PLs are a group of enzymes that cleave uronic acid-containing polysaccharides via a β-elimination mechanism [26]. In order to elucidate the alginolytic potential of *Zobellia*, we calculated the content of the PL genes in these genomes (Figure 2). The heat map shows the frequency of the PL genes dedicated to PL families. Notably, the genomes of *Z. amurskyensis* MAR 2009 138 and *Z. galactanivorans* Dsij^T encode 25 and 24 PLs, while the genomes *Zobellia* sp. Asnod3-E08-A and Asnod2 B07BT encode only 17 and 16 PLs, respectively. Among the identified PLs, the genus-specific ones were PL1, PL6, PL7, PL14, PL17, and PL40. Since all *Zobellia* genomes contain PL genes of families PL6, PL7, PL14, and PL17, this indicates that they all have a functional alginate utilization pathway. PL7 is an important enzyme in the utilization of alginate. Surprisingly, PL7 genes are the most abundant PLs in *Zobellia* genomes, accounting for one to five copies (Figure 2).

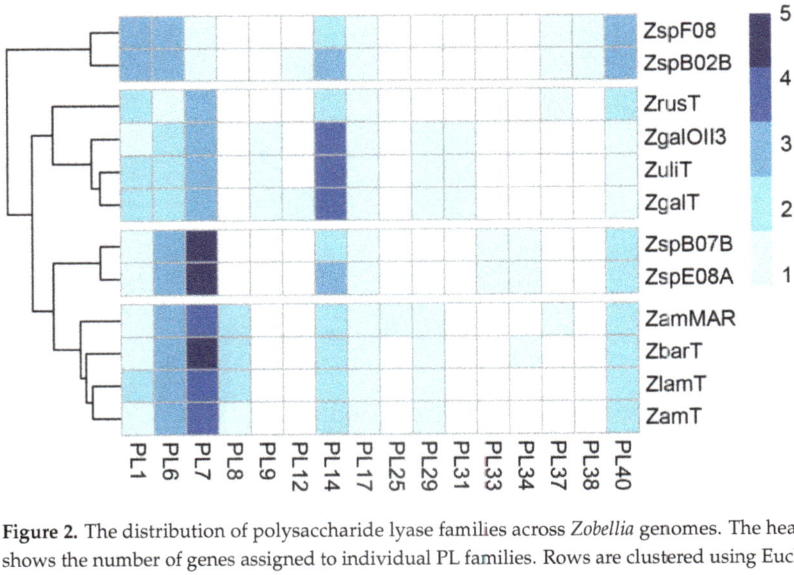

Figure 2. The distribution of polysaccharide lyase families across *Zobellia* genomes. The heat map shows the number of genes assigned to individual PL families. Rows are clustered using Euclidean distances. ZspF08, *Zobellia* sp. Asnod1-F08; ZspB02B, *Zobellia* sp. Asnod2-B02-B; ZrusT, *Z. russellii* KMM 3677^T; ZgOII3, *Z. galactanivorans* OII3; ZulT, *Z. uliginosa* DSM 2061^T; ZgaT, *Z. galactanivorans* Dsij^T; ZspB07B, *Zobellia* sp. Asnod2-B07-B; ZspE08A, *Zobellia* sp. Asnod3-E08-A; Zammar, *Z. amurskyensis* MAR 2009 138; ZbarT, *Z. barbeyroni'* KMM 6746^T; ZlamT, *Z. laminariae* KMM 3676^T; ZamT, *Z. amurskyensis* KMM 3526^T.

2.2. PL7 Phylogenic and Structural Analyses

Alginate lyases from the PL7 family are widely distributed in bacteria and have a typical β-jelly roll fold, which can possess both endolytic and exolytic activities. To date, crystal structures of eleven PL7 algninate lyases have been elucidated, and at least 40 representatives were characterized from the PL7 family (CAZy database, February 2021). Based on the sequence similarity of catalytic domains, the family of PL7 has been subdivided into five subfamilies (SF1-SF5) [26]. Subfamily 6 (SF6) was proposed by Thomas et al. in the extensive study of the PL7 alginate lyases from *Z. galactanivorans* Dsij^T [21]. It has been suggested that PL7 enzymes from SF6 appear to be conserved only in marine representatives of the *Flavobacteriaceae*. Later, alginate lyase Aly7B_Wf from *Wenyingzhuangia funcanilytica* CZ1127^T was characterized and classified as belonging to subfamily SF6 of the PL7 family [27]. Recently, the crystal structure of a novel PL7 alginate lyase AlyC3 from *Psychromonas* sp. C-3 was reported [28]. The AlyC3, along with several other unclassified PL7 alginate lyases, was attributed to subfamily SF6, which implies belonging to the novel subfamily SF7. Thus, despite some confusion in the literature, subfamilies SF6 and SF7 could be distinguished in addition to the well-known subfamilies SF1-SF5.

Initially, to clarify the classification of predicted alginate lyases from *Zobellia* within the PL7 family, a phylogenetic tree was constructed with all the available characterized PL7 alginate lyases derived from CAZy database (data not provided). To avoid redundancy one more phylogenetic tree was obtained, which included only the most representative alginate lyases for each subfamily along with their target sequences (Figure 3).

In accordance with the ML tree, 41 PL7 alginate lyases from *Zobellia* fall strictly into SF3, SF5, and SF6. Although AlyA1 and AlyA5 from *Z. galactanivorans* Dsij^T have been biochemical and structural characterized in detail [21], it is worth noting that close inspections of orthologous and paralogous genes are of great value for investigation of the PL7 enzymes evolution at the genus level. For convenience, paralogs and orthologs in subclades were numbered from one to six on the phylogenetic tree.

It was revealed that only 6 of the 12 *Zobellia* representatives encode PL7 lyases belong to subfamily SF3—namely, *Z. galactanivorans* Dsij^T, *Z. galactanivorans* OII3, *Z. uliginosa*, *Zobellia* sp. Asnod2-B07-B, *Zobellia* sp. Asnod3-E08-A, and '*Z. barbeyronii*'. Five of these, including AlyA1, were clustered together as presumptive orthologs, while ZbarT_PL7sf3_3 was reliably clustered with AlyQ from *Persicobacter* sp. CCB-QB2. We identified that the studied PL7-SF3 lyases had a modular organization and that all of them contain cleaved lipoprotein signals and CBM32 in addition to the catalytic domain. The same was determined earlier for AlyA1 [3]. Furthermore, domains moderately resembling CBM16 and CBM6 were also found in the architectures of AlyQ and ZbarT_PL7sf3_3, respectively.

The AlyA1 is an endolytic guluronate lyase [21], and AlyQ is most active on alginate, although it can also act on polyguluronate (poly-G) and polymannuronate (poly-M) [29]. For putative PL7-SF3 lyases from *Zobellia*, homology modeling based on the AlyA1 and AlyQ crystal structures was carried out (data not provided). The congruence of the phylogeny and structural similarities between these so-called orthologs indicate that they may possess similar activities. Considering that AlyA1 appears to have been acquired via horizontal gene transfer (HGT) from marine *Actinobacteria* [19], it becomes obvious that ZbarT_PL7sf3_3 was laterally acquired from other taxa. It is possible that CBM modules were fused with catalytic domains in ancestral genes before their transfer.

We consider that the PL7 alginate lyases belonging to subfamily SF5 are conserved within the *Zobellia* genus because at least one homolog was found in each genome. According to the phylogenetic tree, a well-supported (bpp = 94) orthologous group (OG) composed of 12 sequences is clearly distinguished and designated as 4 in Figure 3. Genes encoding putative PL7-SF5 lyases were duplicated and presented by paralogous pairs (marked as 5 on Figure 3) in seven *Zobellia* representatives—namely, *Z. russellii*, *Z. laminariae*, '*Z. barbeyronii*', *Zobellia* sp. Asnod2-B07-B, *Zobellia* sp. Asnod3-E08-A, *Z. amurskyensis* KMM 3526^T, and *Z. amurskyensis* MAR 2009 138. It should be noted that *Zobellia* sp. Asnod1-F08 and Asnod2-B02-B encode only one copy of the PL7 alginate lyases attributed to OG of the subfamily SF5. All the predicted PL7-SF5 lyases possess cleaved lipoprotein signals and PL7 catalytic domains. The amino acid sequence identities of PL7 catalytic domains between in-paralogous SF5_4 and SF5_5 were in the range of 65.0% to 67.51%. The identities between AlyA5 from *Z. galactanivorans* Dsij^T and orthologs from OG SF5_4 varied within the range of 83.16% to 98.97%, while they were in the range of 64.98% tc 66.79% across the AlyA5 and out-paralogs from SF5_5 (Table S2).

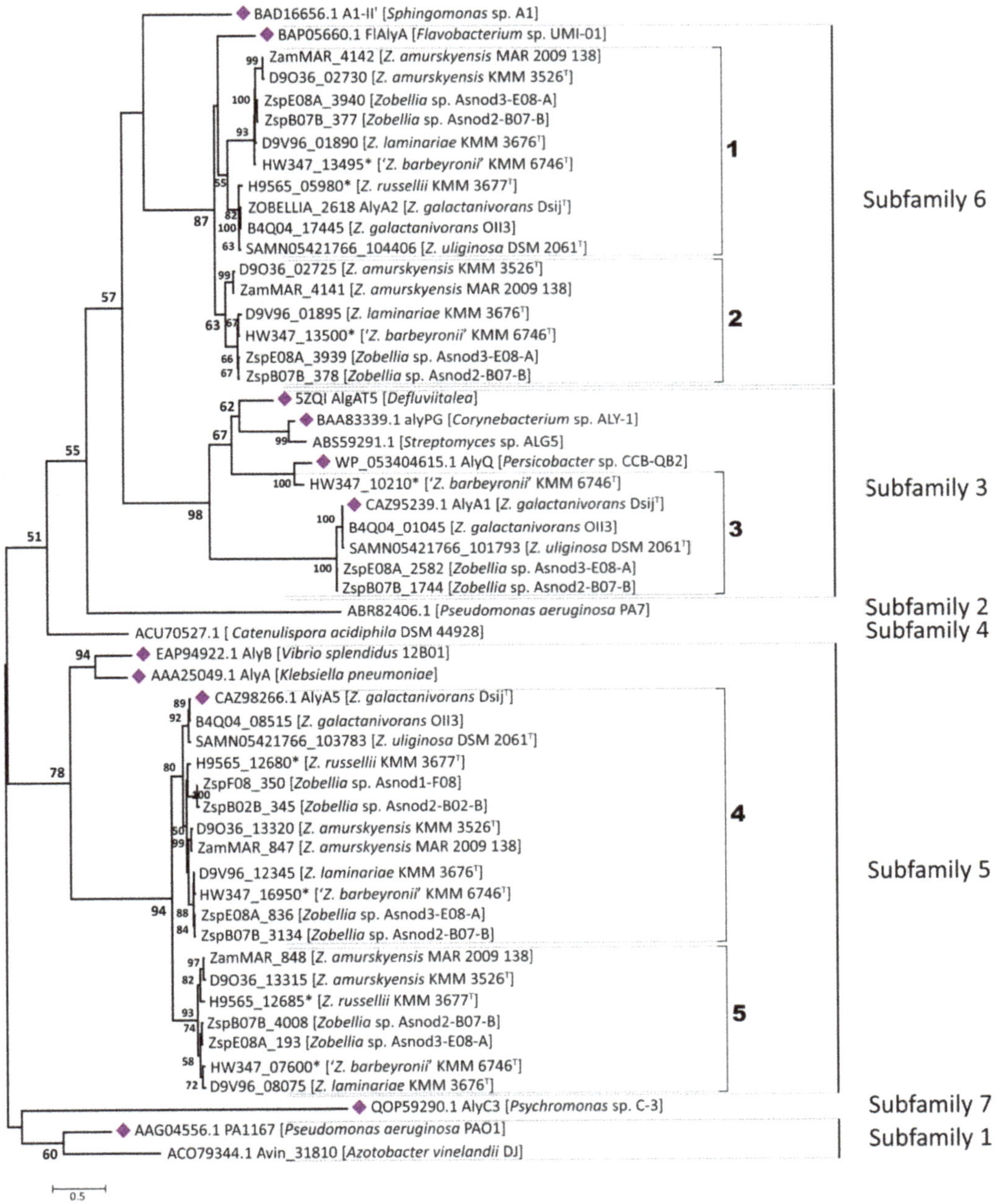

Figure 3. The phylogenetic tree of PL7 family alginate lyases from *Zobellia* and selected characterized representatives of PL7 family. For characterized PL7 family members, the corresponding GenBank accession numbers are given; for PL7 proteins of *Zobellia,* locus tags or RAST ORFs identifiers are listed. The organism names are listed in brackets. Proteins with a clarified crystal structure are marked as violet diamonds, and identifiers from unpublished genomes are marked with asterisks. Bootstrap values lower than 50 are not indicated.

In comprehensive study, it was determined that AlyA5 cleaves monomers from the non-reducing end of oligoalginates in an exolytic fashion [21]. The three-dimensional structures of in-paralogous ZlamT_PL7sf5_4 and ZlamT_ PL7sf5_5 were modelled using AlyA5 (PDB ID: 4BE3) as the template and superimposed. Based on structural alignment (Figure S1), the main divergences between the catalytic domains of the ortholog and paralogs were colored in ribbon representation (Figure 4a). It was revealed that the orthologs from OG SF5_4 shared a closer similarity in 3D structures to each other in contrast to in-paralogs, which confirms the conclusions following from the pairwise sequence identity analysis. Although the overall fold of *Zobellia* representatives PL7-SF5 is mostly similar, particularly in terms of the catalytic groove, there are slight differences in the external loop configurations.

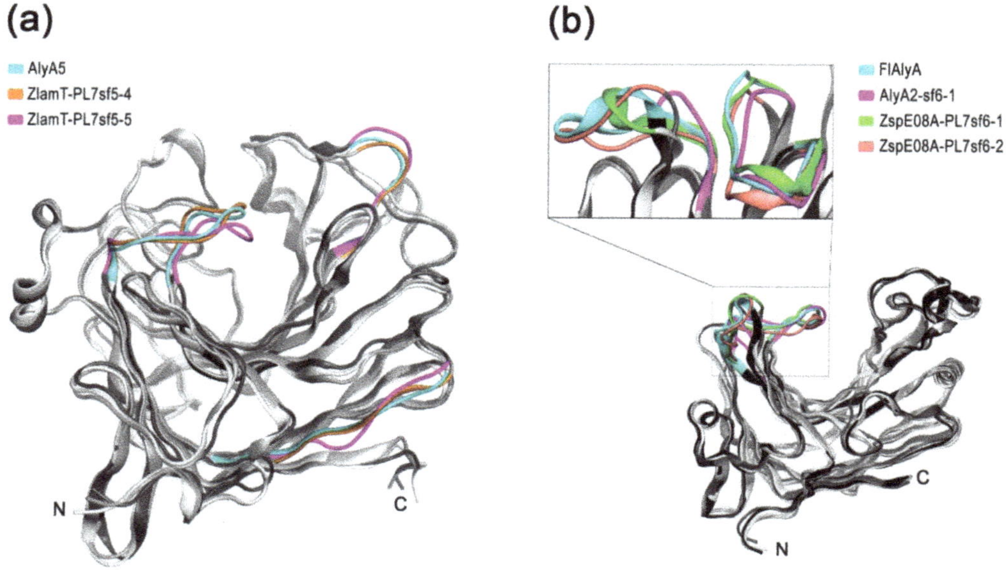

Figure 4. The superimposition of paralogous alginate lyases PL7 from SF5 and SF6 to known structures: (**a**) ribbon representation of the superimposition of the predicted 3D models for paralogous PL7-SF5_4 and SF5_5 from *Z. laminariae* KMM 3676^T onto prototype AlyA5 from *Z. galactanivorans* Dsij^T (PDB code 4BE3); (**b**) ribbon representation of the superimposition of predicted 3D models for paralogous PL7-SF6_1 and SF6_2 from *Zobellia* sp. Asnod3-E08-A and AlyA2 SF6_1 from *Z. galactanivorans* Dsij^T onto prototype FlAlyA from *Flavobacterium* sp. UMI-01 (PDB code 5Y33). Ribbon representations of superimpositions are presented in shades of grey. Differences between spatial structures of the paralogous alginate lyases in types of external loops are shown in color. The color corresponding to each structure is indicated.

The most curious and puzzling insights were obtained regarding the PL7 enzymes belonging to the subfamily SF6. It was revealed that 10 of 12 *Zobellia* representatives encode PL7-SF6 lyases, except *Zobellia* sp. Asnod1-F08 and Asnod2-B02-B. A suppositive OG for subfamily SF6 was distinguished (bpp = 55) and designated as one at the phylogenetic tree (Figure 3). In the same representatives, except *Z. russellii*, these lyases were presented by in-paralogs, as was revealed for subfamily SF5. All putative PL7-SF6 lyases contain only cleaved lipoprotein signals along with PL7 catalytic domains. Pairwise identities, calculated for as in-paralogs (59.92–61.54%), AlyA2 versus alleged orthologs (66.80–100%), and out-paralogs (64.78–68.83%, Table S2), were insufficient for reliable delineation of orthologous group in SF6 because obtained values did not agreed with generally accepted criteria.

To date, among the PL7 alginate lyases from subfamily SF6, only two enzymes have been studied. The Aly7B_Wf from *W. funcanilytica* CZ1127^T was characterized as endo-acting bifunctional alginate lyase and preferably cleaved polyM [27]. The FlAlyA from *Flavobacterium* sp. UMI-01 was first characterized as an endolytic enzyme with a preference for polymannuronate [30], and later, its crystal structure was clarified [31]. The three-dimensional structures of the orthologous AlyA2, and in-paralogous PL7-SF6 from *Zobellia* sp. Asnod3-E08-A, were modeled using FlAlyA (PDB ID: 5Y33) as the template and superimposed. The most significant divergences between 3D structures were colored in ribbon representation (Figure 4b). The overall fold of *Zobellia* PL7-SF6 lyases was mostly matched to prototype structure, but there were moderate differences in external loop configurations, which may imply the sub-functionalization of PL7-SF6. Considering the observed peculiarities of 3D structures, which are reflected in structural alignment (Figure S2), it has become clear that the PL7-SF6 are characterized by high diversification within the *Zobellia* genus.

A detailed exploration of genetic loci containing genes for PL7 lyases may shed light on the debatable issue regarding both the OGs delineation and the role of gene duplication.

2.3. Comparative Analysis of PL7-Containing Loci between Zobellia Genomes

The marine flavobacterium *Z. galactanivorans* Dsij^T constitutes a model organism for studying algal polysaccharide bioconversions, including alginates [14]. For the first time, the Alginate Utilization System (AUS) for marine bacterium has been identified in *Z. galactanivorans* Dsij^T and studied in detail [3]. One more comprehensive investigation regarding alginate utilization loci was carried out for the marine '*Gramella forsetii*' KT0803 [6]. Recently, it was reported that key enzymes for alginate utilization are widespread across 60 strains, which were isolated from marine environments and belong to the phyla *Proteobacteria* and *Bacteroidetes* [7].

According to the literature, in *Z. galactanivorans* Dsij^T the AUS is encoded by two operons and two genes isolated in the genome [14]. The activity of the system is tightly controlled by the presence of alginate in the medium [32] and AusR, a GntR family repressor [33]. As described in [33], the current model of alginate degradation by *Z. galactanivorans* Dsij^T implies stepwise depolymerization of alginate by coherent action of extracellular lyases AlyA1 (PL7) and AlyA7 (PL14) then oligosaccharides, recruited by surface-exposed PKD-containing and SusD-like proteins, are imported to the periplasm via TBDT where they are subjected to further degradation into unsaturated mono-uronic acid by the alginate lyases AlyA2 (PL7), AlyA3 (PL17), AlyA4 (PL6), AlyA5 (PL7) and AlyA6 (PL6). Further conversions in the cytoplasm occur through the consecutive action of KdgF, SDR, and KdgK1. Finally, KDPG (2-keto-3-deoxy-6-phosphogluconate) is eventually assimilated into the central metabolism through the Entner-Doudoroff pathway.

In order to clarify the issues mentioned above, we performed a comparative analysis of loci containing PL7 genes across the *Zobellia* genomes (Figure 5).

It was revealed that PL7 genes are localized in six genetic loci (I–VI), which are part of a complex and evolved AUS, in agreement with previously those defined for the AlyA1, AlyA2, and AlyA5 from *Z. galactanivorans* Dsij^T [14]. Thus, the trend toward the distribution of PL7 alginate lyases genes over separate loci is conserved within the *Zobellia* genus. It is noteworthy that PL7 of subfamilies SF3 and SF6 are encoded strictly in separate loci V or VI, and I, respectively. Whereas, in-paralogous genes from PL7-SF5 might be located adjacently in locus II and separately in loci II, III, and IV.

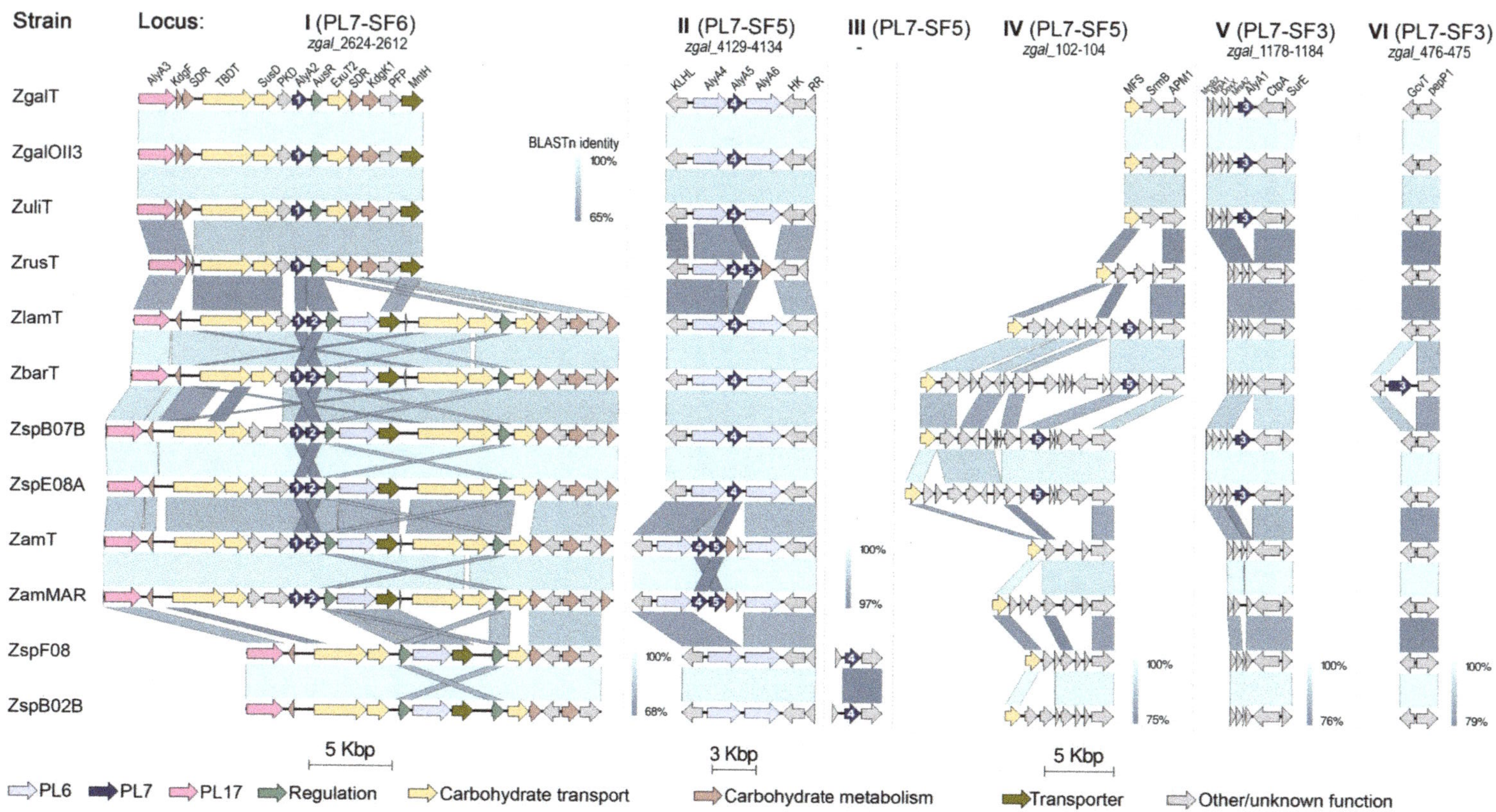

Figure 5. The comparisons of six loci (I–VI) containing PL7 genes among the *Zobellia* representatives. Genes were colored based on their annotation as indicated at the bottom of the figure; ZgaT, *Z. galactanivorans* Dsij^{T}; ZgOII3, *Z. galactanivorans* OII3; ZulT, *Z. uliginosa* DSM 2061^{T}; ZrusT, *Z. russellii* KMM 3677^{T}; ZlamT, *Z. laminariae* KMM 3676^{T}; ZbarT, '*Z. barbeyronii*' KMM 6746^{T}; ZspB07B, *Zobellia* sp. Asnod2-B07-B; ZspE08A, *Zobellia* sp. Asnod3-E08-A; ZamT, *Z. amurskyensis* KMM 3526^{T}; Zammar, *Z. amurskyensis* MAR 2009 138; ZspF08, *Zobellia* sp. Asnod1-F08; ZspB02B, *Zobellia* sp. Asnod2-B02-B.

As presented on the synteny plot, the orthologs PL7-SF3 from *Z. galactanivorans* Dsij^T, *Z. galactanivorans* OII3, *Z. uliginosa*, *Zobellia* sp. Asnod2-B07-B and Asnod3-E08-A were encoded in locus V in the same surroundings, whereas orthologs from '*Z. barbeyronii*' was collocated within another environment in locus VI. This confirms that all of them were laterally acquired, in agreement with phylogenetic analysis points. It can be assumed that such dispersed genes are plastic in genomes and more frequently participate in HGTs processes.

Ten PL7 genes from OG SF5_4 are collocated along with PL6 lyases genes in locus II, which was previously described in *Z. galactanivorans* Dsij^T genome as small operon *zgal*_4130-4132 [3]. Seven PL7-SF5 genes are presented by in-paralogs; among them, three SF5_5 paralogs are collocated with SF5_4 in locus II (in *Z. russellii*, *Z. amurskyensis* KMM 3526^T, *Z. amurskyensis* MAR 2009 138), whereas four other SF5_5 paralogs are in different locus IV (in *Z. laminariae*, '*Z. barbeyronii*', *Zobellia* sp. Asnod2-B07-B and Asnod3-E08-A). Interestingly, SF5_4 genes of *Zobellia* sp. Asnod1-F08 and Asnod2-B02-B are placed in locus III, which is not represented in *Z. galactanivorans* Dsij^T genome. We suppose that the processes of gene duplication and local gene transfer represent the evolution of PL7-SF5 genes at the *Zobellia* genus level. Summarizing insights from phylogenetic and synteny analyses, in the case of SF5 in-paralogs colocation, the SF5_4 from the orthologous group is located upstream. Surprisingly, the pairwise identities of PL7 catalytic domains between SF5_4 and SF5_5 remained about 66% on average, regardless of whether they collocated or not.

Altogether, except genomes of *Zobellia* sp. Asnod1-F08 and Asnod2-B02-B, all the genomes contain the PL7-SF6 orthologs, localized in a large operon (locus I) along with other carbohydrate-related genes. The operon *zgal*_2624-2612 containing AlyA2 from *Z. galactanivorans* Dsij^T was described in detail [3]. Six PL7-SF6 genes are presented by in-paralogs, for which upstream localization is a reliable criterion for belonging to OG PL7-SF6, as was identified for PL7-SF5 and discussed above. A hypothesized sub-functionalization of in-paralogs SF6_1 and SF6_2 is supported by the presence of additional non-paralogous transporters and PKD-containing protein genes, as well as by the duplication of the repressor *AusR*, at the locus I. Our highlight of the variations in PL lyases content is another interesting observation based on synteny analysis of locus I. The first type of locus I contains genes for PL17 and PL7-SF6_1 (strains of *Z. galactanivorans*, *Z. uliginosa*, and *Z. russellii*); the second one comprises PL17, PL7-SF6 in-paralogs, and PL6 (*Z. laminariae*, '*Z. barbeyronii*', *Zobellia* sp. Asnod2-B07-B and Asnod3-E08-A, strains of *Z. amurskyensis*); the third type contains PL17 and PL6 (*Zobellia* sp. Asnod1-F08 and Asnod2-B02-B).

3. Materials and Methods

3.1. Phylogenomic Analysis

Representative genomes for *Zobellia* and related genera were retrieved from GenBank, NCBI. Accession numbers for the genomes used in this study are provided in Table S1. Single-copy marker genes (n = 400) were selected from protein sequences, concatenated, and aligned using the PhyloPhlAn (version 3.0) pipeline [22]. A maximum-likelihood tree was reconstructed by RAxML (version 8.2.12) under the LG+Γ model with non-parametric bootstrapping using 100 replicates [34].

3.2. Annotation of Carbohydrate-Active Enzymes

The carbohydrate-active enzymes and carbohydrate-binding modules were annotated using the dbCAN2 Meta server (version 9.0) with default settings [24]. Predictions by at least one of the three algorithms integrated within the server (DIAMOND, HMMER, and Hotpep) were considered sufficient for CAZy family assignments. The relative abundances of CAZymes and PLs were visualized by stacked bar plots and heat maps using the ggplot2 (version 3.3.3) and pheatmap (version 1.0.12) packages in RStudio (version 1.3.1093) with R (version 4.0.3, R Foundation for Statistical Computing) [35,36].

3.3. Sequence Analyses and Homology Modelling

The PL7 lyases were inspected for the presence of additional functional domains via the Interproscan server (version 83.0) [37]. For phylogenetic analysis protein sequences of 11 alginate lyases with solved structures belonging to the PL7 family were selected in the CAZy database [38]. Multiple sequence alignment of PL7 catalytic domains was performed using ClustalX implemented in MEGA-X (version 10.1.8) and manually corrected [39]. A maximum-likelihood phylogeny with 100 non-parametric bootstrap replicates was calculated using the IQ-TREE web server (version 1.6.12) and the WAG+I+G4 substitution model determined using ModelFinder [40,41].

The pairwise sequence identities were calculated by Clustal (version 2.1) using EMBL-EBI Services [42]. The 3D-structures of paralogous PL7 alginate-lyases from SF3, SF5, and SF6 were generated using homology modeling by the Molecular Operating Environment (MOE, version 2020.09) [43]. The crystal structures of the AlyA1 and AlyA5 from *Z. galactanivorans* Dsij^T, AlyQ from *Persicobacter* sp. CCB-QB2, and FlAlyA from *Flavobacterium* sp. UMI-01 with PDB codes 3ZPY, 4BE3, 5XNR, and 5Y33, respectively, were used as the initial templates. Structure-based sequence alignments were visualized with ESPript 3 [44].

3.4. Comparative Analysis of PL7-Containing Loci

The genomic regions containing PL7 genes were extracted from the GBK files of the *Zobellia* genomes using Geneious Pro software (version 4.8) [45]. Due to the lack of genome annotations for *Zobellia* spp. and *Z. amurskyensis* MAR 2009 138, their genomes were initially annotated by the RAST server [46]. Identifiers for PL7 and adjacent genes included in selected loci are listed in Table S3. Generated GBK files were modified by adding custom color feature qualifiers. Pairwise comparisons of each locus between twelve genomes were carried out using BLASTn (BLAST version 2.11.0+) run in EasyFig (version 2.2.5) [47]. Synteny plots were visualized by Easyfig with the minimum BLAST hit of 680 bp.

4. Conclusions

We revealed that PL7 family alginate lyases are the most abundant among the polysaccharide lyases identified in *Zobellia* genomes. Based on phylogenomic, structural, and comparative analyses, the PL7 lyases belong to SF3, SF5, and SF6 and are involved in local and horizontal gene transfer, as well as gene duplications processes. It is most likely that an individual evolution of PL7 genes may promote Alginate Utilization System variability across the *Zobellia* genus. The PL7 alginate lyases may acquire a sub-functionalization due to diversification between in-paralogs.

Supplementary Materials: Figure S1: Structure-based sequence alignment of PL7 alginate lyases from subfamily SF5; Figure S2: Structure-based sequence alignment of PL7 alginate lyases from subfamily SF6; Table S1: List of accession numbers for genomes used in the phylogenomic analysis; Table S2: Identity matrix of catalytic domains from PL7 alginate lyases identified in *Zobellia* genomes; Table S3: Identifiers of PL7 and adjacent genes included in loci I–VI.

Author Contributions: Conceptualization, M.I. and N.C.; methodology, E.B., O.N., and N.C.; software, N.C. and G.L.; investigation, N.C., E.B., and O.N.; data curation, M.I.; writing—original draft preparation, N.C. and M.I.; writing—review and editing, M.I.; visualization, N.C.; supervision, M.I.; funding acquisition, M.I. All authors have read and agreed to the published version of the manuscript.

Funding: This research received no external funding.

Institutional Review Board Statement: Not applicable.

Informed Consent Statement: Not applicable.

Data Availability Statement: Unpublished genomes are available from the authors.

Acknowledgments: We are grateful to the anonymous reviewers and editors that have improved our manuscript.

Conflicts of Interest: The authors declare no conflict of interest.

References

1. Kirchman, D.L. The ecology of Cytophaga-Flavobacteria in aquatic environments. *FEMS Microbiol. Ecol.* **2002**, *39*, 91–100.
2. Thomas, F.; Hehemann, J.-H.; Rebuffet, E.; Czjzek, M.; Michel, G. Environmental and gut *Bacteroidetes*: The food connection. *Front. Microbiol.* **2011**, *2*, 93. [CrossRef] [PubMed]
3. Thomas, F.; Barbeyron, T.; Tonon, T.; Génicot, S.; Czjzek, M.; Michel, G. Characterization of the first alginolytic operons in a marine bacterium: From their emergence in marine Flavobacteriia to their independent transfers to marine Proteobacteria and human gut Bacteroides. *Environ. Microbiol.* **2012**, *14*, 2379–2394. [CrossRef] [PubMed]
4. Sun, C.; Fu, G.; Zhang, C.; Hu, J.; Xu, L.; Wang, R.; Su, Y.; Han, S.; Yu, X.; Cheng, H.; et al. Isolation and complete genome sequence of *Algibacter alginolytica* sp. nov., a novel seaweed-degrading Bacteroidetes bacterium with diverse putative polysaccharide utilization loci. *Appl. Environ. Microbiol.* **2016**, *82*, 2975–2987.
5. Mann, A.J.; Hahnke, R.L.; Huang, S.; Werner, J.; Xing, P.; Barbeyron, T.; Huettel, B.; Stüber, K.; Reinhardt, R.; Harder, J.; et al. The genome of the alga-associated marine flavobacterium *Formosa agariphila* KMM 3901T reveals a broad potential for degradation of algal polysaccharides. *Appl. Environ. Microbiol.* **2013**, *79*, 6813–6822.
6. Kabisch, A.; Otto, A.; Konig, S.; Becher, D.; Albrecht, D.; Schuler, M.; Teeling, H.; Amann, R.I.; Scheweder, T. Functional characterization of polysaccharide utilization loci in the marine Bacteroidetes 'Gramella forsetii' KT0803. *ISME J.* **2014**, *8*, 1492–1502.
7. Chaq, Q.Q.; Wang, X.J.; Ren, X.B.; Li, D.; Wang, P.; Li, P.Y.; Fu, H.H.; Zhang, X.Y.; Chen, X.L.; Zhang, Y.Z.; et al. Comparison of alginate utilization pathways in culturable bacteria isolated from Arctic and Antarctic marine environments. *Front. Microbiol.* **2021**, *12*, 609393.
8. Wolter, L.A.; Mitulla, M.; Kalem, J.; Daniel, R.; Simon, M.; Wietz, M. CAZymes in *Maribacter dokdonensis* 62-1 from the Patagonian shelf: Genomics and physiology compared to related flavobacteria and a co-occurring *Alteromonas* strain. *Front. Microbiol.* **2020**, *12*, 717.
9. Reintjes, G.; Arnosti, C.; Fuchs, B.M.; Amann, R. An alternative polysaccharide uptake mechanism of marine bacteria. *ISME J.* **2017**, *11*, 1640–1650. [CrossRef]
10. Matos, M.N.; Lozada, M.; Anselmino, L.E.; Musumeci, M.A.; Henrissat, B.; Jansson, J.K.; Mac Cormack, W.F.; Carroll, J.; Sjöling, S.; Lundgren, L.; et al. Metagenomics unveils the attributes of the alginolytic guilds of sediments from four distant cold coastal environments. *Environ. Microbiol.* **2016**, *18*, 4471–4484.
11. Barbeyron, T.; L'Haridon, S.; Corre, E.; Kloareg, B.; Potin, P. *Zobellia galactanovorans* gen. nov., sp. nov., a marine species of Flavobacteriaceae isolated from a red alga, and classification of [Cytophaga] uliginosa (ZoBell and Upham 1944) Reichenbach 1989 as *Zobellia uliginosa* gen. nov., comb. nov. *Int. J. Syst. Evol. Microbiol.* **2001**, *51*, 985–997.
12. Nedashkovskaya, O.I.; Suzuki, M.; Vancanneyt, M.; Cleenwerck, I.; Lysenko, A.M.; Mikhailov, V.V.; Swings, J. *Zobellia amurskyensis* sp. nov., *Zobellia laminariae* sp. nov. and *Zobellia russellii* sp. nov., novel marine bacteria of the family Flavobacteriaceae. *Int. J. Syst. Evol. Microbiol.* **2004**, *54*, 1643–1648.
13. Martin, M.; Portetelle, D.; Michel, G.; Vandenbol, M. Microorganisms living on macroalgae: Diversity, interactions, and biotechnological applications. *Appl. Microbiol. Biotechnol.* **2014**, *98*, 2917–2935.
14. Barbeyron, T.; Thomas, F.; Barbe, V.; Teeling, H.; Schenowitz, C.; Dossat, C.; Goesmann, A.; Leblanc, C.; Glöckner, F.O.; Czjzek, M.; et al. Habitat and taxon as driving forces of carbohydrate catabolism in marine heterotrophic bacteria: Example of the model algae-associated bacterium *Zobellia galactanivorans* Dsij. *Environ. Microbiol.* **2016**, *18*, 4610–4627. [CrossRef]
15. Hehemann, J.H.; Correc, G.; Thomas, F.; Bernard, T.; Barbeyron, T.; Jam, M.; Helbert, W.; Michel, G.; Czjzek, M. Biochemical and structural characterization of the complex agarolytic enzyme system from the marine bacterium *Zobellia galactanivorans*. *J. Biol. Chem.* **2012**, *287*, 30571–30584.
16. Barbeyron, T.; Gerard, A.; Potin, P. The kappa-carrageenase of the marine bacterium Cytophaga drobachiensis. Structural and phylogenetic relationships within family-16 glycoside hydrolases. *Mol. Biol. Evol.* **1998**, *15*, 528–537.
17. Rebuffet, E.; Barbeyron, T.; Jeudy, A.; Jam, M.; Czjzek, M.; Michel, G. Identification of catalytic residues and mechanistic analysis of family GH82 iota-carrageenases. *Biochemistry* **2010**, *49*, 7590–7599. [CrossRef]
18. Harms, H.; Poehlein, A.; Thürmer, A.; König, G.M.; Schäberle, T.F. Draft Genome Sequence of *Zobellia* sp. Strain OII3, Isolated from the Coastal Zone of the Baltic Sea. *Genome Announc.* **2017**, *5*, e00737-17. [CrossRef] [PubMed]
19. Zhu, Y.; Thomas, F.; Larocque, R.; Li, N.; Duffieux, D.; Cladière, L.; Souchaud, F.; Michel, G.; McBride, M.J. Genetic analyses unravel the crucial role of a horizontally acquired alginate lyase for brown algal biomass degradation by *Zobellia galactanivorans*. *Environ. Microbiol.* **2017**, *19*, 2164–2181. [CrossRef]
20. Labourel, A.; Jam, M. The beta-glucanase ZgLamA from *Zobellia galactanivorans* evolved a bent active site adapted for efficient degradation of algal laminarin. *J. Biol. Chem.* **2014**, *289*, 2027–2042.
21. Thomas, F.; Lundqvist, L.C.E.; Jam, M.; Jeudy, A.; Barbeyron, T.; Sandström, C.; Michel, G.; Czjzek, M. Comparative Characterization of Two Marine Alginate Lyases from *Zobellia galactanivorans* Reveals Distinct Modes of Action and Exquisite Adaptation to Their Natural Substrate. *J. Biol. Chem.* **2013**, *288*, 23021–23037. [CrossRef]
22. Asnicar, F.; Thomas, A.M.; Beghini, F.; Mengoni, C.; Manara, S.; Manghi, P.; Zhu, Q.; Bolzan, M.; Cumbo, F.; May, U.; et al. Precise phylogenetic analysis of microbial isolates and genomes from metagenomes using PhyloPhlAn 3.0. *Nat. Commun.* **2020**, *11*, 2500. [CrossRef] [PubMed]

23. Chernysheva, N.; Bystritskaya, E.; Stenkova, A.; Golovkin, I.; Nedashkovskaya, O.; Isaeva, M. Comparative Genomics and CAZyme Genome Repertoires of Marine *Zobellia* Amurskyensis KMM 3526T and *Zobellia* Laminariae KMM 3676T. *Mar. Drugs* **2019**, *17*, 661.
24. Zhang, H.; Yohe, T.; Huang, L.; Entwistle, S.; Wu, P.; Yang, Z.; Busk, P.K.; Xu, Y.; Yin, Y. dbCAN2: A meta server for automated carbohydrate-active enzyme annotation. *Nucleic Acids Res.* **2018**, *46*, W95–W101. [CrossRef] [PubMed]
25. Thomas, F.; Barbeyron, T.; Michel, G. Evaluation of reference genes for real-time quantitative PCR in the marine flavobacterium *Zobellia galactanivorans*. *J. Microbiol. Methods* **2011**, *84*, 61–66.
26. Lombard, V.; Bernard, T.; Rancurel, C.; Brumer, H.; Coutinho, P.M.; Henrissat, B. A hierarchical classification of polysaccharide lyases for glycogenomics. *Biochem. J.* **2010**, *432*, 437–444. [CrossRef] [PubMed]
27. Pei, X.; Chang, Y.; Shen, J. Cloning, expression and characterization of an endo-acting bifunctional alginate lyase of marine bacterium *Wenyingzhuangia fucanilytica*. *Protein Expr. Purif.* **2019**, *154*, 44–51. [CrossRef]
28. Xu, F.; Chen, X.L.; Sun, X.H.; Dong, F.; Li, C.Y.; Li, P.Y.; Ding, H.; Chen, Y.; Zhang, Y.Z.; Wang, P. Structural and molecular basis for the substrate positioning mechanism of a new PL7 subfamily alginate lyase from the arctic. *J. Biol. Chem.* **2020**, *295*, 16380–16392. [CrossRef]
29. Sim, P.F.; Furusawa, G.; Teh, A.H. Functional and structural studies of a multidomain alginate lyase from *Persicobacter* sp. CCB-QB2. *Sci. Rep.* **2017**, *7*, 13656. [CrossRef]
30. Inoue, A.; Takadono, K.; Nishiyama, R.; Tajima, K.; Kobayashi, T.; Ojima, T. Characterization of an Alginate Lyase, FlAlyA, from *Flavobacterium* sp. Strain UMI-01 and Its Expression in *Escherichia coli*. *Mar. Drugs* **2014**, *12*, 4693–4712. [CrossRef] [PubMed]
31. Qin, H.M.; Miyakawa, T.; Inoue, A.; Nishiyama, R.; Nakamura, A.; Asano, A.; Ojima, T.; Tanokura, M. Structural basis for controlling the enzymatic properties of polymannuronate preferred alginate lyase FlAlyA from the PL-7 family. *Chem. Commun. (Camb.)* **2018**, *54*, 555–558. [CrossRef]
32. Thomas, F.; Bordron, P.; Eveillard, D.; Michel, G. Gene expression analysis of *Zobellia galactanivorans* during the degradation of algal polysaccharides reveals both substrate-specific and shared transcriptome-wide responses. *Front. Microbiol.* **2017**, *8*, 1808. [CrossRef] [PubMed]
33. Dudek, M.; Dieudonné, A.; Jouanneau, D.; Rochat, T.; Michel, G.; Sarels, B.; Thomas, F. Regulation of alginate catabolism involves a GntR family repressor in the marine flavobacterium *Zobellia galactanivorans* Dsij. *Nucleic Acids Res.* **2020**, *48*, 7786–7800. [CrossRef] [PubMed]
34. Stamatakis, A. RAxML version 8: A tool for phylogenetic analysis and post-analysis of large phylogenies. *Bioinformatics* **2014**, *30*, 1312–1313. [CrossRef]
35. Wickham, H. ggplot2. *Wiley Interdiscip. Rev. Comput. Stat.* **2011**, *3*, 180–185. [CrossRef]
36. Kolde, R.; Kolde, M.R. Package 'pheatmap'. *R Package* **2015**, *1*, 790.
37. Blum, M.; Chang, H.-Y.; Chuguransky, S.; Grego, T.; Kandasaamy, S.; Mitchell, A.; Nuka, G.; Paysan-Lafosse, T.; Qureshi, M.; Raj, S. The InterPro protein families and domains database: 20 years on. *Nucleic Acids Res.* **2021**, *49*, D344–D354. [CrossRef]
38. Lombard, V.; Golaconda Ramulu, H.; Drula, E.; Coutinho, P.M.; Henrissat, B. The carbohydrate-active enzymes database (CAZy) in 2013. *Nucleic Acids Res.* **2013**, *42*, D490–D495. [CrossRef]
39. Kumar, S.; Stecher, G.; Li, M.; Knyaz, C.; Tamura, K. MEGA X: Molecular evolutionary genetics analysis across computing platforms. *Mol. Biol. Evol.* **2018**, *35*, 1547–1549. [CrossRef]
40. Nguyen, L.-T.; Schmidt, H.A.; von Haeseler, A.; Minh, B.Q. IQ-TREE: A fast and effective stochastic algorithm for estimating maximum-likelihood phylogenies. *Mol. Biol. Evol.* **2015**, *32*, 268–274. [CrossRef] [PubMed]
41. Kalyaanamoorthy, S.; Minh, B.Q.; Wong, T.K.F.; von Haeseler, A.; Jermiin, L.S. Modelfinder: Fast model selection for accurate phylogenetic estimates. *Nat. Methods* **2017**, *14*, 587–589. [CrossRef]
42. Madeira, F.; Park, Y.M.; Lee, J.; Buso, N.; Gur, T.; Madhusoodanan, N.; Basutkar, P.; Tivey, A.R.N.; Potter, S.C.; Finn, R.D.; et al. The EMBL-EBI search and sequence analysis tools APIs in 2019. *Nucleic Acids Res.* **2019**, *47*, W636–W641. [CrossRef]
43. *Molecular Operating Environment (MOE)*; 2020.09; Chemical Computing Group ULC: Montreal, QC, Canada, 2020.
44. Robert, X.; Gouet, P. Deciphering Key Features in Protein Structures with the New ENDscript Server. *Nucleic Acids Res.* **2014**, *42*, W320–W324. [CrossRef] [PubMed]
45. Kearse, M.; Moir, R.; Wilson, A.; Stones-Havas, S.; Cheung, M.; Sturrock, S.; Buxton, S.; Cooper, A.; Markowitz, S.; Duran, C. Geneious Basic: An integrated and extendable desktop software platform for the organization and analysis of sequence data. *Bioinformatics* **2012**, *28*, 1647–1649. [CrossRef]
46. Aziz, R.K.; Bartels, D.; Best, A.A.; DeJongh, M.; Disz, T.; Edwards, R.A.; Formsma, K.; Gerdes, S.; Glass, E.M.; Kubal, M.; et al. The RAST Server: Rapid annotations using subsystems technology. *BMC Genom.* **2008**, *9*, 75. [CrossRef] [PubMed]
47. Sullivan, M.J.; Petty, N.K.; Beatson, S.A. Easyfig: A genome comparison visualizer. *Bioinformatics* **2011**, *27*, 1009–1010. [CrossRef] [PubMed]

MDPI
St. Alban-Anlage 66
4052 Basel
Switzerland
Tel. +41 61 683 77 34
Fax +41 61 302 89 18
www.mdpi.com

Molecules Editorial Office
E-mail: molecules@mdpi.com
www.mdpi.com/journal/molecules

www.ingramcontent.com/pod-product-compliance
Lightning Source LLC
Chambersburg PA
CBHW040840250926
51850CB00007BA/1363
* 9 7 8 3 0 3 6 5 2 5 0 4 4 *